Austin He
100/6 and 3000 Owners Workshop Manual

by J H Haynes
Associate Member of the Guild of Motoring Writers
and B L Chalmers-Hunt
TEng(CEI), AMIMI, AMIRTE, AMVBRA

Models covered

100/6 Mk I and II. 2 and 2/4 seater
3000 Mk I, II and III. 2 and 2/4 seater

ISBN 0 900550 49 X

Printed in England *(049-2N1)*

Haynes Publishing Group
Sparkford Nr Yeovil
Somerset BA22 7JJ England

Haynes Publications, Inc
861 Lawrence Drive
Newbury Park
California 91320 USA

Acknowledgements

Our thanks are due to BLMC for their assistance with technical material and illustrations and to Castrol for lubrication details.

Additionally much invaluable work has been done in preparing this book by experts in their field, the typesetters, photographers, layout, graphic, editorial and print room staffs and we wish to record our thanks.

Photographic captions and cross references

The book is divided into twelve chapters each covering a part of the car. These chapters have sections, numbered in sequence, which are headed in bold type between horizontal lines and each section has serially numbered paragraphs.

There are two types of illustration:

1 Figures which are numbered in sequence for each chapter and all have a caption giving detail, often specifying individual components.

2 Photographs having a number in the bottom left hand corner and referring to the section and paragraph in which the operation is described. There are a few photographs notably at the end of Chapter One, which are grouped together to illustrate several steps in one operation, and these have a caption and no reference number.

So far as possible illustrations are placed close to the relevant text, mostly on the same or facing page, for easy reference.

Procedures once described in the text are not normally repeated but are referred to by starting the chapter, section and paragraph in which the information can be found - for example a reference given as Chapter 1/6.5 means that the operation is described in Chapter One, Section 6, Paragraph 5. Cross references given without reference to a chapter apply to a section within the same chapter.

When the left or right hand side of the car is mentioned it is as if one is looking in the forward direction of travel.

AUSTIN HEALEY 3000 Mk 2

Introduction

This is a manual for the do-it-yourself owner of an Austin Healey 100/6 or 3000. It shows how to maintain these cars in first class condition and how to carry out repairs when components become worn or broken. Regular and careful maintenance is essential if maximum reliability and minimum wear are to be achieved.

Although these cars are hardwearing and robust it is inevitable that their reliability and performance will decrease as they become older. Early models requiring attention are frequently bought by the more impecunious motorist who can least afford garage repairs.

It is in these circumstances that this manual will prove to be of maximum assistance, as it is the ONLY workshop manual written from practical experience.

Manufacturers' official manuals are often good publications which contain a wealth of technical information. Because they are issued primarily to help the manufacturers' authorised dealers and distributors they tend to be written in technical language, and to skip details of some jobs which are of common knowledge to fully trained mechanics. Owners' Workshop Manuals differ in that they are primarily to help the owner. They, therefore, go into many jobs in great detail with extensive photographic support to ensure that everything is properly understood so that the repair can be done properly.

Owners who intend to do their own maintenance and repairs should have a reasonably comprehensive tool kit. Some jobs do require special tools, but in many instances it is possible to avoid their use with a little care and ingenuity.

Throughout, ways of avoiding this use of special equipment and tools are shown. However, in some cases the proper tool must be used. Where this is the case a description of the tool and its correct use is included.

Actual repairs are becoming more and more a case of replacing the defective item with an exchange rebuilt unit. This is excellent practice when a component is thoroughly worn out, but it is a waste of good money when the total component is only half worn, and requires the replacement of but a single small item to effect a complete repair.

A further intent of this manual is to show the owner how to examine malfunctioning parts; determine what is wrong, and then how to make the repair.

Given the time, a certain amount of mechanical do-it-yourself aptitude, and a reasonable collection of tools, this manual will show the ordinary private owner how to maintain and repair his car economically.

Ordering spare parts

Always order genuine British Leyland Unipart spare parts from your nearest BLMC dealer or local garage if you can. BLMC authorised dealers carry a stock of genuine parts and can supply items 'over the counter', however you may find certain genuine parts are no longer available as the parts stock is now exhausted. 'Pirate' replacement parts will be obtainable from Austin Healey specialists.

When ordering spare parts it is essential to give full details of your car to the storeman. He will want to know the chassis and engine numbers. When ordering parts for the gearbox or rear axle it is also necessary to quote the numbers which may be found on the respective units.

Identification

Chassis number – This is located on a plate mounted on the bulkhead, under the bonnet.
Engine number – The engine carries a serial number stamped on a metal plate secured to the left hand (carburettor) side of the cylinder block.
Gearbox number – This is stamped on the right hand side of the gearbox casing.
Rear axle number – This is stamped on the rear of the left hand axle tube adjacent to the rebound rubber.
Body number – Stamped on a plate fixed to the front valance.

When ordering new parts, remember that some assemblies can be exchanged. This is very much cheaper than buying them outright and throwing away the old part.

Contents

Routine maintenance

The maintenance instructions listed below are basically those recommended by the manufacturer. They are supplemented by additional maintenance tasks which, through practical experience, can be recommended.

The additional tasks are indicated by an asterisk and are primarily of a preventative nature in that they will assist in eliminating the unexpected failure of a component due to fair wear and tear.

It should however be realised that all models will be outside any manufacturers servicing schedule now. Such is their age, at least four years old, that the servicing they require will be of a special nature and certainly more frequent than a 'modern' car. Consider this servicing schedule to be of a 'long term' life extension plan for the car and a necessary safety insurance.

Full instructions of how to service the individual components are given in each chapter. This is merely the schedule and some additional illustrations.

Maintenance schedule

Daily

Check oil in engine. Top up if necessary.
Check water in radiator. Top up if necessary.

Weekly

Test tyre pressures and adjust if necessary.
Check battery level and top up if necessary.

Monthly (1000 miles, whichever is sooner)

1 ENGINE
Top up carburettor piston dampers.
Lubricate carburettor controls.
Top up radiator.

2 IGNITION
Give the distributor shaft greaser cap one half-turn.

3 CLUTCH
Check level of fluid in the hydraulic clutch fluid reservoir and top up if necessary.

4 BRAKES
Check brake pedal travel and adjust if necessary.
Make visual inspection of brake lines and pipes.
Check level of fluid in the hydraulic brake fluid reservoir and top up if necessary.

5 HYDRAULIC DAMPERS
Examine all hydraulic dampers for leaks.

6 ELECTRICAL
Check battery cell specific gravity readings and top up to correct level.

OIL FILLER CAP

7 LUBRICATION
Top up engine, gearbox, rear axle, steering box and steering idler oil levels.
8 Lubricate all grease nipples.
9 WHEELS AND TYRES
Check tyre pressures.
Check 'knock-on' wheel caps for tightness.

Bi-monthly (3000 miles, whichever is sooner)

All servicing items in the monthly service, plus:

1 ENGINE
Lubricate carburettor controls.
Clean and re-oil air cleaners.
Check dynamo drive belt tension.
Clean and adjust spark plugs.
2 BRAKES
Check brakes and adjust if necessary.
Inspect disc brake friction pads for unequal wear.
3 BODY
Lubricate door hinges, door locks, bonnet lock, safety catches and operating mechanism.
4 LUBRICATION
Change engine oil.
5 WHEELS AND TYRES
Change road wheels round diagonally, including spare, to regularise tyre wear.
Check tyre pressures.
6 *Adjust brakes. Inspect pads and hoses.
*Inspect all rubber boots on steering.
*Wash bodywork and chrome fittings.
*Clean interior of car.
*Check all lights for correct operation.
*Lubricate dynamo bearings.
*Lubricate all controls.

Six monthly (6000 miles, whichever is sooner)

All servicing items in the monthly and bi-monthly service plus:

1 ENGINE
Lubricate water pump sparingly.
Check valve rocker clearances and adjust if necessary.
Clean carburettor and fuel pump filters (when applicable).
2 IGNITION
Check automatic ignition control lubricating distributor drive shaft (by one half-turn of grease cap), cam and advance mechanism.
Check and adjust, if necessary, distributor contact points.
3 GENERAL
Tighten rear road spring seat bolts.
4 BODY
Check and tighten, if necessary, door hinges and striker plate securing screws.
5 LUBRICATION
Change oil in gearbox, overdrive and rear axle.
Fit new oil filter element.
Top up oil level in steering box and steering idler.
Re-pack front hub caps with grease.
6 WHEELS AND TYRES
Check wheel alignment.
7 *Check condition of all cooling system and heater hoses.
*Lubricate fuel lines and union joints for leaks.
*Adjust carburettor slow running and tune if necessary.
*Examine exhaust system for leaks or holes.
*Wax polish body and chrome plating.
*Balance front wheels.
*Check spokes of wire wheels.

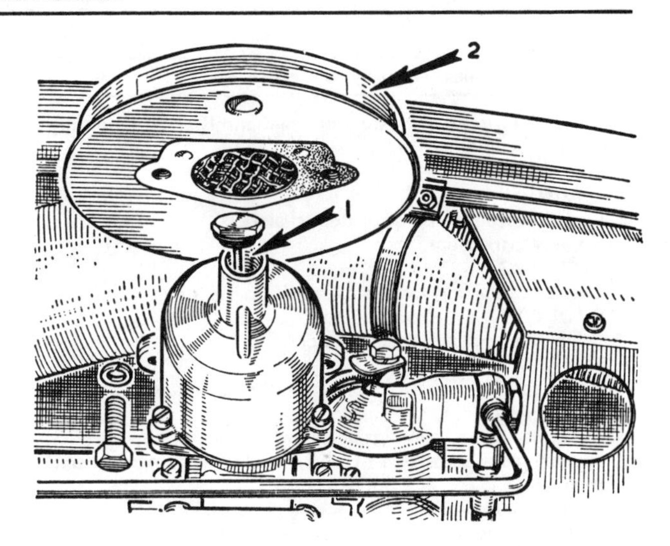

CARBURETTOR DAMPER RESERVOIR (1) AND AIR CLEANER (2)

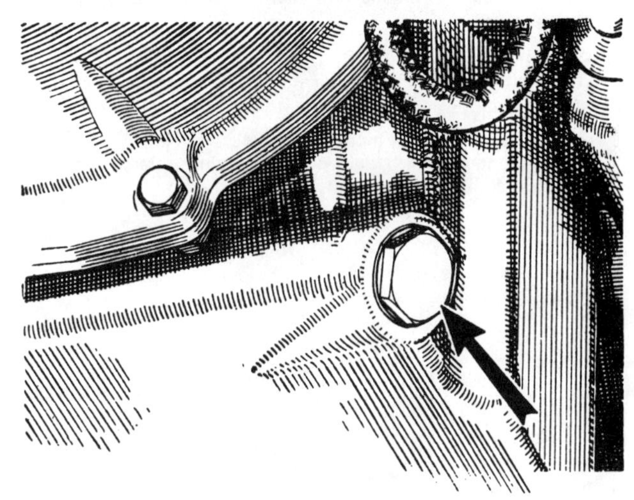

ENGINE SUMP DRAIN PLUG

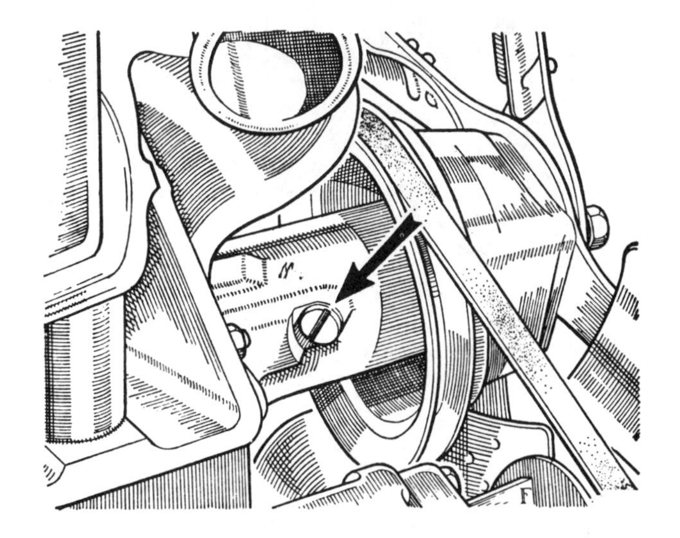

WATER PUMP LUBRICATION PLUG

Yearly (12,000 miles, whichever is sooner)

All servicing items in the monthly, bi-monthly and six monthly service, plus:

1 ENGINE
Clean out from carburettor float chambers.
Fit new spark plugs.
2 STEERING
Check steering and suspension moving parts for wear.
3 RADIATOR
Drain, flush out and refill radiator.
4 LUBRICATION
Grease sparingly speedometer and tachometer cables.
5 HEADLAMPS
Check headlamp beam setting and reset if necessary.
6 *Steam clean underside of body, engine compartment and engine.
*Inspect ignition leads for cracking or perishing.
*Fit new windscreen wiper blades.
*Fit new contact breaker points.
*Check and adjust any loose play in the cam and peg steering.
*Examine ball joints and hub bearings for wear and replace as necessary.
*Check tightness of battery earth lead on bodywork.
*Renew condenser in distributor.
*Test engine cylinder compression; if low determine cause and rectify as necessary.
7 See Section 4 of the bi-monthly service
Remove engine sump and pick-up strainer, clean and re-assemble, filling with fresh oil.

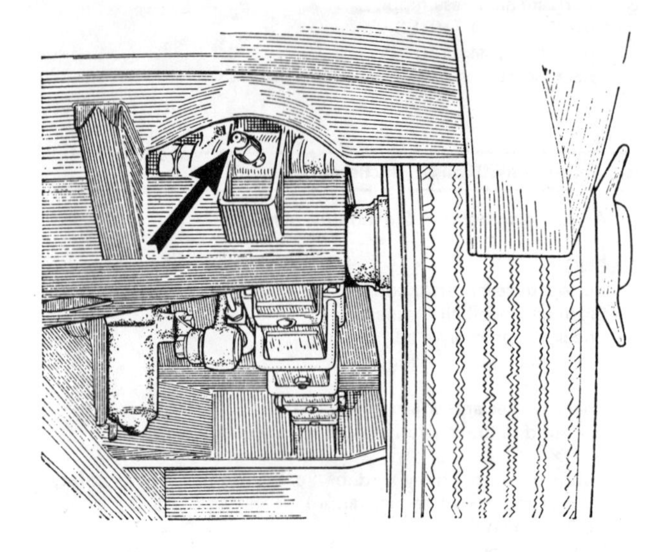

REAR SPRING REAR SHACKLE LUBRICATION POINT

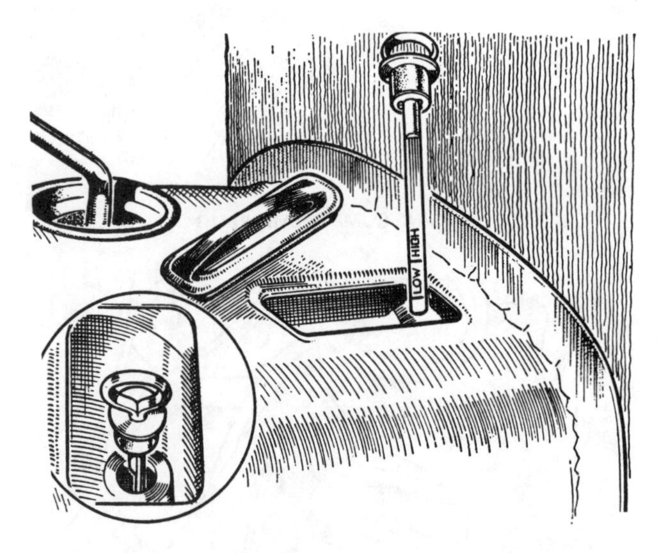

MANUAL GEARBOX COMBINED FILLER PLUG AND DIPSTICK

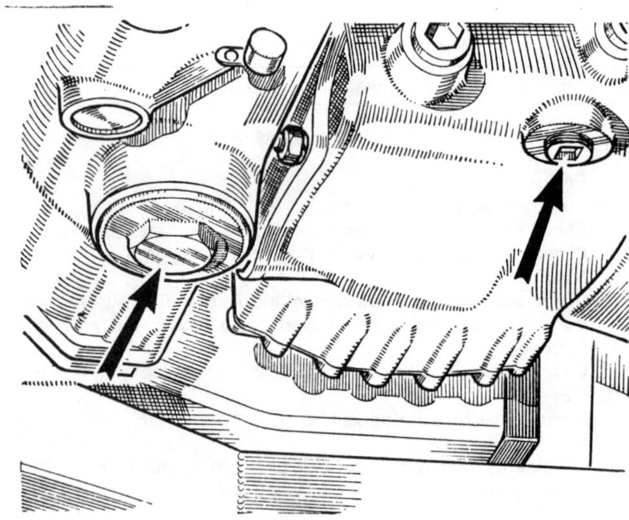

GEARBOX AND OVERDRIVE DRAIN PLUGS

Safety first!

Professional motor mechanics are trained in safe working procedures. However enthusiastic you may be about getting on with the job in hand, do take the time to ensure that your safety is not put at risk. A moment's lack of attention can result in an accident, as can failure to observe certain elementary precautions.

There will always be new ways of having accidents, and the following points do not pretend to be a comprehensive list of all dangers; they are intended rather to make you aware of the risks and to encourage a safety-conscious approach to all work you carry out on your vehicle.

Essential DOs and DON'Ts

DON'T rely on a single jack when working underneath the vehicle. Always use reliable additional means of support, such as axle stands, securely placed under a part of the vehicle that you know will not give way.
DON'T attempt to loosen or tighten high-torque nuts (e.g. wheel hub nuts) while the vehicle is on a jack; it may be pulled off.
DON'T start the engine without first ascertaining that the transmission is in neutral (or 'Park' where applicable) and the parking brake applied.
DON'T suddenly remove the filler cap from a hot cooling system – cover it with a cloth and release the pressure gradually first, or you may get scalded by escaping coolant.
DON'T attempt to drain oil until you are sure it has cooled sufficiently to avoid scalding you.
DON'T grasp any part of the engine, exhaust or catalytic converter without first ascertaining that it is sufficiently cool to avoid burning you.
DON'T allow brake fluid or antifreeze to contact vehicle paintwork.
DON'T syphon toxic liquids such as fuel, brake fluid or antifreeze by mouth, or allow them to remain on your skin.
DON'T inhale dust – it may be injurious to health (see *Asbestos* below).
DON'T allow any spilt oil or grease to remain on the floor – wipe it up straight away, before someone slips on it.
DON'T use ill-fitting spanners or other tools which may slip and cause injury.
DON'T attempt to lift a heavy component which may be beyond your capability – get assistance.
DON'T rush to finish a job, or take unverified short cuts.
DON'T allow children or animals in or around an unattended vehicle.
DO wear eye protection when using power tools such as drill, sander, bench grinder etc, and when working under the vehicle.
DO use a barrier cream on your hands prior to undertaking dirty jobs – it will protect your skin from infection as well as making the dirt easier to remove afterwards; but make sure your hands aren't left slippery. Note that long-term contact with used engine oil can be a health hazard.
DO keep loose clothing (cuffs, tie etc) and long hair well out of the way of moving mechanical parts.
DO remove rings, wristwatch etc, before working on the vehicle – especially the electrical system.
DO ensure that any lifting tackle used has a safe working load rating adequate for the job.
DO keep your work area tidy – it is only too easy to fall over articles left lying around.
DO get someone to check periodically that all is well, when working alone on the vehicle.
DO carry out work in a logical sequence and check that everything is correctly assembled and tightened afterwards.
DO remember that your vehicle's safety affects that of yourself and others. If in doubt on any point, get specialist advice.
IF, in spite of following these precautions, you are unfortunate enough to injure yourself, seek medical attention as soon as possible.

Asbestos

Certain friction, insulating, sealing, and other products – such as brake linings, brake bands, clutch linings, torque converters, gaskets, etc – contain asbestos. *Extreme care must be taken to avoid inhalation of dust from such products since it is hazardous to health.* If in doubt, assume that they *do* contain asbestos.

Fire

Remember at all times that petrol (gasoline) is highly flammable. Never smoke, or have any kind of naked flame around, when working on the vehicle. But the risk does not end there – a spark caused by an electrical short-circuit, by two metal surfaces contacting each other, by careless use of tools, or even by static electricity built up in your body under certain conditions, can ignite petrol vapour, which in a confined space is highly explosive.

Always disconnect the battery earth (ground) terminal before working on any part of the fuel or electrical system, and never risk spilling fuel on to a hot engine or exhaust.

It is recommended that a fire extinguisher of a type suitable for fuel and electrical fires is kept handy in the garage or workplace at all times. Never try to extinguish a fuel or electrical fire with water.

Fumes

Certain fumes are highly toxic and can quickly cause unconsciousness and even death if inhaled to any extent. Petrol (gasoline) vapour comes into this category, as do the vapours from certain solvents such as trichloroethylene. Any draining or pouring of such volatile fluids should be done in a well ventilated area.

When using cleaning fluids and solvents, read the instructions carefully. Never use materials from unmarked containers – they may give off poisonous vapours.

Never run the engine of a motor vehicle in an enclosed space such as a garage. Exhaust fumes contain carbon monoxide which is extremely poisonous; if you need to run the engine, always do so in the open air or at least have the rear of the vehicle outside the workplace.

If you are fortunate enough to have the use of an inspection pit, never drain or pour petrol, and never run the engine, while the vehicle is standing over it; the fumes, being heavier than air, will concentrate in the pit with possibly lethal results.

The battery

Never cause a spark, or allow a naked light, near the vehicle's battery. It will normally be giving off a certain amount of hydrogen gas, which is highly explosive.

Always disconnect the battery earth (ground) terminal before working on the fuel or electrical systems.

If possible, loosen the filler plugs or cover when charging the battery from an external source. Do not charge at an excessive rate or the battery may burst.

Take care when topping up and when carrying the battery. The acid electrolyte, even when diluted, is very corrosive and should not be allowed to contact the eyes or skin.

If you ever need to prepare electrolyte yourself, always add the acid slowly to the water, and never the other way round. Protect against splashes by wearing rubber gloves and goggles.

When jump starting a car using a booster battery, for negative earth (ground) vehicles, connect the jump leads in the following sequence: First connect one jump lead between the positive (+) terminals of the two batteries. Then connect the other jump lead first to the negative (–) terminal of the booster battery, and then to a good earthing (ground) point on the vehicle to be started, at least 18 in (45 cm) from the battery if possible. Ensure that hands and jump leads are clear of any moving parts, and that the two vehicles do not touch. Disconnect the leads in the reverse order.

Mains electricity

When using an electric power tool, inspection light etc, which works from the mains, always ensure that the appliance is correctly connected to its plug and that, where necessary, it is properly earthed (grounded). Do not use such appliances in damp conditions and, again, beware of creating a spark or applying excessive heat in the vicinity of fuel or fuel vapour.

Ignition HT voltage

A severe electric shock can result from touching certain parts of the ignition system, such as the HT leads, when the engine is running or being cranked, particularly if components are damp or the insulation is defective. Where an electronic ignition system is fitted, the HT voltage is much higher and could prove fatal.

Recommended lubricants and fluids

REFERENCE ON CHART	COMPONENT	CASTROL PRODUCT
1	ENGINE	Castrol GTX
3	GEARBOX AND OVERDRIVE	Castrol GTX
4	REAR AXLE	Castrol Hypoy
5	STEERING GEARBOX	Castrol Hypoy
6	STEERING IDLER	Castrol Hypoy
7	STEERING JOINTS	Castrol LM Grease
8	PROPELLER SHAFT	Castrol LM Grease
9	HANDBRAKE CABLE	Castrol LM Grease
10	HANDBRAKE COMPENSATOR	Castrol LM Grease
11	HANDBRAKE LINKAGE	Castrol GTX or Everyman
12	REAR SPRING SHACKLES	Castrol LM Grease
13	CARBURETTORS	Castrol GTX
14	AIR CLEANERS	Castrol GTX
15	STEERING COLUMN	Castrol GTX
19	DISTRIBUTOR	Cam - Castrol LM Grease Advance Mechanism and CB Pivot - Castrol GTX
20	DYNAMO	Castrol GTX
22	WATER PUMP	Castrol LM Grease

Additionally Castrol Everyman Oil can be used to lubricate door, boot and bonnet hinges and locks, pivots etc.

LUBRICATION CHART

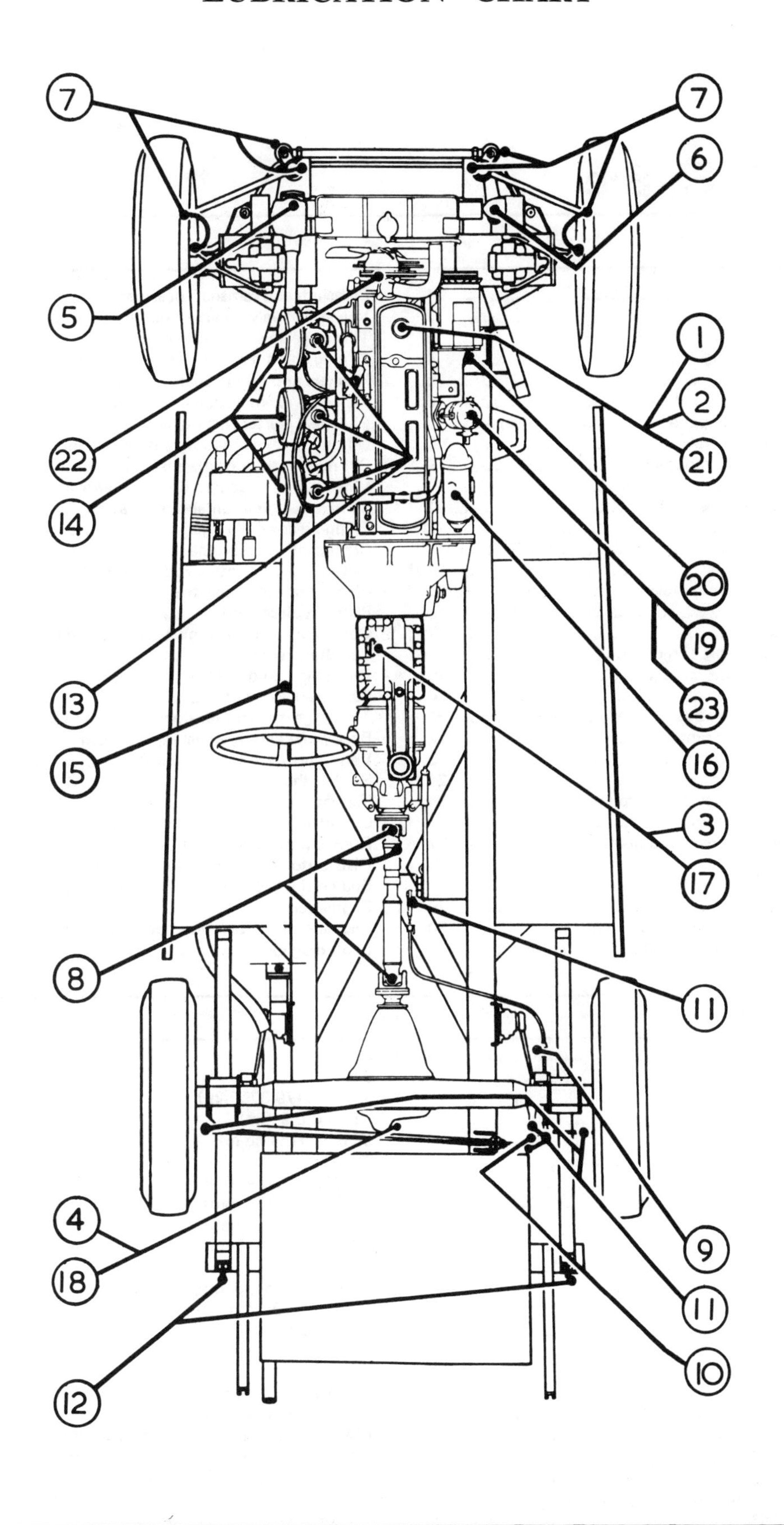

Chapter 1 Engine

Contents

Specifications

Engine type ... BN4
Number of cylinders ... 6
Bore ... 3.125 in (79.375 mm)
Stroke ... 3.5 in (88.9 mm)
Capacity ... 2639 cc (161 cu in)
Compression ratio ... 8.25:1
Bore, 1st oversize ... + 0.010 in (.254 mm)
Bore, 2nd oversize ... + 0.020 in (.508 mm)
Bore, 3rd oversize ... + 0.030 in (.762 mm)
Bore, 4th oversize ... + 0.040 in (1.016 mm)
Firing order ... 1 5 3 6 2 4
Cooling ... Thermo-siphon; pump; fan and thermostat
Torque ... 142 lb ft (19.77 kg/m) at 2,400 rpm
B M E P ... 139 lb/sq in (9.77 kg/cm^2)

Valves
Position ... Overhead, push-rod operated
Lift ... 0.3145 in (8.054 mm)
Diameter: Head: Inlet ... 1.693 to 1.683 in (42.99 to 42.75 mm)
Exhaust ... 1.420 to 1.415 in (36.07 to 35.94 mm)
Stem: Inlet ... 0.34175 to 0.34225 in (8.68 to 8.69 mm)
Exhaust ... 0.34175 to 0.34225 in (8.68 to 8.69 mm)
Stem/guide clearance: Inlet ... 0.0025 to 0.0015 in (0.063 to 0.038 mm)

Stem/guide clearance: Exhaust	0.002 to 0.001 in (0.051 to 0.025 mm)
Valve rocker clearance	0.012 in (0.3 mm) hot
Seat angle: Inlet	30°
Exhaust	45°
Seat face width: Inlet	0.091 to 0.097 in (2.31 to 2.46 mm)
Exhaust	0.198 to 0.217 in (5.03 to 5.51 mm)
Valve guides	
Length: Inlet	2.266 in (57.55 mm)
Exhaust	2.578 in (65.49 mm)
Valve springs	
Free length: Inner	1.969 in (50 mm)
Outer	2.047 in (51.99 mm)
Fitted length: Inner	1.517 in (38.53 mm) load 25.3 lb (11.476 kg)
Outer	1.607 in (40.82 mm) load 53.2 lb (24.58 kg)
Tappets	
Type	Cylindrical, spherical foot
Diameter	0.93725 in (23.81 mm)
Length	2.548 in (64.72 mm)
Rockers	
Bushes	Steel and white metal
Outside diameter (before fitting)	0.913 in (23.17 mm)
Inside diameter (reamed in position)	0.8115 to 0.9125 in (20.62 to 20.65 mm)
Clearance	0.0025 to 0.0005 in (0.063 to 0.012 mm)
Bore of arm	0.909 to 0.910 in (23.076 to 23.101 mm)
Piston	
Material	Low expansion aluminium alloy
Clearance at skirt (right angles to gudgeon pin)	0.0226 to 0.0008 in (0.066 to 0.020 mm)
Width of ring groove: Compression	0.0952 to 0.0962 in (2.410 to 2.436 mm)
Oil	0.189 to 0.190 in (4.81 to 4.83 mm)
Oversizes	+ 0.010 in + 0.020 in + 0.030 in + 0.040 in
Piston rings	
Number	3 compression (2 taper), 1 oil control
Width: Compression	0.0938 to 0.0928 in (2.383 to 2.357 mm)
Oil	0.1865 to 0.1875 in (4.737 to 4.762 mm)
Clearance in groove: Compression	0.0034 to 0.0014 in (0.086 to 0.036 mm)
Oil	0.0015 to 0.0035 in (0.038 to 0.088 mm)
Ring gap (compression and oil)	0.009 to 0.014 in (0.23 to 0.35 mm)
Gudgeon pins	
Type	Clamped in rod. Fully floating from Engine No 40501
Fit	Selective; push in piston
Diameter	0.8748 to 0.8750 in (22.215 to 22.220 mm)
Crankshaft	
Journal diameter	2.3742 to 2.3747 in (60.305 to 60.317 mm)
Crankpin diameter	2.0000 to 2.0005 in (50.80 to 50.8127 mm)
Undersizes (journals and crankpins)	– 0.010 in – 0.020 in – 0.030 in – 0.040 in (– 0.254 mm – 0.508 mm – 0.762 mm – 1.016 mm)
End-float	Taken on thrust washer at front middle (No 2) main bearing; 0.0025 to 0.0055 in (0.063 to 0.140 mm)
Thrust washer: Standard	0.091 to 0.093 in (2.315 to 2.366 mm)
+ 0.0025 in (+ 0.063 mm)	0.0935 to 0.0955 in (2.378 to 2.429 mm)
+ 0.005 in (+ 0.127 mm)	0.0960 to 0.0980 in (2.442 to 2.493 mm)
+ 0.0075 in (+ 0.190 mm)	0.0985 to 0.1005 in (2.505 to 2.556 mm)
+ 0.010 in (+ 0.254 mm)	0.1010 to 0.1030 in (2.569 to 2.620 mm)
Main bearings	
Number	4
Type	White-metalled steel shell
Length	1.495 to 1.506 in (37.973 to 38.227 mm)
Running clearance	0.0013 to 0.0028 in (0.033 to 0.071 mm)
Sizes for reground journals	0.010 in U/S; 0.020 in U/S; 0.030 in U/S; 0.040 in U/S. (0.254 mm 0.508 mm 0.762 mm 1.016 mm)
Connecting rods	
Length (centres)	6.601 to 6.605 in (167.66 to 167.76 mm)
Side clearance	0.007 to 0.004 in (0.18 to 0.10 mm)

Big-ends: Type	White-metalled steel shells
Diametrical clearance	0.005 to 0.002 in (0,0127 to 0.051 mm)
Small-end bush	0.8750 to 0.8755 in (22.225 to 22.328 mm)
Sizes for reground crankpins	0.010 in U/S; 0.020 in U/S; 0.030 in U/S; 0.040 in U/S; (0.254 mm 0.508 mm 0.762 mm 1.016 mm)

Camshaft

Journal diameters: Front	1.78875 to 1.78925 in (45.434 to 45.447 mm)
Middle front	1.76875 to 1.76925 in (44.926 to 44.939 mm)
Middle rear	1.74875 to 1.74925 in (44.418 to 44.431 mm)
Rear	1.72875 to 1.72925 in (43.910 to 43.923 mm)
End-float	Taken on thrust plate at front end: 0.003 to 0.006 in (0.076 to 0.152 mm)

Camshaft bearings

Number and type	4 thin-wall rolled bush
Outside diameter (before fitting):	
Front	1.9205 in (48.780 mm)
Middle front	1.9005 in (48.272 mm)
Middle rear	1.8805 in (47.762 mm)
Rear	1.8605 in (47.252 mm)
Inside diameter (remade in position):	
Front	1.79025 to 1.79075 in (45.472 to 45.485 mm)
Middle front	1.77025 to 1.77075 in (44.964 to 44.977 mm)
Middle rear	1.75025 to 1.75075 in (44.456 to 44.469 mm)
Rear	1.73025 to 1.73075 in (43.946 to 43.959 mm)
Clearance	0.002 to 0.001 in (0.051 to 0.025 mm)

Valve timing

Marking	Adjoining gear teeth are marked
Chain pitch and number of pitches	0.375 in (9.525 mm) 62
Rocker clearance for valve	0.0234 in (0.610 mm)
Inlet valve: Opens	5° B T D C
Closes	45° A B D C
Exhaust valve: Opens	40° B B D C
Closes	10° A T D C

Lubrication

System	Pressure
Pump type	Rotor
External filter	Full flow; Tecalemit
Oil pressures: Running	55 to 60 lb/sq in (3.9 to 4.2 kg/cm^2)
Idling	25 to 30 lb/sq in (1.758 to 2.109 kg/cm^2)
Release valve spring, free length	2.562 in (65.09 mm)
Release valve: number of coils	13
Diameter	$0.484\ in\ ^{+0.000\ in}_{-0.015\ in}$ $(12.30\ mm\ ^{+0.000\ mm}_{-0.381\ mm})$

Flywheel

Diameter	12.8125 in (325.45 mm)
Number of teeth on starter ring	106

Engine type BN6

The following are the differing details of vehicles fitted with the 6 port cylinder head engine and should therefore be used in conjunction with the preceding specification.

Torque	149 lb ft (20.7 kg m) at 3000 rpm
Compression ratio	8.7:1

Piston rings:

1st ring	Taper
Connecting rod, type of bearing	Steel backed lead indium
Standard journal diameter	2.3742 in (60.305 mm) to 2.3747 in (60.32 mm)

Exhaust valve:

Throat diameter	1.3125 in (33.34 mm)
Head diameter	1.5625 in (39.69 mm) to 1.5575 in (39.56 mm)

Inlet valve:

Throat diameter	1.5 in (38.1 mm)
Head diameter	1.750 (44.45 mm) to 1.745 in (44.32 mm)

Valve seat angles:

Inlet	45°

Exhaust	45°
Engine type	Mk I 29D Mk II 29E BJ7 29F Mk III 29K
Number of cylinders	6
Bore	3.281 in (83.34 mm)
Stroke	3.5 in (88.9 mm)
Capacity	2912 cc (177.7 cu in)
B M E P	142 lb/sq in (9.98 kg/cm^2) at 2,700 rpm
Torque	167 lb ft (23.09 kg m) at 2,700 rpm
Compression ratio	9:1
Cranking pressure	175 lb/sq in (12.25 kg cm^2)
Engine idle speed (approx)	500 rpm
Firing order	1 5 3 6 2 4
Bore, 1st oversize	+ 0.010 in (0.25 mm)
Bore, 2nd oversize	+ 0.020 in (0.50 mm)
Bore, 3rd oversize	+ 0.030 in (0.76 mm)
Bore, 4th oversize	+ 0.040 in (1.02 mm)
Valves	
Position	Overhead push-rod operated
Lift	0.3145 in (8.054 mm) Mk II and III 0.368 in (9.36 mm)
Diameter; Head: Inlet	1.750 to 1.745 in (44.45 to 44.32 mm)
Exhaust	1.5625 to 1.5575 in (39.69 to 39.56 mm)
Stem: Inlet	0.34175 to 0.34225 in (8.68 to 8.69 mm)
Exhaust	0.34175 to 0.34225 in (8.68 to 8.69 mm)
Stem/guide clearance: Inlet	0.0025 to 0.0015 in (0.063 to 0.038 mm)
Exhaust	0.002 to 0.001 in (0.051 to 0.025 mm)
Valve rocker clearance	0.012 in (0.30 mm) cold
Competition (Mk II and III)	0.015 in (0.34 mm) cold
Seat angle: Inlet	45°
Exhaust	45°
Seat face width: Inlet	0.091 to 0.097 in (2.31 to 2.46 mm)
Exhaust	0.198 to 0.217 in (5.03 to 5.51 mm)
Valve guides	
Length: Inlet	2.266 in (57.55 mm)
Exhaust	2.578 in (65.49 mm)
Fitted height above head: Inlet	1.348 in (34.23 mm)
Exhaust	1.036 in (26.32 mm)
Valve springs	
Free length: Inner	1.969 in (50 mm)
Outer	2.047 in (51.99 mm) Mk II and III 2.031 in (51.5 mm)
Fitted length: Inner	1.504 in (38.2 mm) Load 26 lb (11.8 kg)
Outer	1.594 in (40.49 mm) Load 55.7 lb (25.2 kg) Mk II and III 67.5 lb (30.63 kg)
Tappets	
Type	Cylindrical, spherical foot
Diameter	0.937 in (23.81 mm)
Length	2.548 in (64.72 mm)
Rockers	
Bushes	Steel and white metal
Outside diameter (before fitting)	0.913 in (23.17 mm)
Inside diameter (reamed in position)	0.8115 to 0.8125 in (20.62 to 20.65 mm)
Clearance	0.0025 to 0.0005 in (0.063 to 0.012 mm)
Bore of arm	0.909 to 0.910 (23.076 to 23.101 mm)
Pistons	
Material	Low expansion aluminium alloy
Clearance at skirt: Top	0.0032 to 0.0043 in (0.081 to 0.109 mm)
Bottom	0.0010 to 0.0016 in (0.025 to 0.040 mm)
Width of ring groove: Compression	0.1417 to 0.1482 in (3.599 to 3.764 mm)
Oil	0.1567 to 0.1632 in (3.98 to 4.137 mm)
Oversizes	+ 0.010 in + 0.020 in + 0.030 in + 0.040 in (+ 0.25 mm + 0.50 mm + 0.76 mm + 1.02 mm)
Piston rings	
Number	3 compression (2 taper), 1 oil control
Width: Compression and oil	3.2055 to 3.3832 mm
Clearance in groove: Compression and oil	0.0015 to 0.0035 in (0.038 to 0.088 mm)
Ring gap: Compression and oil	0.013 to 0.018 in (0.33 to 0.46 mm)

Gudgeon pins

Type	Fully floating
Fit	Selective; push in piston
Diameter	0.8748 to 0.8750 in (22.219 to 22.225 mm)
Location	Circlips in piston

Crankshaft

Journal diameter	2.3742 to 2.3747 in (60.305 to 60.317 mm)
Crankpin diameter	2.0000 to 2.0005 in (50.80 to 50.813 mm)
Undersizes (journals and crankpins)	- 0.010 in - 0.020 in (- 0.254 mm - 0.254 mm - 0.508 mm)
End float	Taken on thrust washer at front middle (No 2) main bearing; 0.0025 to 0.0055 in (0.063 to 0.140 mm)
Thrust washer: Standard	0.091 to 0.093 in (2.315 to 2.366 mm)
+ 0.0025 in (+ 0.063 mm)	0.0935 to 0.0955 in (2.378 to 2.429 mm)
+ 0.005 in (+ 0.127 mm)	0.0960 to 0.0980 in (2.442 to 2.493 mm)
+ 0.0075 in (+ 0.190 mm)	0.0985 to 0.1005 in (2.505 to 2.556 mm)
+ 0.010 in (+ 0.254 mm)	0.1010 to 0.1030 in (2.569 to 2.620 mm)

Main bearings

Number	4
Type	White metalled steel shell, lead indium plated
Length	1.495 to 1.505 in (37.973 to 38.227 mm)
Running clearance	0.0013 to 0.0028 in (0.033 to 0.071 mm)
Sizes for reground journals	0.010 in U/S; 0.020 in U/S; (0.254 mm 0.508mm U/S)

Connecting rods

Length (centres)	to 6.605 in (167.66 to 167.76 mm)
Side clearance	0.005 to 0.009 in (0.13 to 0.23 mm)
Big end bearings: Type	White metalled steel shells, lead indium plated
Diametrical clearance	0.002 to 0.0035 in (0.051 to 0.089 mm)
Small end bush	0.8749 to 0.8751 in (22.22 to 22.23 mm)
Sizes for reground crankpins	0.010 in U/S; 0.020 in U/S (0.254 mm 0.508 mm U/S)

Camshaft

Journal diameters: Front	1.78875 to 1.78925 in (45.434 to 45.447 mm)
Middle front	1.76875 to 1.76925 in (44.926 to 44.939 mm)
Journal diameters: Middle rear	1.74875 to 1.74925 in (44.418 to 44.431 mm)
Rear	1.72875 to 1.72925 in (43.910 to 43.923 mm)
End float	Taken on thrust plate at front end: 0.003 to 0.006 in (0.076 to 0.152 mm)

Camshaft bearings

Number and type	4 thin-wall rolled bush
Outside diameter (before fitting):	
Front	1.9205 in (48.780 mm)
Middle front	1.9005 in (48.272 mm)
Middle rear	1.8805 in (47.762 mm)
Rear	1.8605 in (47.252 mm)
Inside diameter (reamed in position):	
Front	1.79025 to 1.79075 in (45.472 to 45.485 mm)
Middle front	1.77025 to 1.77075 in (44.964 to 44.977 mm)
Middle rear	1.75025 to 1.75075 in (44.456 to 44.469 mm)
Rear	1.73025 to 1.73075 in (43.946 to 43.959 mm)
Clearance	0.002 to 0.001 in (0.051 to 0.025 mm)

Valve timing

Marking	Adjoining gear teeth are marked
Chain pitch and number of pitches	0.375 in (9.525 mm) 62
Rocker clearance for valve timing	0.030 in (0.76 mm)

	Mk I	Mk II To 29F/2285	Mk II BJ7 From 29F/2286	Mk III
Inlet valve: Opens	5° B T D C	5° B T D C	10° B T D C	16° B T D C
Closes	45° A B D C	45° A B D C	50° A B D C	56° A B D C
Exhaust valve: Opens	40° B B D C	51° B B D C	45° B B D C	51° B B D C
Closes	10° A T D C	21° A T D C	15° A T D C	21° A T D C

Lubrication

Capacity (including filter)	12¾ pints 15.3 U.S. pints (7.25 litres)
System	Pressure
Pump type	Gear
External filter	Full flow; Tecalemit or Purolator
Oil pressure: Running	50 lb/sq in (3.52 kg/cm^2) at 40 mph
Idling	20 lb/sq in (1.4 kg/cm^2) at 600 rpm

Release valve spring, free length	2.687 in (68.26 mm)
Release valve: number of coils	13
Diameter	0.484 in $^{+\ 0.000\ in}_{-\ 0.015\ in}$ (12.30 mm $^{+\ 0.000\ mm}_{-\ 0.318\ mm}$)
Flywheel	
Diameter	12.8125 in (325.45 mm)
Number of teeth on starter ring	106
Torque wrench settings	
Cylinder head studs	400 lb in (4.6 kg.m)
Cylinder head nuts	900 lb in (10.4 kg.m)
Main bearing nuts	900 lb in (10.4 kg.m)
Connecting rod set screws	600 lb in (6.91 kg.m)
Front cover screws	7/16 in (11.11 mm) less than 150 lb in (1.73 kg.m) ½ in (12.70 mm) less than 150 lb in (1.73 kg.m)
Front mounting plate screws	200 lb in (2.30 kg.m)
Rear mounting plate screws	600 lb in (6.91 kg.m)
Flywheel bolts	600 lb in (6.91 kg.m)
Rocker shaft bracket nuts	300/324 lb in (3.45/3.72 kg.m)
Bell housing bolts	420 lb in (4.83 kg.m)
Clutch to flywheel	300 lb in (3.45 kg.m)

1 General description

The engines are basically identical with minor differences such as increased engine cubic capacity and the fitting of a crankshaft vibration damper on the later produced engines.

The engine is a six cylinder in line overhead valve design, water cooled and mounted in unit with the clutch and gearbox. The cylinder block and cylinder head are of cast iron.

The overhead valves are mounted vertically in steel valve guides within the cylinder head. Double valve springs are fitted and oil seals are used on the valve stems to control stem lubrication. The pushrods operate the tappets (cam followers) which are lifted by a roller chain driven camshaft that is positioned on the right hand side of the engine. Valve rockers are in contact with the upper end of the camshaft and are adjustable so as to allow correct clearance setting.

The front mounted roller chain is automatically tensioned by a spring loaded adjuster incorporating a synthetic rubber slipper which is lubricated through its centre.

On later engines coinciding with the introduction of the 3000 Mk II power unit, the timing chain vibration damper was fitted in addition to the automatic chain tensioner already in use. The damper comprises an angle bracket bolted with a single setscrew to the cylinder block and located on the engine front plate by a dowel. An oil resistant rubber pad bonded to the bracket maintains light rubbing contact with the timing chain and dampens chain vibration under light running conditions.

To assist further smooth running of the engine a vibration damper incorporated in the fan belt pulley was fitted to the end of the crankshaft on later produced engines.

The oil pump and distributor are driven by a skew gear machined on the camshaft. The oil pump fitted to early engines was of the Hobourn Eaton rotary vane type whilst on later engines a gear type pump was fitted.

The camshaft runs in four white metal steel backed bearings, the end thrust being taken by a plate which is screwed to the front end of the cylinder block.

The pistons are made of tin plated aluminium alloy and are of the solid skirt design. Four piston rings are used, three compression and one single slotted oil control ring. On early engines the gudgeon pin was clamped in the small end of the connecting rod by a screw and spring washer whereas on later produced engines a fully floating gudgeon pin was fitted. In both cases they are offset in the piston towards the thrust side. Renewable steel backed shell bearings are fitted to the connecting rod big ends.

The forged steel counterbalance crankshaft runs in four renewable steel backed bearings with the end thrust taken by split thrust washers located on number 2 main bearing housing.

A centrifugal impeller type water pump with a fan mounted on the front is located at the front of the cylinder block.

Attached to the end of the crankshaft by four bolts and one dowel is the flywheel to which is bolted the clutch. Attached to the engine backplate is the gearbox bellhousing.

On later engines an entirely new cylinder head together with a detachable induction manifold was fitted. New features of this modified cylinder head included redesigned combustion chambers, larger inlet and exhaust valve head diameters and reshaped inlet and exhaust ports to provide an even more efficient gas flow. At the same time the engine compression ratio was raised from 8.25 : 1 to 8.71 : 1. Full details of the dimensional and power changes brought about by these modifications will be found in the Specifications at the beginning of this Chapter.

2 Routine maintenance

1 Daily remove the dipstick and check the engine oil level which should be at the 'MAX' mark. Top up the oil in the sump with Castrol GTX. On no account allow the oil to fall below the 'MIN' mark on the dipstick.

2 See Routine Maintenance for frequency then run the engine until it is at its normal operating temperature and place a container with a capacity of 13 pints under the drain plug in the sump. Undo and remove the drain plug and allow the oil to drain out for 10 minutes. While the oil is draining out wash the oil filler cap in petrol, shake dry and refit. DO NOT oil.

3 Clean the drain plug, ensure that the washer is in place and refit the plug to the sump. Tighten the plug fully. Refill the sump with 11¾ pints of Castrol GTX. Note that on engines produced after 1964 the oil capacity is 11 pints.

4 Every 6000 miles the oil filter element should be renewed. Full details of this operation will be found in Section 24 of this Chapter.

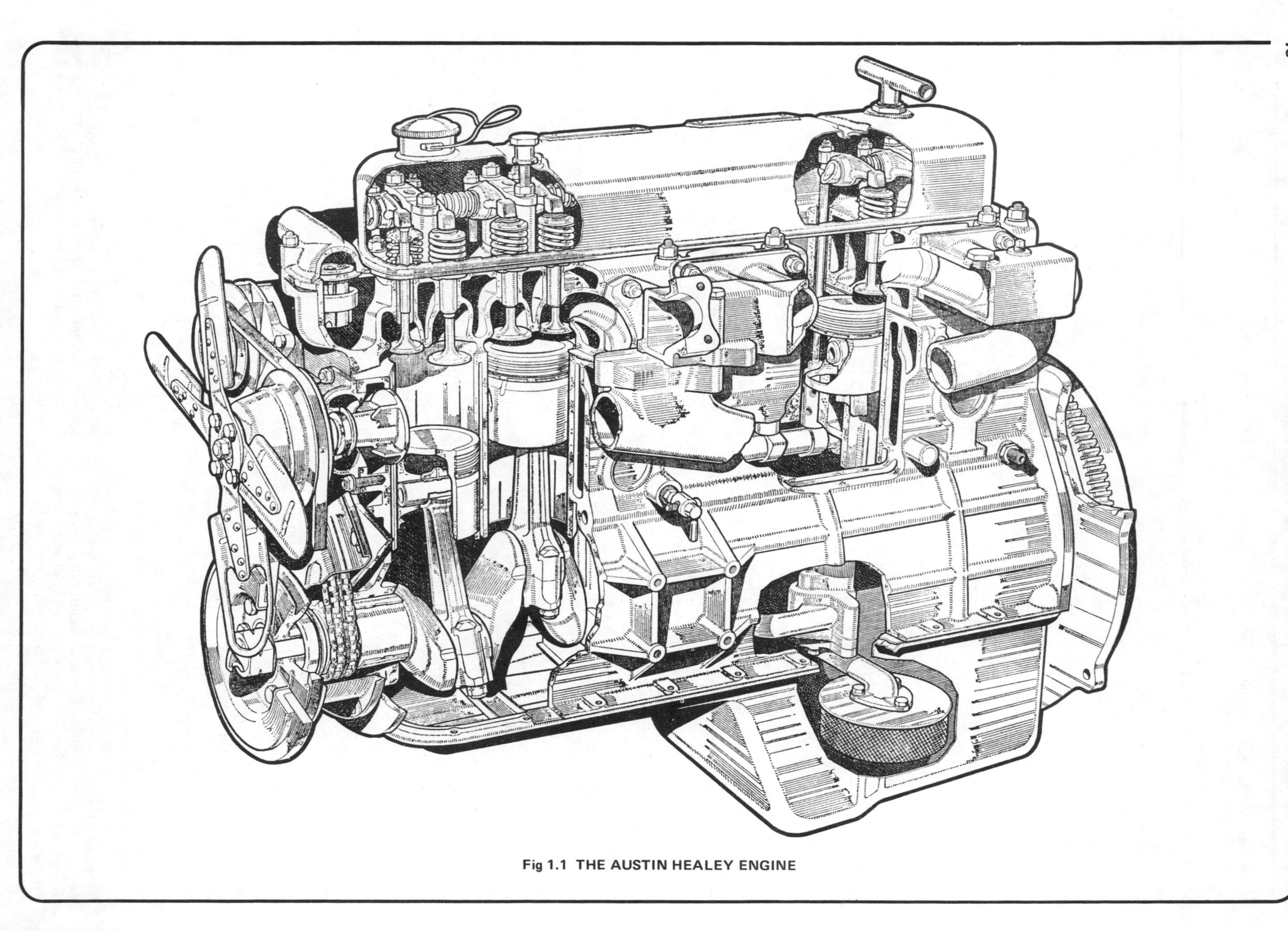

Fig 1.1 THE AUSTIN HEALEY ENGINE

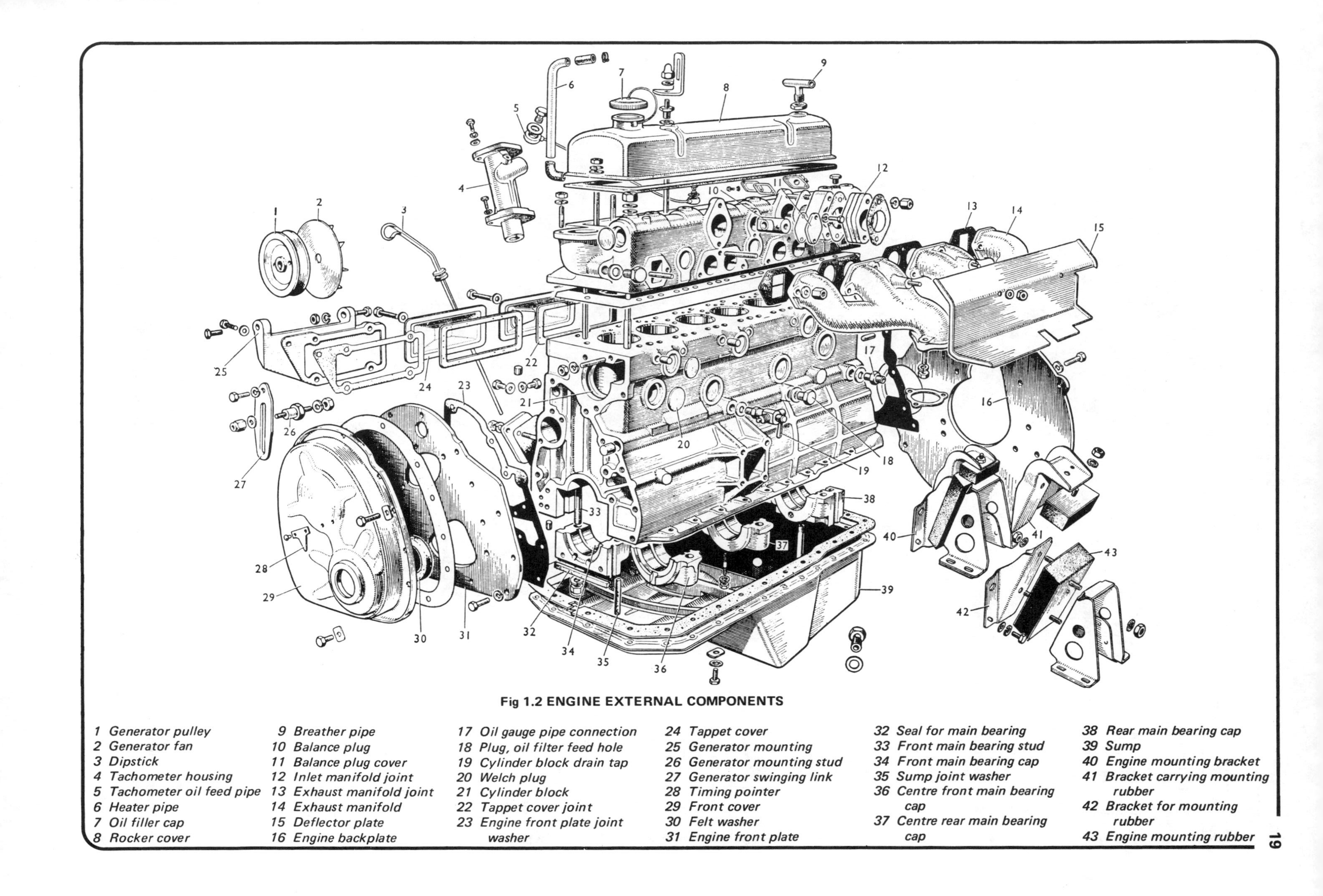

Fig 1.2 ENGINE EXTERNAL COMPONENTS

1 Generator pulley
2 Generator fan
3 Dipstick
4 Tachometer housing
5 Tachometer oil feed pipe
6 Heater pipe
7 Oil filler cap
8 Rocker cover
9 Breather pipe
10 Balance plug
11 Balance plug cover
12 Inlet manifold joint
13 Exhaust manifold joint
14 Exhaust manifold
15 Deflector plate
16 Engine backplate
17 Oil gauge pipe connection
18 Plug, oil filter feed hole
19 Cylinder block drain tap
20 Welch plug
21 Cylinder block
22 Tappet cover joint
23 Engine front plate joint washer
24 Tappet cover
25 Generator mounting
26 Generator mounting stud
27 Generator swinging link
28 Timing pointer
29 Front cover
30 Felt washer
31 Engine front plate
32 Seal for main bearing
33 Front main bearing stud
34 Front main bearing cap
35 Sump joint washer
36 Centre front main bearing cap
37 Centre rear main bearing cap
38 Rear main bearing cap
39 Sump
40 Engine mounting bracket
41 Bracket carrying mounting rubber
42 Bracket for mounting rubber
43 Engine mounting rubber

3 Major operations with engine in place

The following major operations may be carried out to the engine with it in the car:

1 Removal and replacement of the cylinder head
2 Removal and replacement of the sump
3 Removal and replacement of the big end bearings
4 Removal and replacement of the pistons and connecting rods
5 Removal and replacement of the timing chain and gears and the timing chain cover oil seal
6 Removal of the camshaft
7 Removal and replacement of the oil pump

4 Major operations with the engine removed

The following major operations can be carried out with the engine out and on the bench or floor:

1 Removal and replacement of the main bearings
2 Removal and replacement of the crankshaft
3 Removal and replacement of the flywheel

5 Engine - removal

1 There are two basic methods of removing the engine and in each case it has to be lifted up through the engine compartment. The first method is to remove the cylinder head, separate the engine from the gearbox and remove the part assembled engine. This is the method recommended for it is more manageable in this form for removal and refitting using the equipment that will probably be available or easily borrowed.
2 The second method is to remove the complete engine and gearbox as a unit. Substantial lifting equipment is needed as the overall weight is approximately 728 lbs. Both methods are described.

6 Engine - removal with gearbox

1 Switch off the battery master switch located within the boot.
2 Open the bonnet and secure in the open position using the bonnet stay.
3 Mark the outline of the hinges relative to the hinge bracket on the bonnet so that the bonnet hinge may be refitted in its original position.
4 Place an old blanket under the top edge of the lid and spread over the rear of the wings and under the windscreen to avoid scratching the paintwork.
5 A second person is needed to assist in taking the weight of the bonnet and lifting over the engine compartment.
6 Undo the bonnet securing bolts from each bracket (photo).
7 Carefully lift the bonnet over the front of the car (photo). Place the bonnet in a safe place.
8 Remove the filler cap from the radiator by turning anti-clockwise.
9 If antifreeze is in the radiator drain it into a clean bowl having a capacity of at least 18 pints (19 pints with heater) for re-use.
10 Open the two drain taps. When viewed from the front, the radiator drain tap is on the bottom right hand side of the radiator and the engine drain tap is halfway down the right hand side of the cylinder block under the exhaust manifold. These taps are shown in Figs 2.3A and 2.3B. A short length of rubber tubing over the radiator tap nozzle will assist draining the coolant without splashing.
11 Slacken the top radiator hose clip at the thermostat housing end of the hose and carefully ease the hose from the elbow (photo).

6.6

6.7

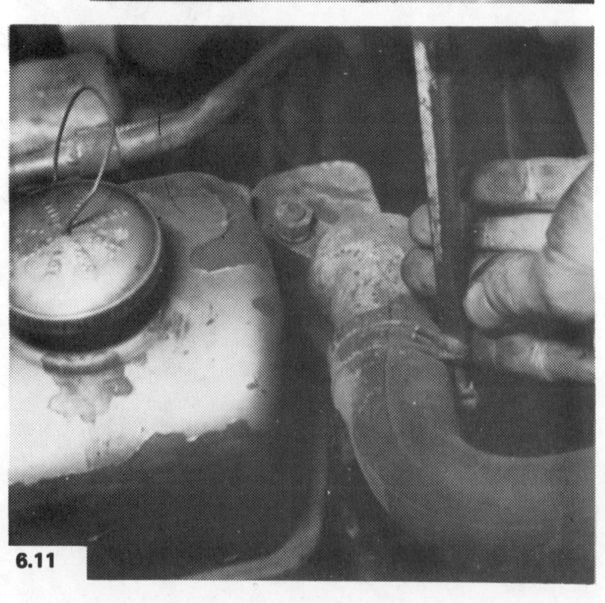
6.11

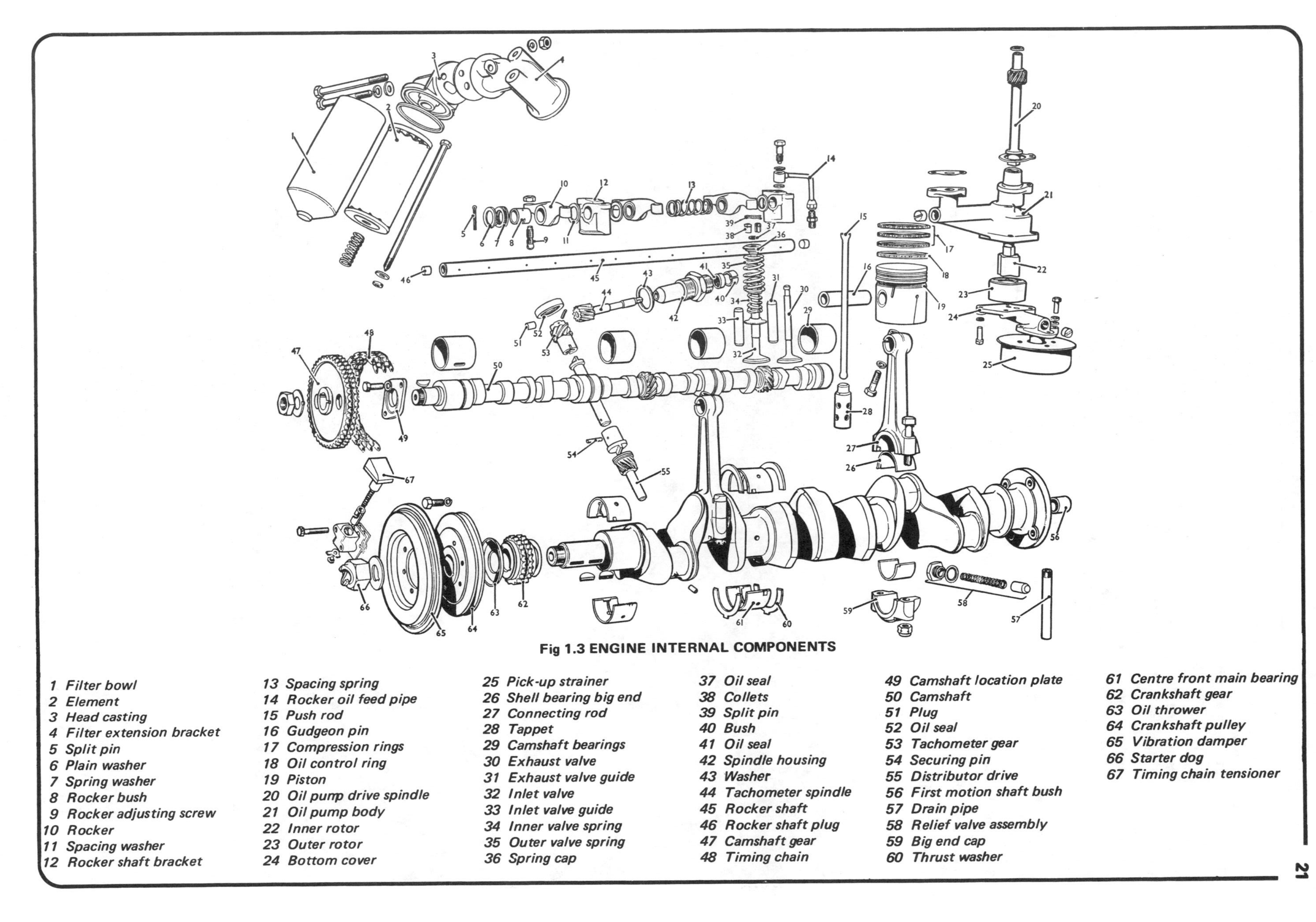

Fig 1.3 ENGINE INTERNAL COMPONENTS

1 Filter bowl
2 Element
3 Head casting
4 Filter extension bracket
5 Split pin
6 Plain washer
7 Spring washer
8 Rocker bush
9 Rocker adjusting screw
10 Rocker
11 Spacing washer
12 Rocker shaft bracket
13 Spacing spring
14 Rocker oil feed pipe
15 Push rod
16 Gudgeon pin
17 Compression rings
18 Oil control ring
19 Piston
20 Oil pump drive spindle
21 Oil pump body
22 Inner rotor
23 Outer rotor
24 Bottom cover
25 Pick-up strainer
26 Shell bearing big end
27 Connecting rod
28 Tappet
29 Camshaft bearings
30 Exhaust valve
31 Exhaust valve guide
32 Inlet valve
33 Inlet valve guide
34 Inner valve spring
35 Outer valve spring
36 Spring cap
37 Oil seal
38 Collets
39 Split pin
40 Bush
41 Oil seal
42 Spindle housing
43 Washer
44 Tachometer spindle
45 Rocker shaft
46 Rocker shaft plug
47 Camshaft gear
48 Timing chain
49 Camshaft location plate
50 Camshaft
51 Plug
52 Oil seal
53 Tachometer gear
54 Securing pin
55 Distributor drive
56 First motion shaft bush
57 Drain pipe
58 Relief valve assembly
59 Big end cap
60 Thrust washer
61 Centre front main bearing
62 Crankshaft gear
63 Oil thrower
64 Crankshaft pulley
65 Vibration damper
66 Starter dog
67 Timing chain tensioner

12 Slacken the bottom radiator hose clip at the water pump elbow end and carefully ease the hose from the elbow (photo).
13 If the car is fitted with a radio this should next be removed. The method of removal will depend on the type fitted and method of attachment.
14 Unclip the carpeting and felt from the short gearbox tunnel and remove from the inside of the car (photo).
15 Unscrew and remove the twelve setscrews and plain washers that secure the tunnel to the body of the car (photo).
16 Undo and remove the six setscrews (three on each side) which secure the carpet covered bulkhead and remove the bulkhead.
17 Unscrew the speedometer cable knurled nut from the rear of the instrument and withdraw the cable (photo).
18 Slacken the gear change lever knob locknut and unscrew the knob (photo).
19 Ease the gear change lever grommet for the moulding on the tunnel and draw the grommet over the gear change lever (photo).
20 Lift away the tunnel by lifting it up over the gear change lever (photo).
21 Disconnect the thermometer element for the radiator heater tank.
22 Undo and remove the six nuts and spring washers that secure the radiator to the mounting flanges (photo).
23 Slacken the radiator top and bottom hoses from the radiator and remove the two hoses. Carefully lift the radiator up and out of the engine compartment (photo).
24 Using a pair of circlip pliers remove the circlip retaining the gear change lever spring to the turret (photo). Lift away the washer and spring over the top of the gear change lever. Undo and remove the gear change lever securing bolt located in the side of the turret (photo). Lift away the gear change lever (photo).
25 Make a note of the electrical cables to the switch in the side of the gearbox turret and detach the two cables (photo).
26 Slacken the heater hose clip securing the rubber hose to the metal pipe on the left hand side of the engine and pull off the hose (photo).
27 Unscrew the speedometer drive cable knurled nut from the drive on the side of the gearbox or overdrive unit (photo).
28 Slacken the clip from the vacuum hose on the induction manifold and withdraw the hose (photo).

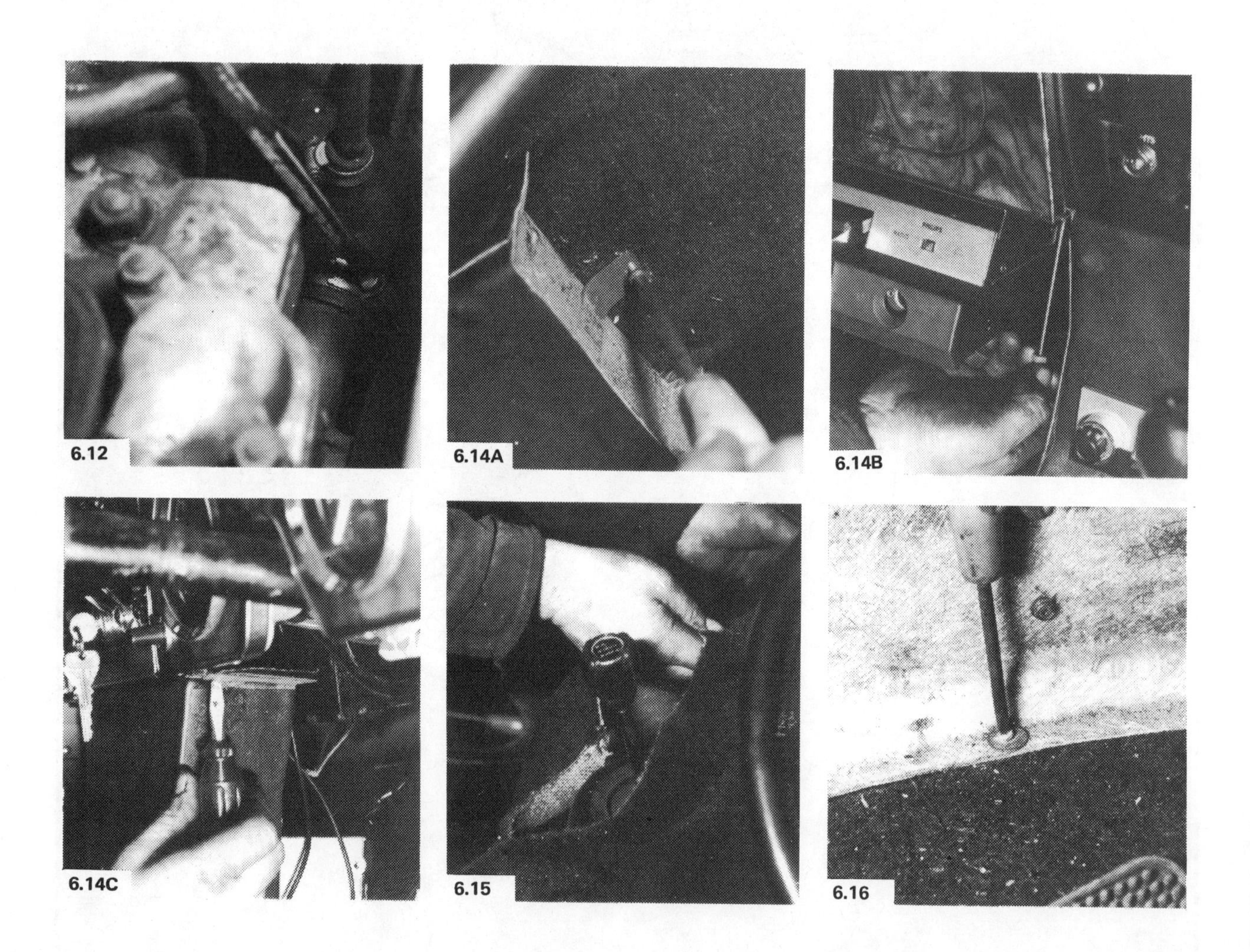

6.12 6.14A 6.14B

6.14C 6.15 6.16

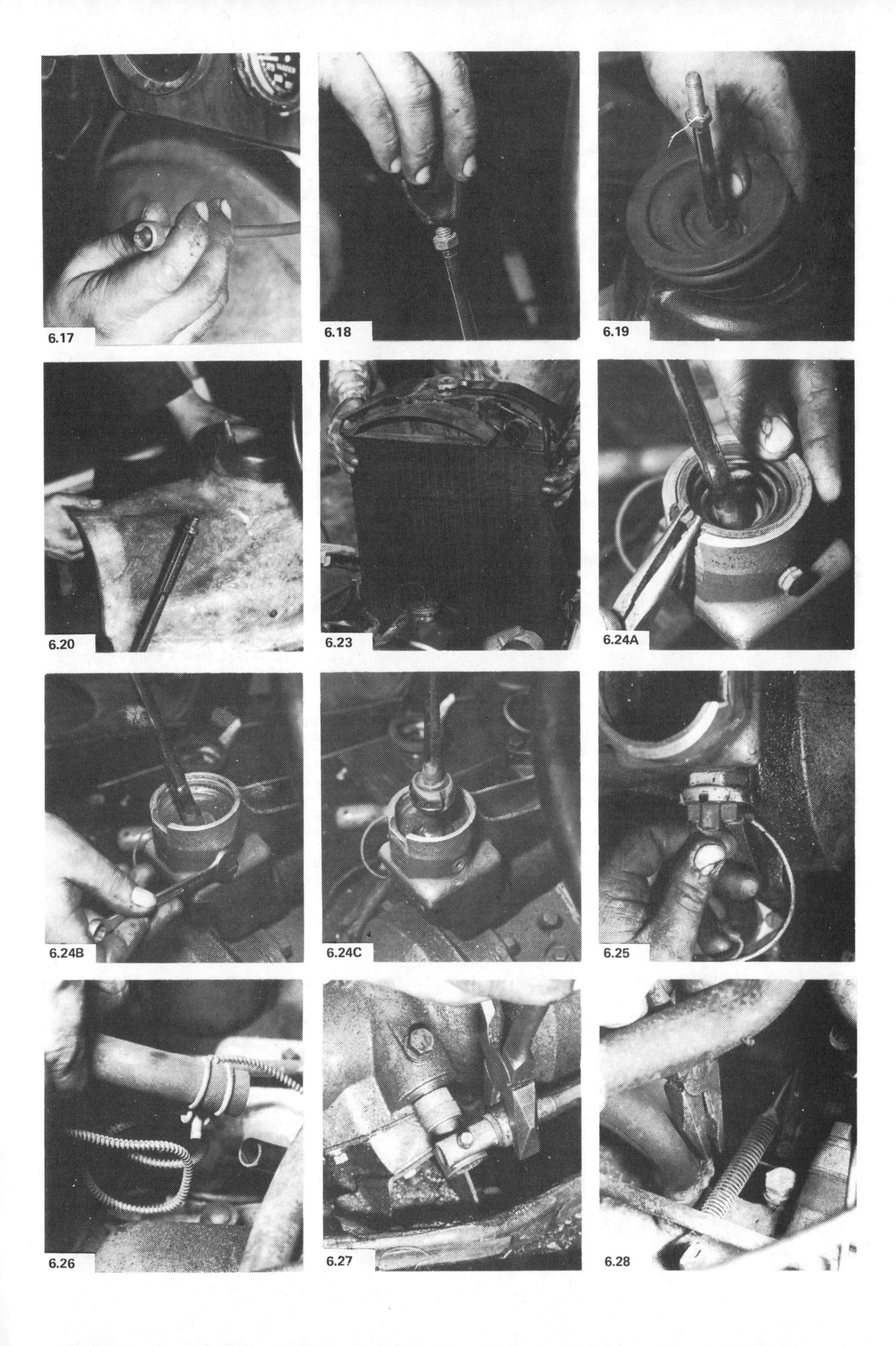
6.17
6.18
6.19
6.20
6.23
6.24A
6.24B
6.24C
6.25
6.26
6.27
6.28

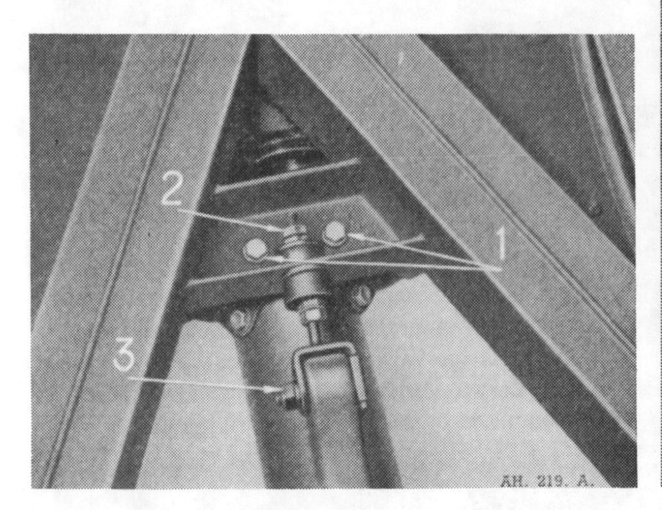

Fig 1.4 GEARBOX LOWER SECURING POINTS

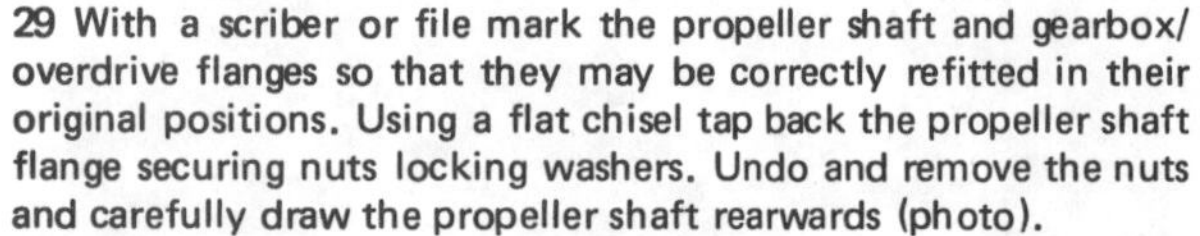
1 Bolts *3 Securing pin*
2 Stabliiser adjusting nut

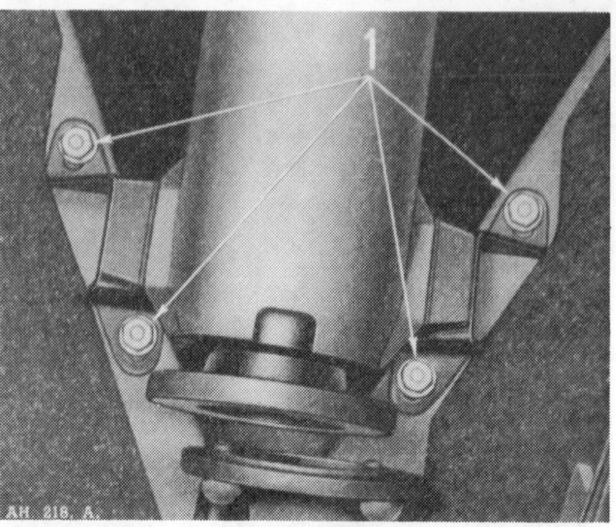

Fig 1.5 GEARBOX REAR UPPER SECURING BRACKET

1 Bolts

29 With a scriber or file mark the propeller shaft and gearbox/ overdrive flanges so that they may be correctly refitted in their original positions. Using a flat chisel tap back the propeller shaft flange securing nuts locking washers. Undo and remove the nuts and carefully draw the propeller shaft rearwards (photo).
30 Place a jack or support stand under the gearbox or overdrive unit. Unscrew and remove the two bolts with plain and spring washers securing the left hand gearbox mounting to the chassis frame (photo). See also Fig 1.4 and 1.5. Unscrew and remove the two bolts with plain and spring washers securing the right hand gearbox mounting to the chassis frame (photo).
31 Undo and remove the two nuts with spring washers securing each air cleaner to the carburettor air intake flange (photo) and lift away each air cleaner.
32 Slacken the hose clip securing the rubber hose to the union in the right hand side of the cylinder head and pull off the rubber hose (photo).
33 Note the electric cables to the ignition coil so that they may be refitted the correct way round and disconnect the two cables (photo).
34 Mark each spark plug lead in turn so that they may be refitted to the correct spark plugs and disconnect the HT leads from the spark plugs (photo).
35 Unscrew the knurled moulded nut securing the HT lead to the centre of the ignition coil. Spring back the two distributor cap retaining clips and lift away the distributor cap and HT leads.
36 Note the electric cables on the rear of the dynamo and disconnect the two cables (photo). Detach the LT cable from the side of the distributor and place the complete harness on one side.
37 Slacken the dynamo adjustment link nuts and bolts (photo). Then slacken the dynamo mounting bracket nuts and bolts (photo). Push the dynamo towards the cylinder block.
38 Detach the throttle linkage ball joint from the relay bracket in the side of the cylinder head by springing back the clip and pulling the cup from the ball (photo).
39 Detach the throttle linkage bracket from the side of the cylinder head behind the carburettor installation (photo). Lift away the throttle operating rod from the rear of the carburettor installation (photo).
40 Undo the two crosshead bolts securing the cable crosshead bracket to the bulkhead (photo). Then undo the crosshead bolt

6.29

6.30A

6.30B

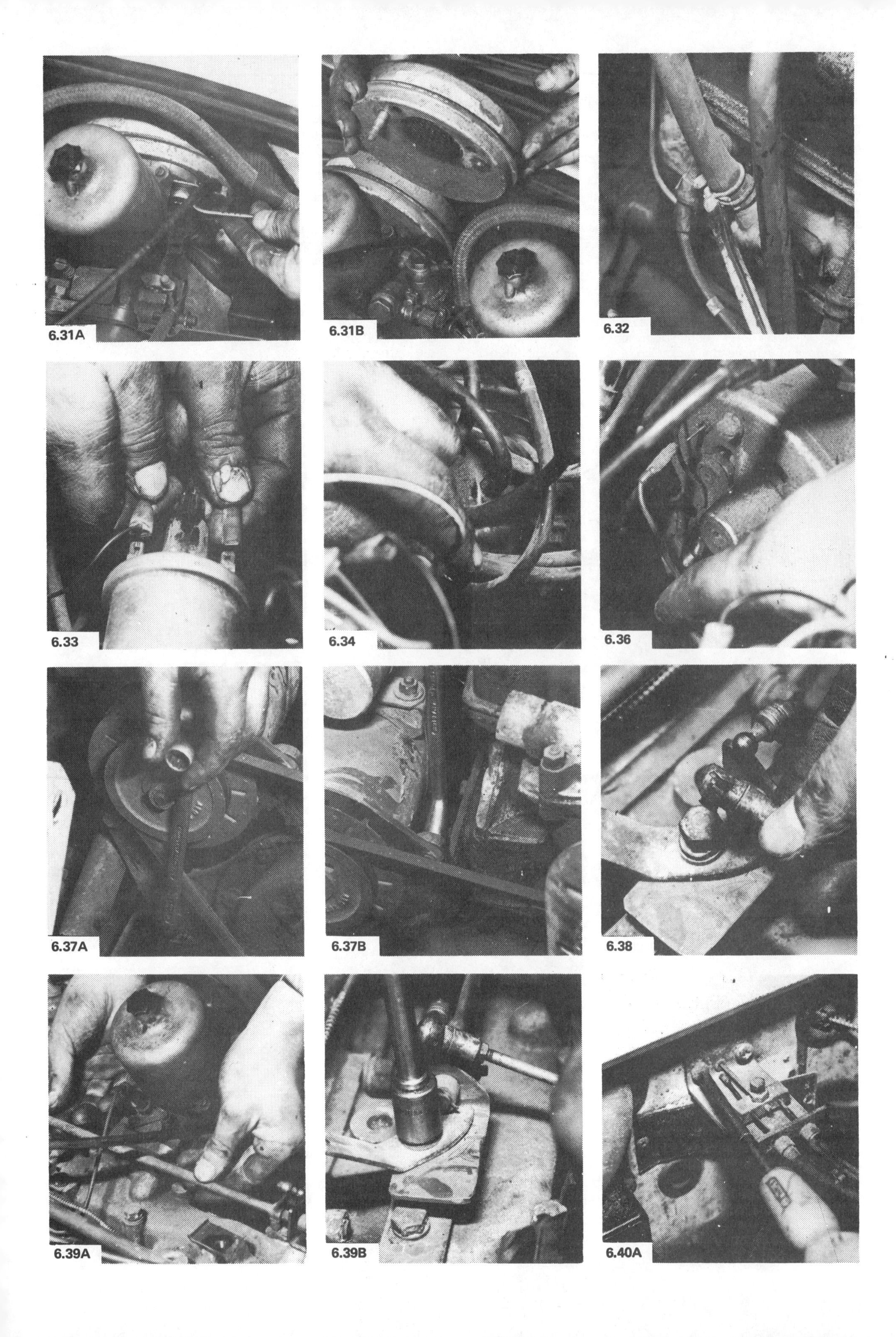

6.31A 6.31B 6.32
6.33 6.34 6.36
6.37A 6.37B 6.38
6.39A 6.39B 6.40A

and draw the bracket away from the bulkhead and single operating cable (photo).

41 If the temperature indicator is connected to the underside of the thermostat housing undo the securing union nut and draw away the indicator and cable from the thermostat housing (photo).

42 Slacken the fuel feed pipe hose to the carburettor float chamber union and pull off the flexible hose (photo). Plug the end with a pencil to stop petrol syphoning out.

43 This photo shows two of the four engine mounting bracket to chassis frame securing bolts. See also Fig 1.6.

44 Undo and remove the eight engine mounting bracket to chassis frame securing bolts (photo).

45 Position the saddle of a hydraulic jack under the front crossmember (photo).

46 Slacken the breather pipe flexible hose clip from the T union at the rear rocker cover securing point (photo).

47 Undo and remove the rocker cover securing bolts (photo) and lift the rocker cover and gasket from the top of the cylinder head (photo). Undo the nut on the front and rear rocker cover securing studs (photo).

48 Fit lifting brackets to the rocker cover securing studs and secure with the two nuts and spring washers previously removed (photo).

49 Place a metal bar with a piece of tube about 6 inches long through the two mounting bracket holes in preparation for slinging the complete power unit (photo).

50 Undo the four bolts and spring washers securing the fan blades to the pulley hub (photo) and lift away the fan blades (photo).

51 Undo the external oil filter centre bolt and lift away the filter housing and element. Be prepared to catch a little oil which may issue from the housing.

52 Remove the heavy duty cable to the starter motor at the solenoid (photo).

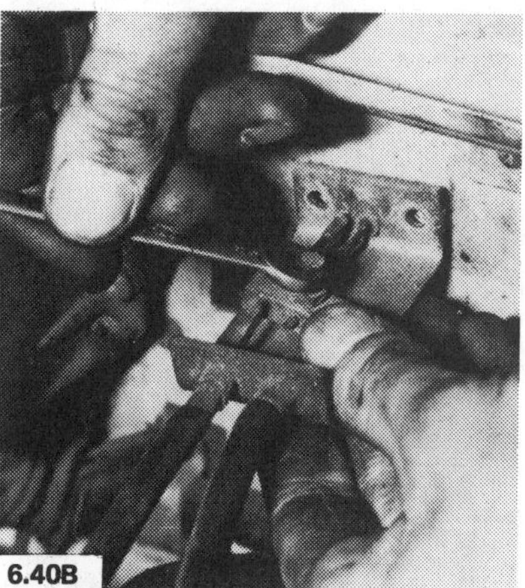

6.40B

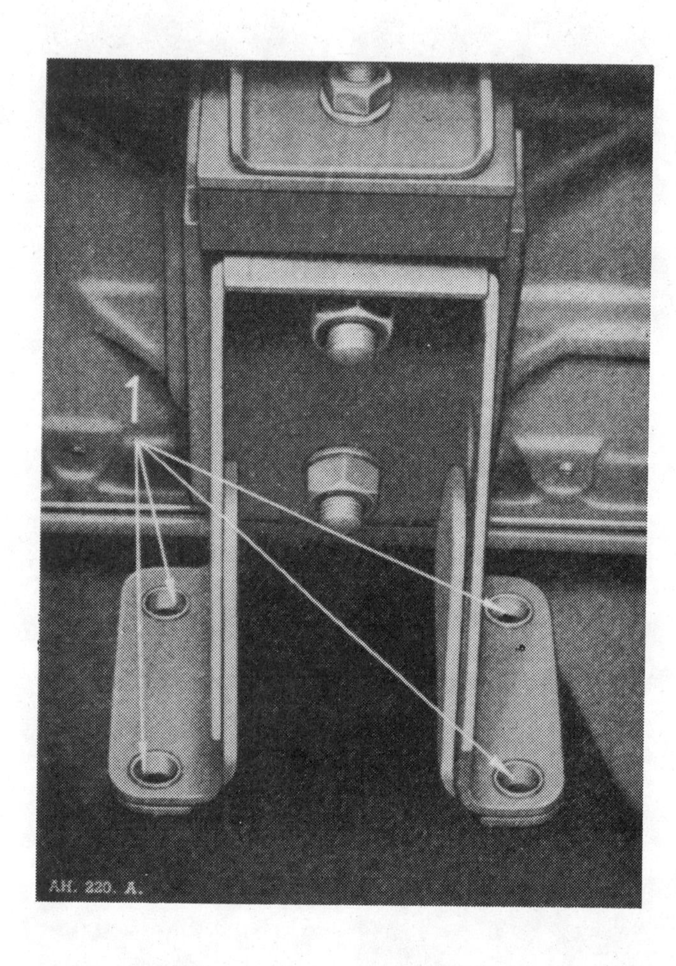

Fig 1.6 ENGINE LEFT HAND FRONT MOUNTING BRACKET

1 Bolt holes

6.41

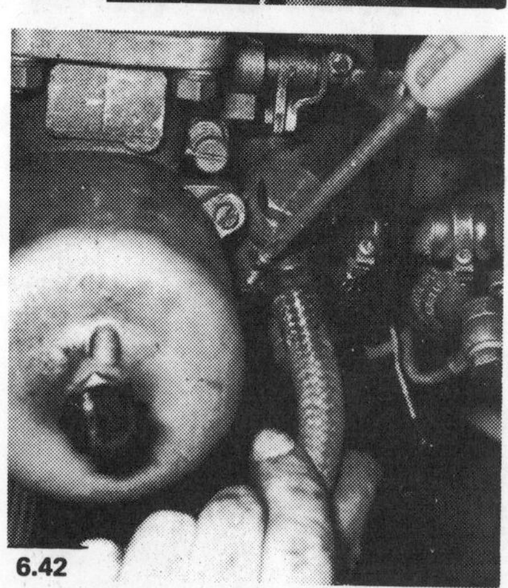

6.42

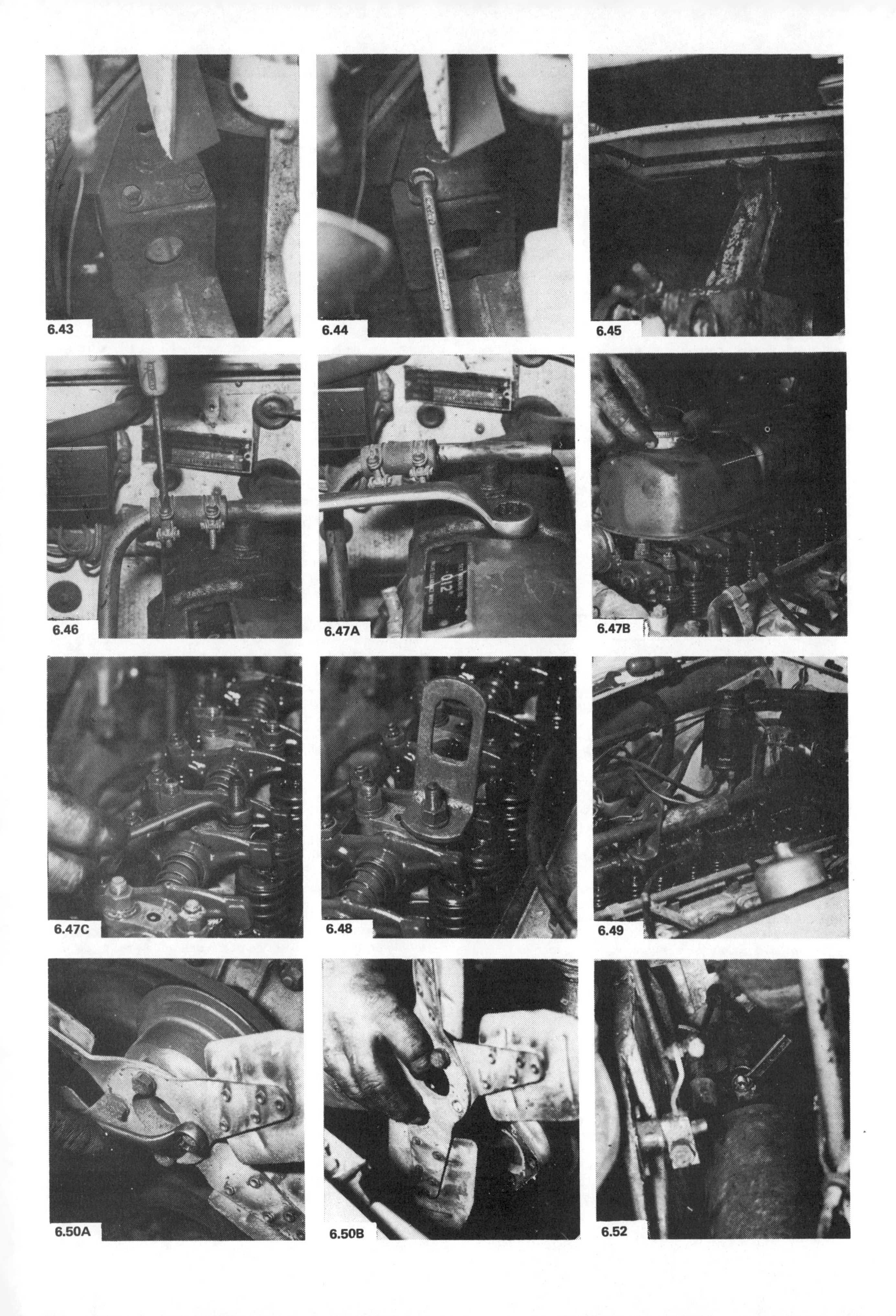

6.43 6.44 6.45 6.46 6.47A 6.47B 6.47C 6.48 6.49 6.50A 6.50B 6.52

53 Unscrew and remove the six brass nuts securing the exhaust downpipe to the exhaust manifolds and pull the downpipe away from the manifold studs (photo). Recover the two asbestos gaskets.
54 Refer to Fig 1.4 and remove the gearbox stabiliser securing pin (3).
55 Extract the split pin and remove the clevis pin so as to release the pushrod from the clutch fork and lever.
56 Undo and remove the two set bolts and spring washers that secure the slave cylinder to the clutch housing.
57 By means of a hoist with the hook around the metal tube on the metal rod, commence lifting the complete power unit, tilting it slightly as shown in this photo.
58 Continue lifting the power unit until the crankshaft pulley is just above the front grille (photo).
59 Undo and remove the four right hand engine mounting to cylinder block securing bolts and spring washers and lift away the mounting bracket (photo).
60 Detach the oil pressure gauge pipe from the main oil gallery on the left hand side of the cylinder block.
61 Check that there are no control or electric cables still attached to the engine or gearbox that could hinder removal.
62 Continue lifting the power unit tilting it as shown in this photo.
63 Whilst the power unit is being lifted take care that the rear of the gearbox/overdrive unit does not swing and damage the bodywork. When sufficient height has been achieved draw the power unit away from the body over the front of the car. Lower the unit to the ground.

7 Engine removal - partially stripped, without manual gearbox

1 Refer to Section 6 and follow the instructions from paragraph 1 to 23 inclusive with the exception of paragraph 17.
2 Refer to Section 11 and remove the cylinder head.
3 Place a jack or support stand under the gearbox to support its weight.
4 Note the electric cables to the ignition coil so that they may be refitted the correct way round and disconnect the two cables.
5 Unscrew the knurled moulded nut securing the HT lead to the centre of the ignition coil. Spring back the two distributor cap retaining clips and lift away the distributor cap and HT leads.
6 Refer to Section 6 and follow the instructions in paragraphs 36 and 37.
7 Undo the two crosshead bolts securing the cable crosshead bracket to the bulkhead (photo).
8 Undo the crosshead bolt and draw the bracket away from the bulkhead and single operating cable (photo).
9 Undo and remove the eight engine mounting bracket to chassis frame securing bolts (photos 6.43 and 6.44)
10 Position the saddle of a hydraulic jack under the front crossmember (photo 6.45)
11 Place a rope sling around the engine in such a manner that the rope cannot slip off when the engine is at an angle.
12 Follow the instructions in paragraphs 50, 51 and 42, Section 6.
13 Undo and remove the two set bolts and spring washers that secure the clutch slave cylinder to the clutch housing and support the weight of the slave cylinder with string or wire.
14 Undo and remove the nuts, bolts and washers that secure the gearbox clutch housing flange to the engine backplate.
15 Check that there are no cables or pipes still left connected to the engine and then carefully draw the engine forwards away from the front of the gearbox.
16 Tilt the front of the engine upwards and lift it up out of the engine compartment.
17 Draw the engine forwards over the front of the car and lower

8 Dismantling the engine - general

1 It is best to mount the engine on a dismantling stand, but it can be stood on a strong bench at a comfortable working height. Failing this, it can be stripped down on the floor.
2 During the dismantling process the greatest care should be taken to keep the exposed parts free from dirt. To achieve this, thoroughly clean down the outside of the engine, removing all traces of oil and congealed dirt.
3 A good grease solvent such as Gunk will make the job much easier. Allow it to stand for a time then wash off all the solvent and filth with water. If the dirt is thick and deeply embedded, work the solvent in with a strong stiff paintbrush.
4 Finally wipe down the exterior of the engine with a rag and only then, when it is quite clean, should the dismantling process begin. As the engine is stripped, clean each part in a bath of paraffin or petrol.
5 Never immerse parts with oilways in paraffin, ie, the crankshaft, but clean with a petrol dampened rag. Oilways can be cleaned out with a nylon pipe cleaner. If an airline is present all parts can be blown dry and the oilways blown through.
6 Always use new gaskets throughout. Re-use of old gaskets is false economy and can give rise to oil and water leaks, if nothing worse.
7 Do not throw away the old gaskets as they can be useful as a template if a replacement is not instantly available. Hang up the gaskets as they are removed, on a hook.
8 Work from the top down when stripping the engine. The crankcase provides a firm base on which the engine can be supported in an upright position. When the stage where the crankshaft must be removed is reached, the engine can be turned on its side and all other work carried out with it in this position.
9 Wherever possible, replace nuts, bolts and washers finger tight from wherever they were removed. This helps avoid loss and muddle later. If they cannot be replaced then lay them out in such a fashion that it is clear from whence they came (photo).

9 Removing ancillary engine components

Before basic engine dismantling begins, it is necessary to strip off ancillary components and these are as follows:

Dynamo
Distributor
Thermostat
Inlet manifold and carburation system
Exhaust manifold
Starter motor

It is possible to strip all these items with the engine in the car if it is merely the individual items which require attention. Presuming that the engine is out of the car and on the bench, starting on the right hand side of the unit, follow the procedure described next:

Slacken off the retaining bolts and remove the dynamo from the side of the engine (photo).
2 To remove the distributor first disconnect the vacuum advance/retard pipe from the carburettor installation. Unscrew the clamp bolt at the base of the distributor and lift the distributor away from its baseplate and drive shaft (photo).
3 Remove the thermostat cover by releasing the two nuts and spring washers which hold it in position and then remove the gasket and thermostat unit (photo).
4 Moving to the left hand side of the engine remove the carburettor installation if it is still in place. Further information on this operation will be found in Chapter 3.
5 Remove the exhaust manifold after unscrewing the brass nuts and washers holding the manifold to the side of the cylinder head (photos).

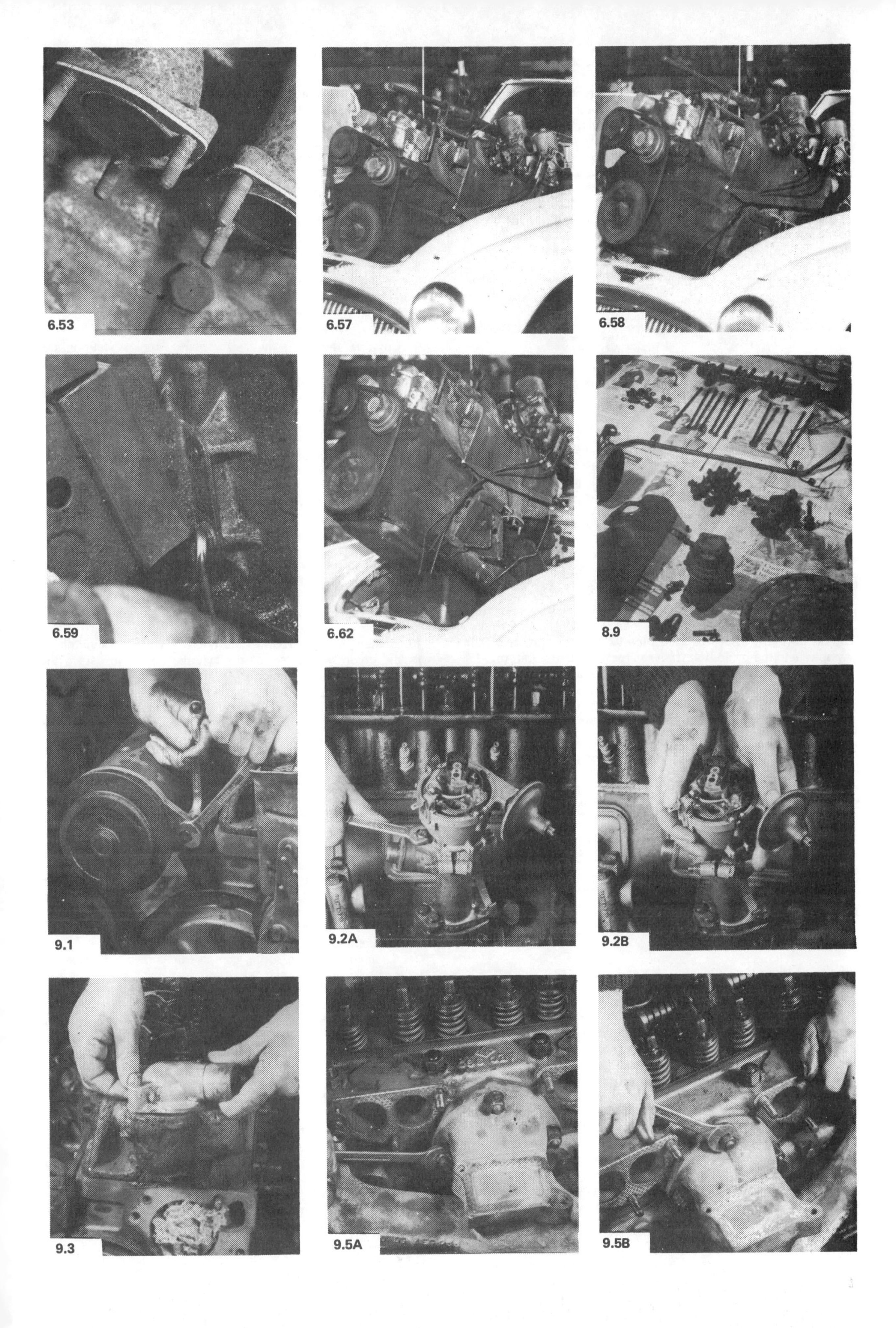
6.53
6.57
6.58
6.59
6.62
8.9
9.1
9.2A
9.2B
9.3
9.5A
9.5B

6 If still fitted, undo the two bolts which hold the starter motor in place and lift away the starter motor.
7 The engine is now stripped of all ancillary components and is ready for major dismantling to begin.

10 Cylinder head removal - engine on bench

1 With the engine out of the car and standing upright on the bench or on the floor, remove the cylinder head as follows:
2 Unscrew the two rocker cover shaped securing nuts and lift the nuts and washers. Lift the rocker cover and gasket away.
3 Unscrew the twelve nuts and plain washers securing the rocker pedestals to the top of the cylinder head (photo).

10.3

4 Remove the rocker shaft oil feed pipe banjo bolt from number 4 rocker pedestal and then undo the oil feed pipe union nut. Lift away the feed pipe with union nut, two fibre washers and the banjo bolt from the rocker pedestal (photo).
5 Remove the rocker assembly complete and place it on one side.
6 Unscrew the sixteen cylinder head nuts half a turn at a time in the order shown in Fig 1.7. When all the nuts are no longer

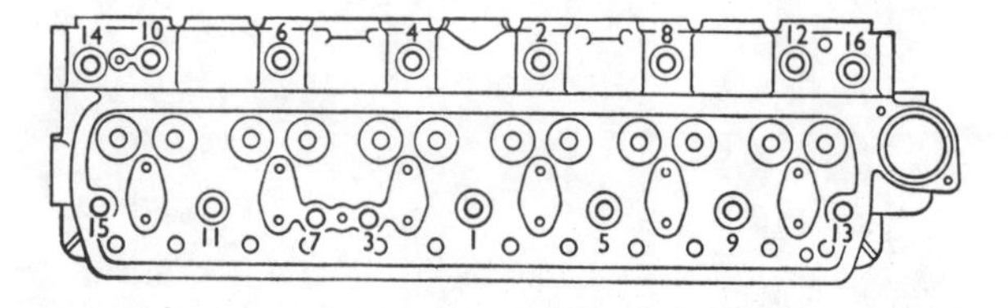

Fig 1.7 CYLINDER HEAD RETAINING NUT REMOVAL/ TIGHTENING SEQUENCE

under tension they may be removed from the cylinder head one at a time (photo).
7 Remove the pushrods, keeping them in the same order in which they were removed. The easiest way to do this is to push them through a sheet of card in the correct sequence (photo).
8 The cylinder head can be removed by straight lifting upwards (photo). If the cylinder head is jammed, try to rock it to break the seal. Under no circumstances try to prise it apart from the block with a screwdriver or cold chisel, as damage may be done to both the faces of the head and the block. If the cylinder head will not free readily, turn the engine over by the flywheel as the compression in the cylinder will often break the cylinder joint. If this fails to work, tap the head sharply with a plastic or wooden headed hammer, or with a metal hammer with an interposed piece of wood to cushion the blow. Under no circumstances hit the head directly with a metal hammer as this will cause the iron casting to fracture. Several sharp taps with the hammer, at the same time pulling upwards, should free it. Lift off the head and place on one side.

11 Cylinder head removal - engine in car

To remove the cylinder head with the engine still in the car the following additional procedures to those mentioned in the previous section must be carried out BEFORE them:

1 Disconnect the battery.
2 Drain the water by turning on the taps at the base of the radiator and at the side of the cylinder block below the exhaust manifold.
3 Slacken the top radiator hose securing clamp at the radiator top tank and disconnect the top hose from the radiator.
4 Unscrew the three nuts and plain washers securing the thermostat housing to the cylinder head and lift the housing and paper gasket away from the top of the cylinder head. If the housing is difficult to remove, gently tap with a soft faced hammer. Lift out the thermostat, if necessary lifting the lip with a screwdriver.
5 Disconnect the cable from the temperature gauge sender unit located at the bottom of the thermostat housing.
6 Refer to Chapter 3, Section 12 and remove the carburettor installation.
7 Undo and remove the two nuts and plain washers securing the heat shield to the two studs on the exhaust manifold. Lift away the heat shield.
8 Undo and remove the ten brass nuts and plain washers securing the two exhaust manifolds to the side of the cylinder head. Lift away the two manifolds.
9 Disconnect the distributor automatic advance/retard pipe from the connection on the union at the rear left hand side at the top of the cylinder head.
10 Slacken the clip securing the brake servo unit vacuum hose to the union at the top of the cylinder head and detach the vacuum hose from the union.
11 Slacken the heater hose securing clip on the take off pipe at the right hand side of the engine and detach the heater hose.
12 Refer to Section 10 and thereafter follow the instructions from the beginning of Paragraph 2 right through to the end of the Section.

12 Valve removal

1 The valves can be removed from the cylinder head by the following method: With a pair of pliers remove the spring circlips holding the two halves of the split tapered collets together. Compress each spring in turn with a valve spring compressor until the two halves of the collets can be removed. Release the compressor and remove the inner and outer valve springs, valve spring cap, valve spring collar - lower, valve stem grommet and finally the valve.
2 If, when the valve spring compressor is screwed down the valve spring retaining cap refuses to free and expose the split collet, do not continue to screw down on the compressor as there is a likelihood of damaging it.
3 Gently tap the top of the tool directly over the cap with a hammer. This will free the cap. To avoid the compressor jumping off the valve retaining cap when it is tapped, hold the compressor firmly in position with one hand.
4 Slide the rubber oil control grommet off the top of each valve stem and then drop out each valve through the combustion chamber.
5 It is essential that the valves are kept in their correct sequence unless they are so badly worn that they have to be renewed. If they are going to be kept and used again, place them in a sheet of card having twelve holes numbered 1 to 12 corresponding with the relative valve locations when fitted. Also keep the valve springs, cap, lower collars and rubber grommets in the correct order.

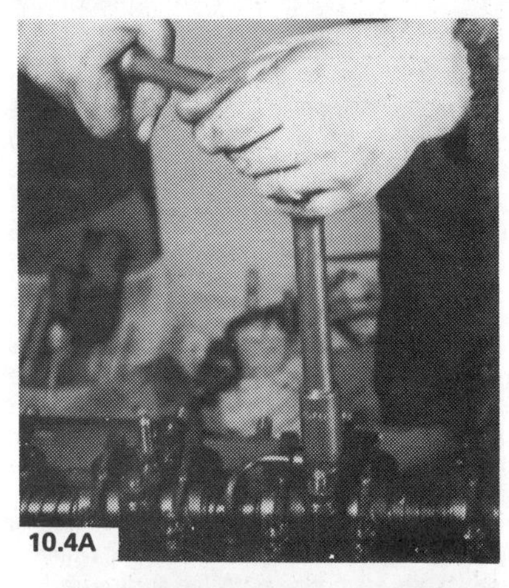
10.4A

10.4B

10.4C

10.6

10.7

10.8

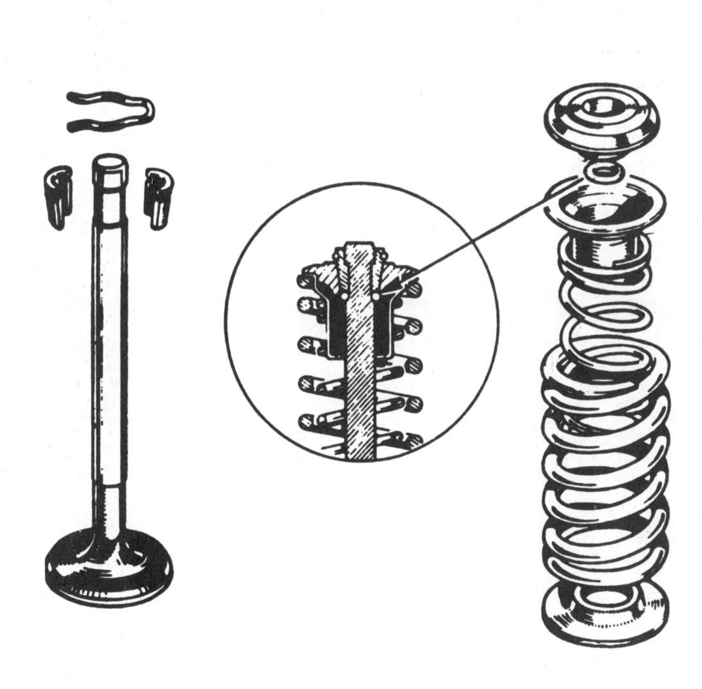
Fig 1.8 VALVE ASSEMBLY

13 Valve guide removal

If it is wished to remove the valve guides they can be removed from the cylinder head in the following manner: Place the cylinder head with the gasket face on the bench and with a suitable hard steel punch drift the guide out of the cylinder head. See Section 41.

14 Rocker assembly - dismantling

1 To dismantle the rocker assembly the locating screw in the form of the oil feed pipe banjo bolt will have already been removed. Now remove the split pins, flat washers and spring washers from each end of the shaft and slide from the shaft the pedestals, rocker arms and rocker spacing washer.
2 From the end of the shaft undo the plug which gives access to the inside of the rocker which can now be cleaned of sludge etc. Ensure the rocker arm lubricating holes are clear.

15 Timing cover, gears and chain - removal

The timing cover, gears and chain can be removed with the engine in the car provided that the radiator and the fan belt have been removed. Otherwise the procedure for removing the timing cover, gears and chain is the same irrespective of whether the engine is in the car or on the bench. It is as follows:

1 Bend back the locking tab of the crankshaft pulley retaining nut and with a large spanner remove the bolt and locking washer (Fig 1.9 and photos).
2 Place two large screwdrivers behind the crankshaft pulley wheel at 180° to each other, and carefully lever off the wheel. It is preferable to use a proper extractor if this is available, but large screwdrivers or tyre levers are quite suitable, providing care is taken not to damage the pulley flange.
3 Remove the Woodruff key from the crankshaft nose with a pair of pliers and note how the channel in the pulley is designed to fit over it. Place the Woodruff key in a glass jam jar as it is very small and can be easily lost.
4 Unscrew the bolts holding the timing cover to the block. Note that different size bolts are used and in some instances that each bolt makes use of a large flat washer.
5 Pull off the timing cover and paper gasket.
6 With the timing cover off, take off the oil thrower. Note: The concave side faces forwards. Take out the bottom plug of the timing chain tensioner, fit an 1/8 inch Allen key in the cylinder and turn the key clockwise until the slipper head is pulled right back and locked behind the limit head. On later produced engines a timing chain vibration damper was fitted in addition to the automatic chain tensioner and this is shown in Fig 1.11. It will be seen that the damper comprises an angled bracket which is secured to the cylinder block with one bolt and spring washer. Accurate location is achieved by using a dowel on the front mounting plate. An oil resistant rubber pad is bonded to the angled bracket and maintains light rubbing contact with the timing chain and dampens chain vibration under light load running conditions. Because of this additional item the cylinder block front mounting plate and gasket were modified.
7 Bend back the locking tab on the washer under the camshaft retaining nut, noting how the locking tab fits in the camshaft gear wheel keyway. Remove the nut using a large ring spanner or socket wrench.
8 To remove the camshaft and crankshaft timing wheel complete with chain, ease each wheel forward a little at a time, levering behind each gear wheel in turn with two large screwdrivers at 180° to each other. If the gear wheels are solid then it will be necessary to use the proper gear wheel and pulley extractor and if one is available this should be used anyway, in preference to a screwdriver. With both gear wheels safely off, remove both Woodruff keys from the crankshaft and camshaft with a pair of pliers and place them in the same jam jar for safe keeping. Note the number of very thin packing washers behind the crankshaft gear wheel and remove them very carefully.

15.1A

15.1B

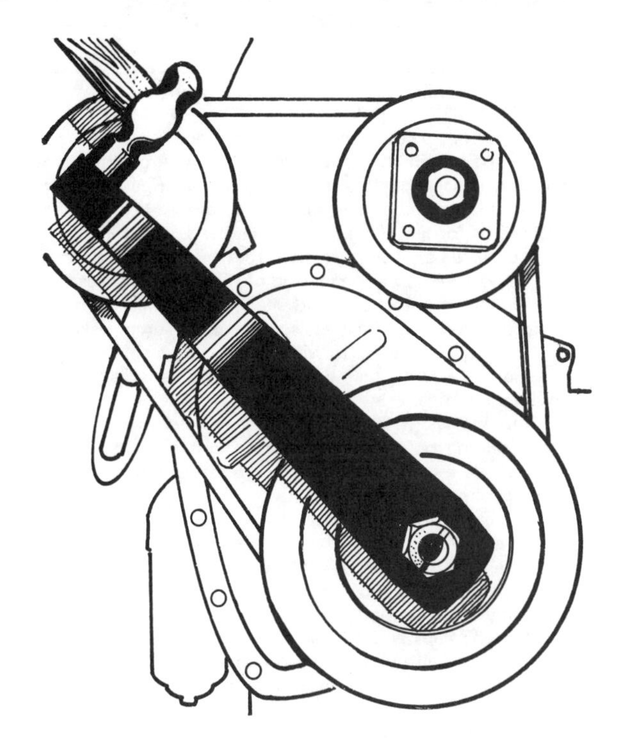

Fig 1.9 REMOVAL OF STARTER DOG NUT

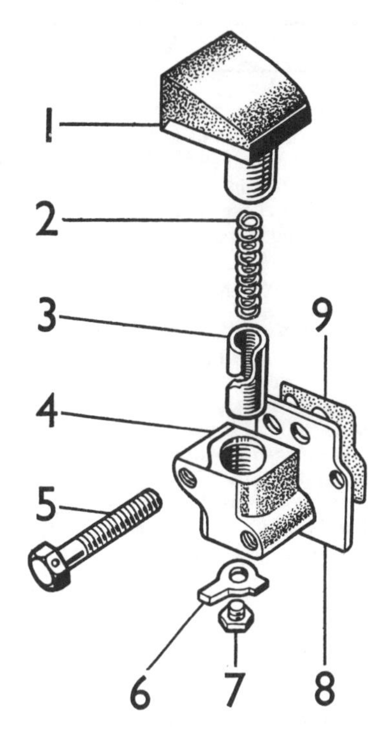

Fig 1.10 COMPONENT PARTS OF TIMING CHAIN TENSIONER

1 *Slipper head*
2 *Spring*
3 *Locating sleeve*
4 *Body*
5 *Set pin*
6 *Lockwasher*
7 *Plug*
8 *Backplate*
9 *Joint washer*

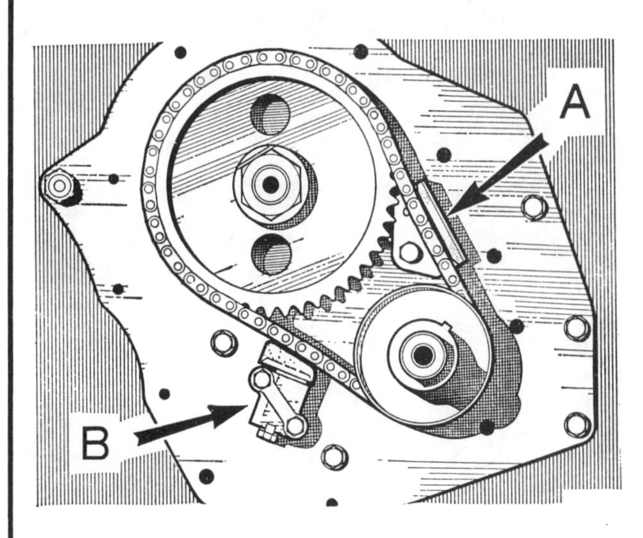

Fig 1.11 LATER TYPE ENGINE SHOWING VIBRATION DAMPER

A Vibration damper *B Tensioner*

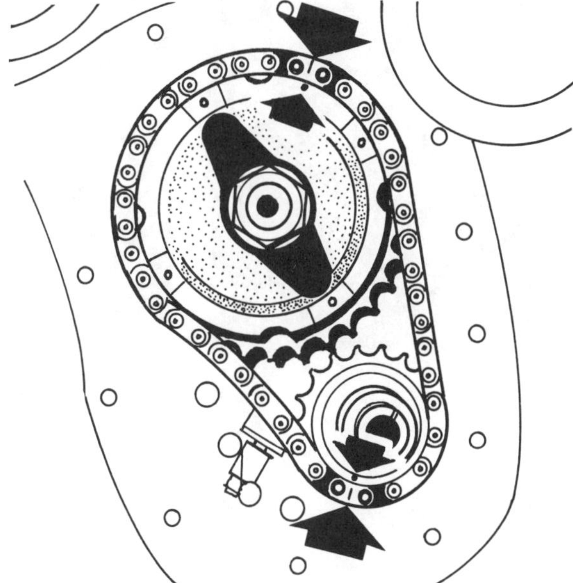

Fig 1.12 BEFORE REMOVING THE TIMING GEARS NOTE POSITIONS OF THE DOTS ON THE GEARS AND BRIGHT LINKS OF THE CHAIN

16 Camshaft removal

The camshaft cannot be removed with the engine in place in the car. With the engine on the bench remove the timing cover, gears and chain as described in Section 15. Then remove the distributor drive gear as described in Section 17. With the drive gear out of the way, proceed in the following manner:

1 Undo and remove the two bolts and spring washers which hold the camshaft locating plate to the cylinder block. The bolts are normally covered by the camshaft gear wheel.
2 Remove the plate. Undo and remove the two long bolts with spring washers securing the cylinder side covers to the side of the cylinder block. Lift away the side covers and felt gaskets (photo).
3 Using a piece of wire and a screwdriver carefully extract the valve tappets and place them in order so that they may be refitted in their original positions.
4 The camshaft can now be withdrawn. Take great care to remove the camshaft gently and in particular ensure that the cam peaks do not damage the camshaft bearings as the shaft is pulled forwards.

17 Distributor drive removal

1 Undo and remove the setscrew together with shakeproof washer that secures the distributor housing to the cylinder block and carefully withdraw the housing (photo).
2 Screw in a 5/16 inch UNF bolt into the distributor drive gear and withdraw the gear complete with the distributor drive coupling from the side of the cylinder block (Fig 1.14).
3 Should it be necessary to separate the coupling from the distributor drive gear, extract the circlip and then withdraw the coupling.

18 Piston, connecting rod and big end removal

1 The sump, pistons and connecting rods can be removed with the engine still in the car or with the engine on the bench. If in the car, proceed as for removing the cylinder head with the engine in the car, as described in Section 11. If on the bench proceed as for removing the cylinder head with the engine in this position as described in Section 10. The pistons and connecting rods are drawn up out of the top of the cylinder bores.
2 Remove the 25 bolts and shaped plain washers holding the sump in position. Remove the sump and the sump gasket (photo).
3 Undo the three bolts with spring and plain washers which hold the oil strainer pick-up pipe to the oil sump and remove the strainer and pipe.
4 Knock back with a screwdriver the locking tabs on the big end retaining bolts and remove the bolts and locking tabs. On some engines self-locking nuts and bolts were used in which case locking tab washers are not fitted.
5 Remove the big end caps one at a time, taking care to keep them in the right order and the correct way round (photo). Also ensure that the shell bearings are also kept with their correct connecting rods and caps unless they are to be renewed. Normally, the numbers 1 to 6 are stamped on adjacent sides of the big end caps and connecting rods, indicating which cap fits on which rod and which way round the cap fits. If no numbers or lines can be found then with a sharp screwdriver scratch mating marks across the joint from the rod to the cap. One line for connecting rod No 1, two for connecting rod No 2 and so on. This will ensure that there is no confusion later as it is most important that the caps go back in the correct position on the connecting rods from which they were removed.
6 If the big end caps are difficult to remove they may be gently tapped with a soft hammer.
7 To remove the shell bearings, press the bearing opposite the groove in both the connecting rod and the connecting rod caps

16.2A

16.2B

17.1

18.2A

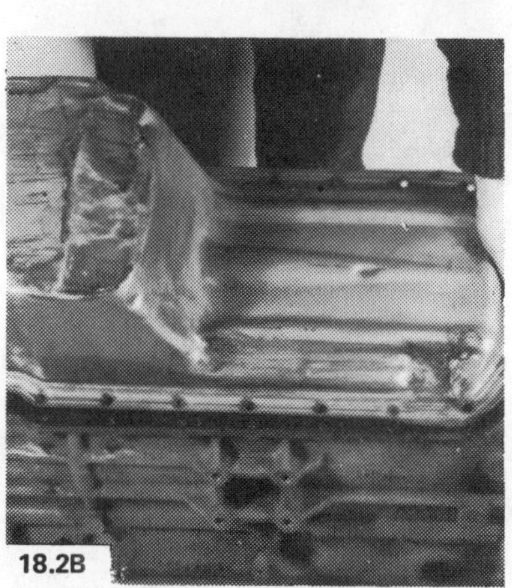
18.2B

18.5

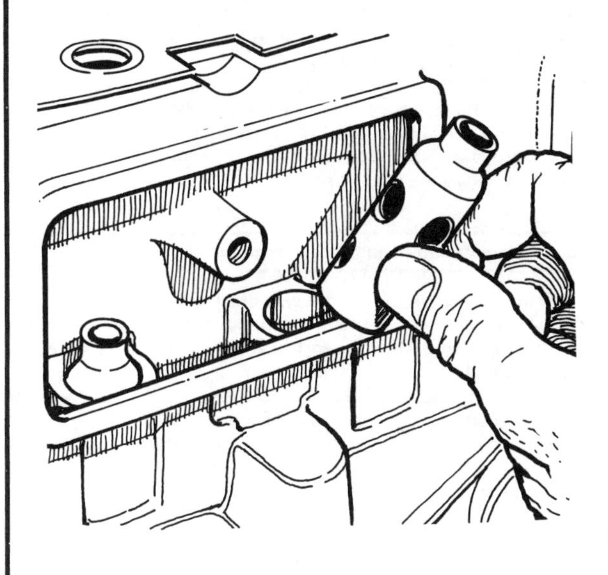

Fig 1.13 REMOVAL OF TAPPETS FROM SIDE CHESTS

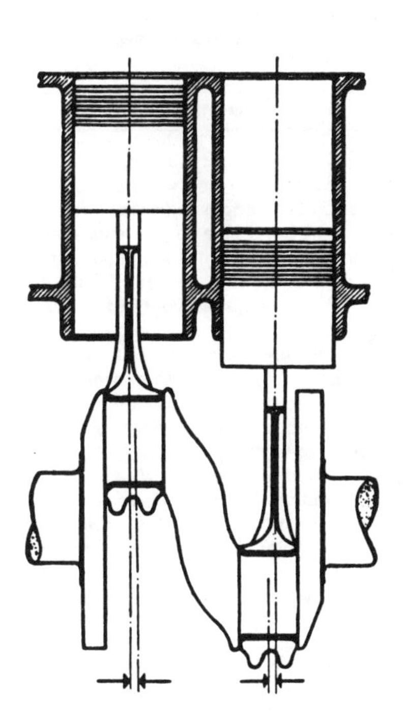

Fig 1.15 THE OFFSET OF THE CONNECTING RODS ON ADJACENT CRANKS

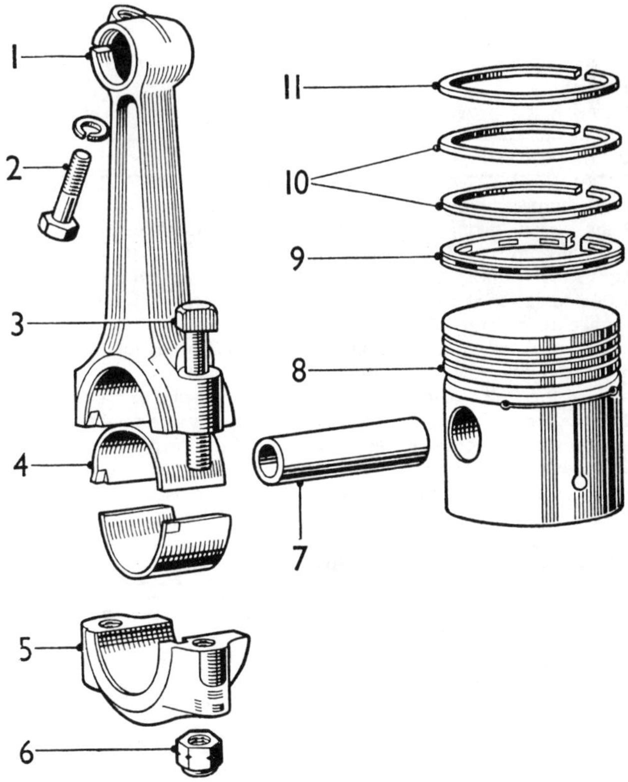

Fig 1.14 USING A BOLT TO REMOVE DISTRIBUTOR DRIVE SHAFT

Fig 1.16 PISTON AND CONNECTING ROD ASSEMBLY

1 Connecting rod
2 Small end clamping screw
3 Big end bolt
4 Shell bearing
5 Big end cap
6 Big end nut
7 Gudgeon pin
8 Piston
9 Oil control ring
10 Taper compression rings
11 Plain compression ring

and the bearings will slide out easily.
8 Withdraw the pistons and connecting rods upwards and ensure they are kept in the correct order for replacement in the same bore. Refit the connecting rod caps and bearings to the rods if the bearings do not require renewal to minimise the risk of getting the caps and rods muddled.

19 Gudgeon pin

On early engines the gudgeon pin was retained in position in the little end by a pinch bolt. This was subsequently modified from engine number 40501 to that of a fully floating gudgeon pin, this being retaining by circlips.

Early engines
1 To remove the gudgeon pin to free the piston from the connecting rod, undo the pinch bolt and remove together with the spring washer.

Later engines
2 To remove the gudgeon pin, using a pair of circlip pliers extract one of the circlips from the end of the gudgeon pin bore.
3 Gently tap out the gudgeon pin using a soft metal drift.
4 If the gudgeon pin shows reluctance to move, then on no account force it out as this could damage the piston. Immerse the piston in a pan of boiling water for about three minutes. On removal the expansion of the aluminium should allow the gudgeon pin to slide out easily.
5 Make sure that the pins are kept in the same piston for ease of refitting.

20 Piston ring removal

1 To remove the piston rings, slide them carefully over the top of the piston, taking care not to scratch the aluminium alloy. Never slide them off the bottom of the piston skirt. It is very easy to break the iron piston rings if they are pulled off roughly so this operation should be done with extreme care. It is helpful to make use of an old 0.020 feeler gauge (photo).
2 Lift one end of the piston ring to be removed out of its groove and insert the end of the feeler gauge under it.
3 Turn the feeler gauge slowly round the piston and as the ring comes out of its groove apply slight upward pressure so that it rests on the one above. It can then be eased off the piston with the feeler gauge stopping it from slipping into any empty groove if it is any but the top piston ring that is being removed.

21 Flywheel and backplate - removal and replacement

Having removed the clutch (see Chapter 5, Section 5) the flywheel and engine backplate can now be removed. It is only possible for this operation to be carried out with the engine out of the car.
1 Bend back the locking tabs from the four bolts which hold the flywheel to the flywheel flange on the rear of the crankshaft (photo).
2 Unscrew the bolts and remove them, complete with the two locking plates (photo).
3 Lift the flywheel away from the crankshaft flange. Note: Some difficulty may be experienced in removing the bolts by the rotation of the crankshaft. Lock the crankshaft in position with a wooden wedge inserted between the crankshaft web, not the journal, and the side of the block inside the crankcase.
4 The engine backplate is held in position by a number of bolts and spring washers of various size. Release the bolts noting where different sizes fit, and place them together to ensure none of them become lost. Lift away the backplate from the block complete with the paper gasket.
5 Flywheel replacement is described in Section 53 of this Chapter. To replace the engine backplate, first fit a new paper gasket in place on the rear of the cylinder block and hold it in place with jointing compound.
6 Remember to fit into its groove the square section cork which will fit into the bottom of the rear main bearing cap. If this is not done the plate will have to be removed later.
7 Replace the backplate, insert the securing bolts and pull up the tabs of any lockwashers used.

22 Crankshaft and main bearing removal

With the engine out of the car, drain the engine oil, remove the timing gears and remove the sump, the oil gauze filter and suction pipe, and the big end bearings, pistons, flywheel and engine backplate as have already been described in Sections 15, 18 and 21. Removal of the crankshaft can only be attempted with the engine on its side on the bench or on the floor.
1 Undo by one turn the self-locking nuts which hold the four main bearing caps in place (photo).
2 Unscrew the nuts and remove them, keeping them in their correct order so that they may be refitted in their original positions.
3 Remove the two bolts and tab washer which hold the front main bearing cap against the engine front plate.
4 Remove the main bearing caps and the bottom half of each shell, taking care to keep the bearing shells in the right cap.
5 When removing the front centre bearing cap, NOTE the bottom semi-circular halves of the thrust washers, one half lying on either side of the main bearing. Lay them with the centre bearing along the correct side (photo).
6 Slightly rotate the crankshaft to free the upper half of the bearing shells and thrust washers which should now be extracted and placed over the correct bearing cap.
7 Remove the crankshaft by lifting it away from the crankcase.

23 Lubrication system - description

1 A forced feed system of lubrication is fitted with oil circulated round the engine from the sump below the block. The level of engine oil in the sump is indicated on the dipstick which is fitted on the right hand side of the engine. It is marked to indicate the optimum level which is the maximum mark.
2 The level of oil in the sump, ideally, should not be above or below this line. Oil is replenished via the filler cap on the front of the rocker cover.
3 The gear type oil pump is bolted in the left hand side of the crankcase and is driven by a short shaft from the skew gear on the camshaft which also drives the distributor shaft.
4 The pump is the non-draining variety to allow rapid pressure build up when starting from cold.
5 Oil is drawn from the sump through a gauze screen in the oil strainer and is sucked up the pick-up pipe and drawn into the oil pump. From the oil pump it is forced under pressure along a gallery on the right hand side of the engine and through drillings to the big end, main and camshaft bearings. A small hole in each connecting rod allows a jet of oil to lubricate the cylinder wall with each revolution.
6 From the camshaft front bearing oil is fed through drilled passages in the cylinder block and head to the front rocker pedestal where it enters the hollow rocker shaft. Holes drilled in the shaft allow for the lubrication of the rocker arms and the valve stems and pushrod ends.
7 This oil is at a reduced pressure to the oil delivered to the crankshaft bearings. Oil from the front camshaft bearing also lubricates the timing gears and the timing chain. Oil returns to the sump by various passages, the tappets being lubricated by oil returning via the pushrod drillings in the block.
8 On all models a full-flow oil filter is fitted, and all oil passes through this filter before it reaches the main oil gallery. The oil is passed directly from the oil pump across the block to an external pipe on the right hand side of the engine which feeds into the filter head.

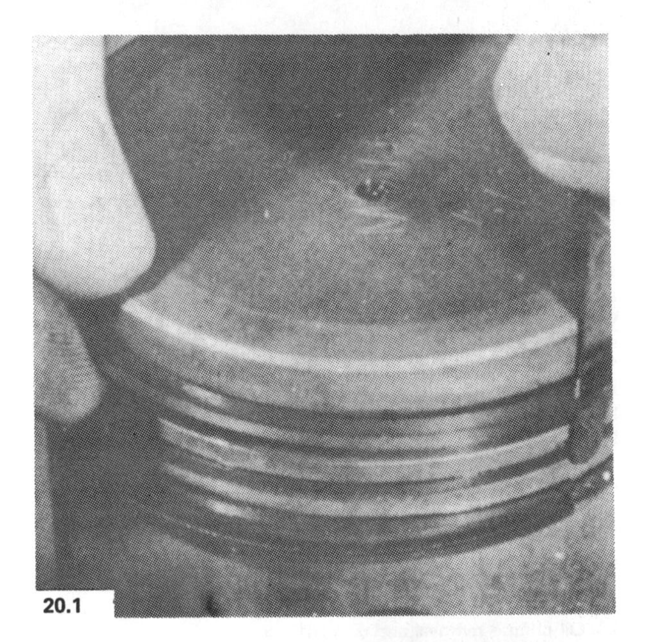
20.1

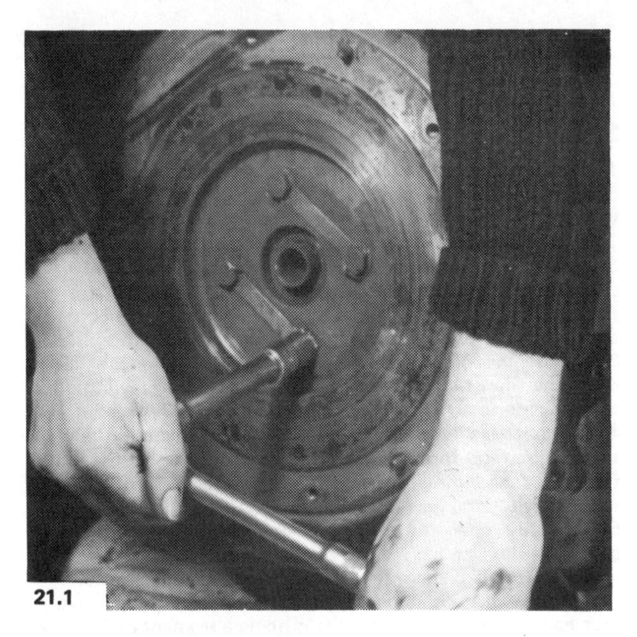
21.1

21.2

22.1

22.5

24 Oil filter - removal and replacement

1 The full flow oil filter fitted to all engines is located three-quarters of the way down the right hand side of the engine towards the front.
2 It is removed by unscrewing the long centre bolt which holds the filter bowl in place. With the bolt released (use a 9/16 AF spanner) carefully lift away the filter bowl which contains the filter, it will be full of oil. It is helpful to have a large basin under the filter body to catch the oil which is bound to spill.
3 Throw the old filter element away and thoroughly clean down the filter bowl, the bolts and associated parts with petrol and when perfectly clean wipe dry with a non-fluffy rag.
4 A rubber sealing ring is located in a groove round the head of the oil filter and forms an effective leakproof joint between the filter head and the filter bowl. A new rubber sealing ring is supplied with each new filter element.
5 Carefully prise out the oil sealing ring from the locating groove. If the ring has become hard and is difficult to move take great care not to damage the sides of the sealing ring groove.
6 With the old ring removed, fit the new ring in the groove at four equidistant points and press it home a segment at a time as shown. Do not insert the ring at just one point and work round the groove pressing it home as, using this method, it is easy to stretch the ring and be left with a small loop of rubber which will not fit into the locating groove (photo).
7 Reassemble the oil filter assembly by first passing up the bolt through the hole in the bottom of the bowl, and with a steel washer under the bolt head and a rubber or felt washer on top of the steel washer and next to the filter bowl
8 Slip the spring over the bolt inside the bowl as shown in the photograph.
9 Then fit the other steel washer and the remaining rubber or felt washer to the centre bolt as shown (photo).
10 Fit the sealing plate over the centre bolt with the concave side facing the bottom of the bowl as shown (photo).
11 Then slide the new element into the oil filter bowl (photo).
12 With the bolt pressed hard up against the filter body (to avoid leakage) three-quarter fill the bowl with engine oil.
13 Offer up the bowl to the rubber sealing ring and before finally tightening down the centre bolt, check that the lip of the filter bowl is resting squarely on the rubber sealing ring and is not offset and off the ring. If the bowl is not seating properly, rotate it until it is. Run the engine and check the bowl for leaks.

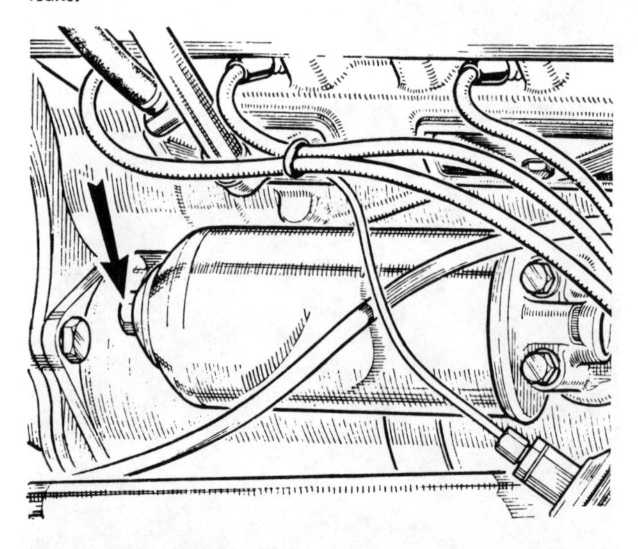

Fig 1.17 THE ENGINE OIL FILTER SHOWING CENTRE SECURING BOLT

25 Oil pressure relief valve - removal and replacement

1 To prevent excessive oil pressure - for example when the engine is cold - an oil pressure relief valve is built into the crank-case just below the oil filter housing.
2 The relief valve is identified externally by a large 9/16 inch domed hexagon nut. To dismantle the unit unscrew the nut and remove it, complete with the two fibre or copper sealing washers. The relief spring and the relief spring cup can then be easily extracted.
3 In position, the metal cup fits over the opposite end of the relief valve spring resting in the dome of the hexagon nut, and bears against a machining in the block. When the oil pressure exceeds 55 lb/sq in (depending on the model) the cup is forced off its seat and the oil bypasses it and returns to the sump.
4 Check the tension of the spring by measuring its length. If it is shorter than 2.87 inch it should be replaced by a new spring. Reassembly of the relief valve unit is a reversal of the dismantling procedure.

26 Oil pump - removal and dismantling

On early engines a Hobourn Eaton rotary vane type pump was fitted as shown in Fig 1.18 but on later engines a gear type pump was fitted. This type is shown in Fig 1.19.
1 Remove the sump as described in Section 18 of this Chapter.
2 Undo and remove the three nuts and spring washers from the studs that secure the oil pump assembly to the crankcase. Lift away the oil pump together with the two gaskets (photo).
3 If the oil pump is being removed whilst the engine is still in the car, the driveshaft will be free to disengage from the cam-shaft. Take care that it does not fall out. Upon reference to Fig 1.19 it will be seen that there is a thrust washer (1) on the end of the driveshaft above the gear.
4 Undo the four bolts and spring washers holding the oil pump cover to the pump body and separate the two parts (photo).
5 Carefully draw the inner and outer rotor (early type) or driven and driving gears from the inside of the pump body.
6 Information on examination and renovation of the oil pump will be found in Section 39 of this Chapter (photo).

27 Timing chain tensioner - removal and dismantling

1 Remove the cover from the timing gears as described in Section 15 of this Chapter and lock the rubber tensioner in its fully retracted position as described in Section 2. Knock back the tabs of the joint lockwasher and undo the two bolts which hold the tensioner and its backplate to the engine.
3 Pull the rubber slipper together with the spring and plunger from the tensioner body. Fit the Allen key to its socket in the cylinder and, holding the slipper and plunger firmly, turn the key clockwise to free the cylinder and spring from the plunger. The component parts of the timing chain tensioner may be seen in Fig 1.10.

28 Engine - examination and renovation - general

With the engine stripped down and all parts thoroughly cleaned, examine everything for wear. Items should be checked and where necessary renewed or renovated as described in the following sections.

29 Crankshaft, big end and main bearings - examination and renovation

1 Examine the crankpin and main journal surfaces for signs of scoring or scratches. Check the ovality of the crankpins in different positions with a micrometer. If more than 0.001 inch

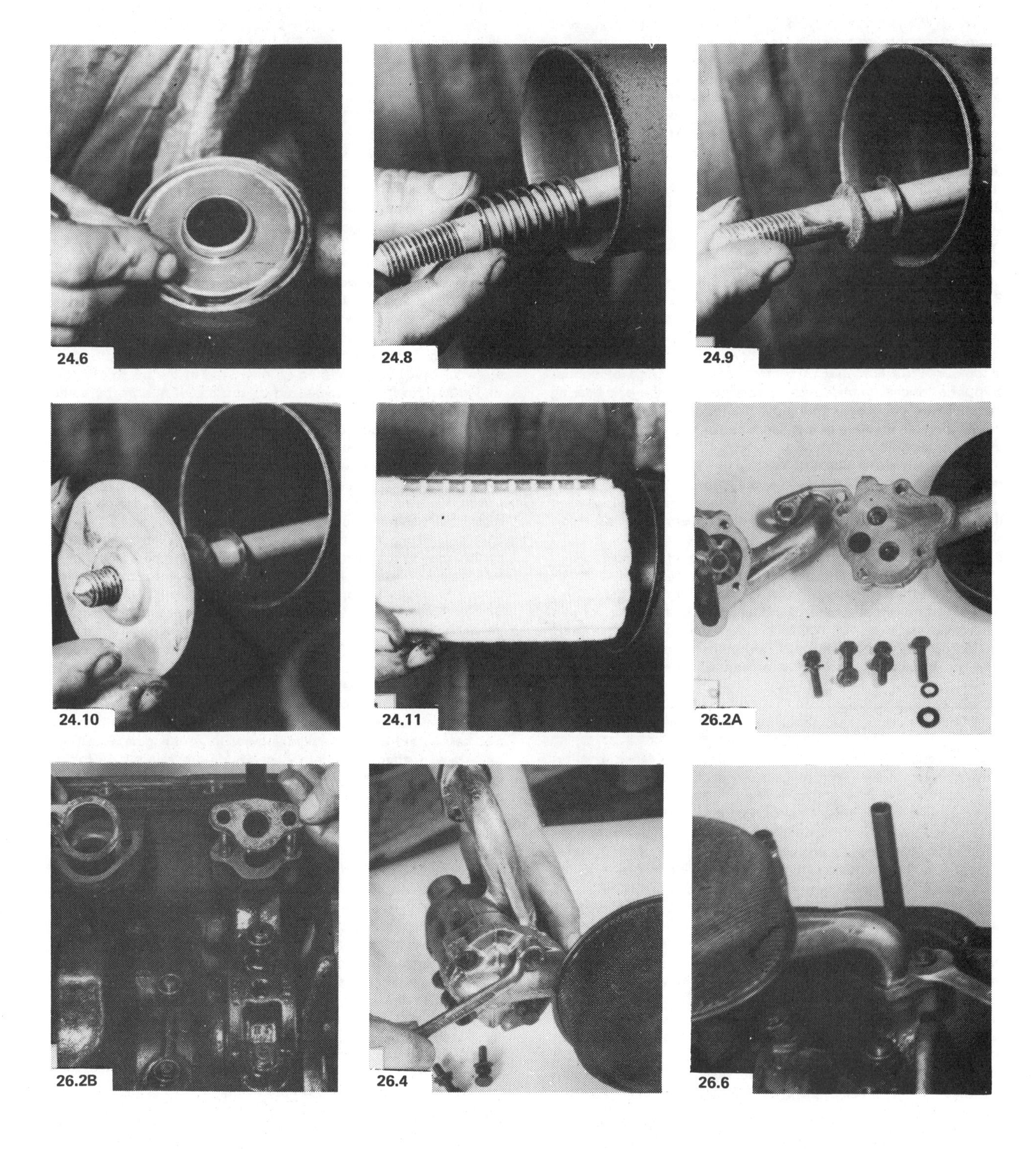

24.6

24.8

24.9

24.10

24.11

26.2A

26.2B

26.4

26.6

out of round the crankpins will have to be reground. It will also have to be reground if there are any scores or scratches present. Also check the journals in the same fashion. The centre main bearings have been known to break up. This is not always apparent immediately, but slight vibration in an otherwise normally smooth engine and a very slight drop in oil pressure under normal conditions are clues. If the centre main bearings are suspected of failure it should be immediately investigated by dropping the sump and removing the centre main bearing cap. Failure to do this will result in a badly scored centre main journal. If it is necessary to regrind the crankshaft and fit new bearings the BLMC garage or engineering works will be able to decide how much metal to grind off and the correct undersize shells to fit.

2 Big end bearing failure is accompanied by a noisy knocking from the crankcase, and a slight drop in oil pressure. Main bearing failure is accompanied by vibration which can be quite severe as the engine speed rises and falls and a drop in oil pressure.

3 Bearings which have not broken up, but are badly worn, will give rise to low oil pressure and some vibration. Inspect the big ends, main bearings, and thrust washers for signs of general wear, scoring, pitting and scratches. The bearings should be matt grey in colour. With lead-indium bearings should a trace of copper colour be noticed, the bearings are badly worn as the lead bearing material has worn away to expose the indium underlay. Renew the bearings if they are in this condition or if there is any sign of scoring or pitting.

4 The undersizes available are designed to correspond with the regrind sizes, ie: 0.010 bearings are correct for a crankshaft reground 0.010 undersize. The bearings are in fact slightly more than the stated undersize as running clearances have been allowed for during their manufacture.

5 Long engine life can be achieved by changing big end bearings at intervals of 30,000 miles and main bearings at intervals of 50,000 miles, irrespective of bearing wear. Normally, crankshaft wear is infinitesimal and regular changes of bearing may ensure mileages of between 100,000 to 120,000 miles before crankshaft regrinding becomes necessary. Crankshafts normally have to be reground because of scoring due to bearing failure.

30 Cylinder bores - examination and renovation

1 The cylinder bores must be examined for taper, ovality, scoring and scratches. Start by carefully examining the top of the cylinder bores. If they are worn at all a very slight ridge will be found on the thrust side. This marks the top of the piston ring travel. The owner will have a good indication of the bore wear prior to dismantling the engine, or removing the cylinder head. Excessive oil consumption accompanied by blue smoke from the exhaust is a sure sign of worn cylinder bores and piston rings.

2 Measure the bore diameter just under the ridge with a micrometer and compare it with the diameter at the bottom of the bore, which is not subject to wear. If the difference between the two measurements is more than 0.006 inch it will be necessary to fit special piston rings or to have the cylinders rebored and fit oversize pistons and rings. If no micrometer is available, remove the rings from a piston and place the piston in each bore in turn about ¾ inch below the top of the bore. If an 0.010 feeler gauge can be slid between the piston and the cylinder wall on the thrust side of the bore then remedial action must be taken. Oversize pistons are available in the following sizes:

+ 0.010 inch (0.254 mm)
+ 0.020 inch (0.508 mm)
+ 0.030 inch (0.762 mm)
+ 0.040 inch (1.016 mm)

These are accurately machined to just below these measurements so as to provide correct running clearances in bores bored out to the exact oversize dimensions.

3 If the bores are slightly worn but not so badly worn as to justify reboring them, special oil control rings can be fitted to the existing pistons which will restore compression and stop the engine burning oil. Several different types are available and the manufacturers instructions concerning their fitting must be followed closely.

31 Pistons and piston rings - examination and renovation

1 If the old pistons are to be refitted, carefully remove the piston rings and then clean them thoroughly. Take particular care to clean out the piston ring groove. At the same time do not scratch the aluminium in any way. If new rings are to be fitted to the old pistons then the top ring should be stepped so as to clear the ridge left above the previous top ring. If a normal but oversize new ring is fitted, it will hit the ridge and break, because the new ring will not have worn in the same way as the old, which will have worn in unison with the ridge.

2 Before fitting the rings on the pistons each should be inserted approximately 3 inches down the cylinder bore and the gap measured with a feeler gauge. This should be within the limits given in the Specifications at the beginning of this Chapter. It is essential that the gap should be measured at the bottom of the ring travel for if it is measured at the top of a worn bore and gives a perfect fit, it could easily seize at the bottom. If the ring gap is too small rub down the ends of the ring with a very fine file until the gap, when fitted, is correct. To keep the rings square in the bore for measurement line each up in turn by inserting an old piston in the bore upside down, and use the piston to push the ring down about 3 inches. Remove the piston and measure the piston ring gap.

3 When fitting new pistons and rings to a rebored engine the piston ring gap can be measured at the top of the bore as the bore will not now taper. It is not necessary to measure the side clearance in the piston ring grooves with the rings fitted as the groove dimensions are accurately machined during manufacture. When fitting new oil control rings to old pistons it may be necessary to have the grooves widened by machining to accept the new wider rings. In this instance the manufacturers representative will make this quite clear and will supply the address to which the pistons must be sent of matching.

4 When new pistons are fitted, take great care to fit the exact size best suited to the particular bores in your engine. BLMC go one stage further than merely specifying one size piston for all standard bores. Because of very slight differences in cylinder machining during production it is necessary to select just the right piston for the bore. Five different sizes are available for the standard bore as well as the four oversize dimensions already shown.

5 Examination of the cylinder block face will show next to each bore a small diamond shaped box with a number stamped in the metal. Careful examination of the piston crown will show a matching diamond and number. These are the standard piston sizes and will be the same for all six bores. If standard pistons are to be refitted or standard low compression pistons changed to standard high compression pistons, then it is essential that only pistons with the same number in the diamond are used. With larger pistons, the amount of oversize is stamped in an ellipse in the piston crown.

6 On engines with tapered second and third compression rings, the top narrow side of the ring is marked with a 'T'. Always fit this side uppermost and carefully examine all rings for this mark before fitting.

32 Camshaft and camshaft bearings - examination and renovation

1 Carefully examine the camshaft bearings for wear. If the bearings are obviously worn or pitted or the metal underlay is showing through, then they must be renewed. This is an operation for your local BLMC dealer or the local engineering works as it

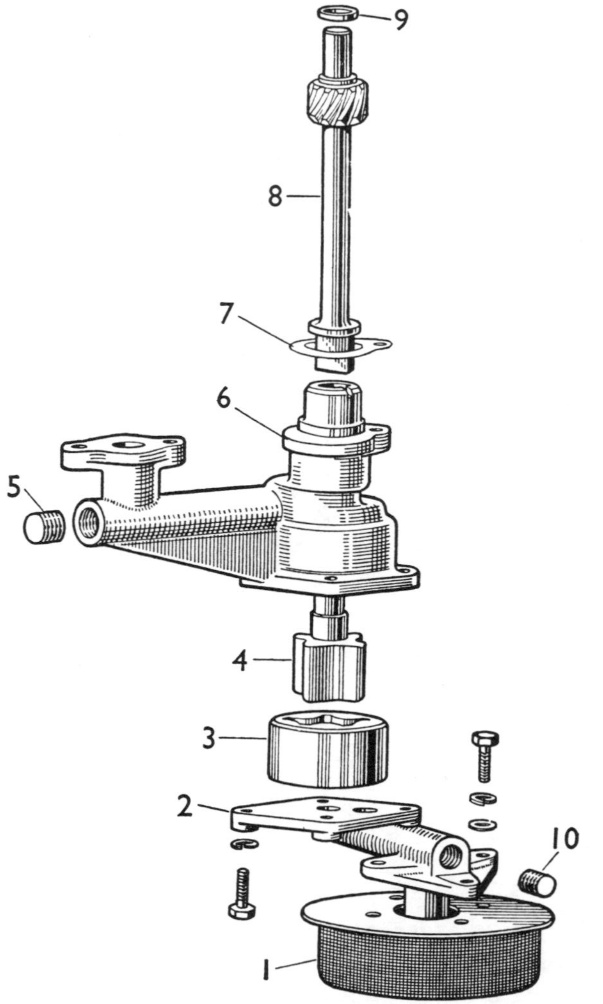

Fig 1.18 COMPONENT PARTS OF ROTARY TYPE PUMP

1 *Pick up strainer*	6 *Pump body*
2 *Bottomlower plate*	7 *Joint washer*
3 *Outer rotor*	8 *Drive spindle*
4 *Inner rotor*	9 *Drive spindle thrust washer*
5 *Screw plug*	10 *Screw plug*

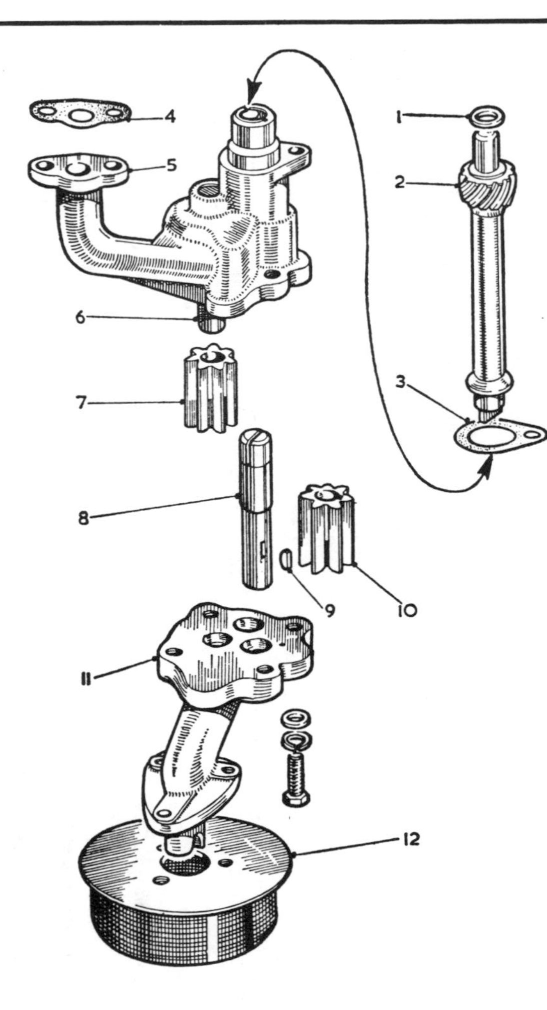

Fig 1.19 COMPONENT PARTS OF GEAR TYPE PUMP

1 *Thrust washer*	7 *Driven gear*
2 *Drive spindle*	8 *Drive gear spindle*
3 *Joint washer*	9 *Key*
4 *Joint washer*	10 *Drive gear*
5 *Pump body*	11 *Pick up*
6 *Driven gear spindle*	12 *Pick up strainer*

Fig 1.20 PISTON CROWN MARKINGS FOR STANDARD AND OVERSIZE BORES

demands the use of specialised equipment. The bearings are removed with a special drift after which new bearings are pressed in, care being taken to ensure the oil holes in the bearings line up with those in the block. The bearings are then reamed in position with a special tool.

2 The camshaft itself should show no signs of wear, but if very slight scoring on the cams is noticed, the score marks can be removed by very gentle rubbing down with a very fine emery cloth. The greatest care should be taken to keep the cam profiles smooth.

3 Fit the retaining plate and then the chain wheel to the end of the camshaft while it is out on the bench, and measure the end float between the thrust face of the camshaft front journal and the retaining plate. If more than 0.007 inch renew the retaining plate.

33 Valves and valve seats - examination and renovation

1 Examine the heads of the valves for pitting and burning, especially the heads of the exhaust valves. The valve seatings should be examined at the same time. If the pitting on valve and seat is very slight the marks can be removed by grinding the seats and valves together with coarse, and then fine, valve grinding paste. Where bad pitting has occurred to the valve seats it will be necessary to recut them and fit new valves. If the valve seats are so worn that they cannot be recut, then it will be necessary to fit new valve seat inserts. These latter two jobs should be entrusted to the local BLMC agent or engineering works. In practice it is very seldom that the seats are so badly worn that they require renewal. Normally it is the valve that is too badly worn for replacement and the owner can easily purchase a new set of valves and match them to the seats by valve grinding.

2 Valve grinding is carried out as follows: Smear a trace of coarse carborundum paste on the seat face and apply a suction grinding tool to the valve head. With a semi-rotary motion, grind the valve head to its seat, lifting the valve occasionally to redistribute the grinding paste. When a dull matt even surface finish is produced on both the valve seat and the valve, wipe off the paste and repeat the process with fine carborundum paste, lifting and turning the valve to redistribute the paste as before. A light spring placed under the valve head will greatly ease this operation. When a smooth unbroken ring of light grey matt finish is produced, on both valve and valve seat faces, the grinding operation is completed.

3 Scrape away all carbon from the valve head and the valve stem. Carefully clean away every trace of grinding compound, taking great care to leave none in the ports or in the valve guides. Clean the valves and valve seats with a paraffin soaked rag then with a clean rag, and finally, if an airline is available, blow the valves, valve guides and valve ports clean.

34 Timing gears and chain - examination and renovation

1 Examine the teeth on both the crankshaft gearwheel and the camshaft gearwheel for wear. Each tooth forms an inverted V with the gearwheel periphery and if worn the side of each tooth under tension will be slightly concave in shape when compared with the other side of the tooth, ie: one side of the inverted V will be concave when compared with the other. If any sign of wear is present the gearwheels must be renewed.

2 Examine the links of the chain for side slackness and renew the chain if any slackness is noticeable when compared with a new chain. It is a sensible precaution to renew the chain at about 30,000 miles and at a lesser mileage if the engine is stripped down for a major overhaul. The actual rollers on a very badly worn chain may be slightly grooved.

35 Timing chain tensioner - examination and renovation

1 Thoroughly clean the component parts in petrol and clean out the oil holes in the slipper and spigot. If either of these holes become blocked slipper wear will increase considerably. After high mileages the slipper head is bound to be worn and must be renewed together with the cylinder assembly.

2 Check the bore of the adjuster body for ovality. If the diameter is more than 0.003 inch (0.076 mm) out of round at the bore mouth, a new adjuster unit must be fitted.

3 Upon introduction of the 3000 Mk II power unit inspect the timing chain vibration damper rubber pad for wear or excessive grooving and, if evident, obtain a new vibration damper.

36 Rockers and rocker shaft - examination and renovation

1 Remove the threaded plug with a screwdriver from the end of the rocker shaft and thoroughly clean out the shaft. As it acts as the oil passage for the valve gear, also ensure that the oil holes in it are quite clear after having cleaned them out. Check the shaft for straightness by rolling it on the bench. It is most unlikely that it will deviate from normal, but, if it does, then a judicious attempt must be made to straighten it. If this is not successful purchase a new shaft. The surface of the shaft should be free from any worn ridges caused by the rocker arms. If any wear is present, renew the shaft. Wear is only likely to have occurred if the rocker shaft oil holes have become blocked.

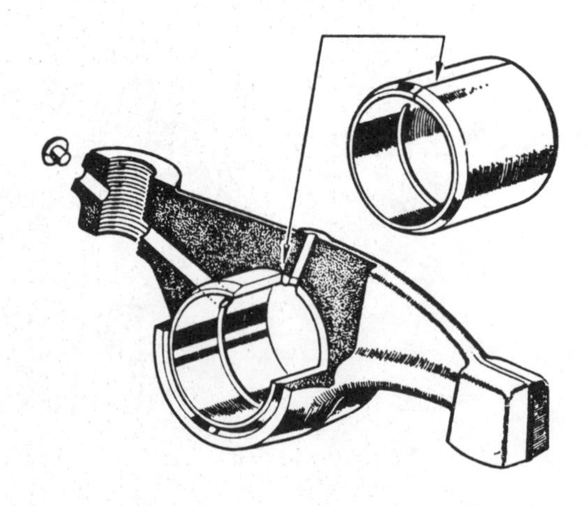

Fig 1.21 CORRECT POSITIONING OF ROCKER BUSH TO ALIGN OILWAYS. ARROW SHOWS PROPER LOCATION OF BUSH JOINT

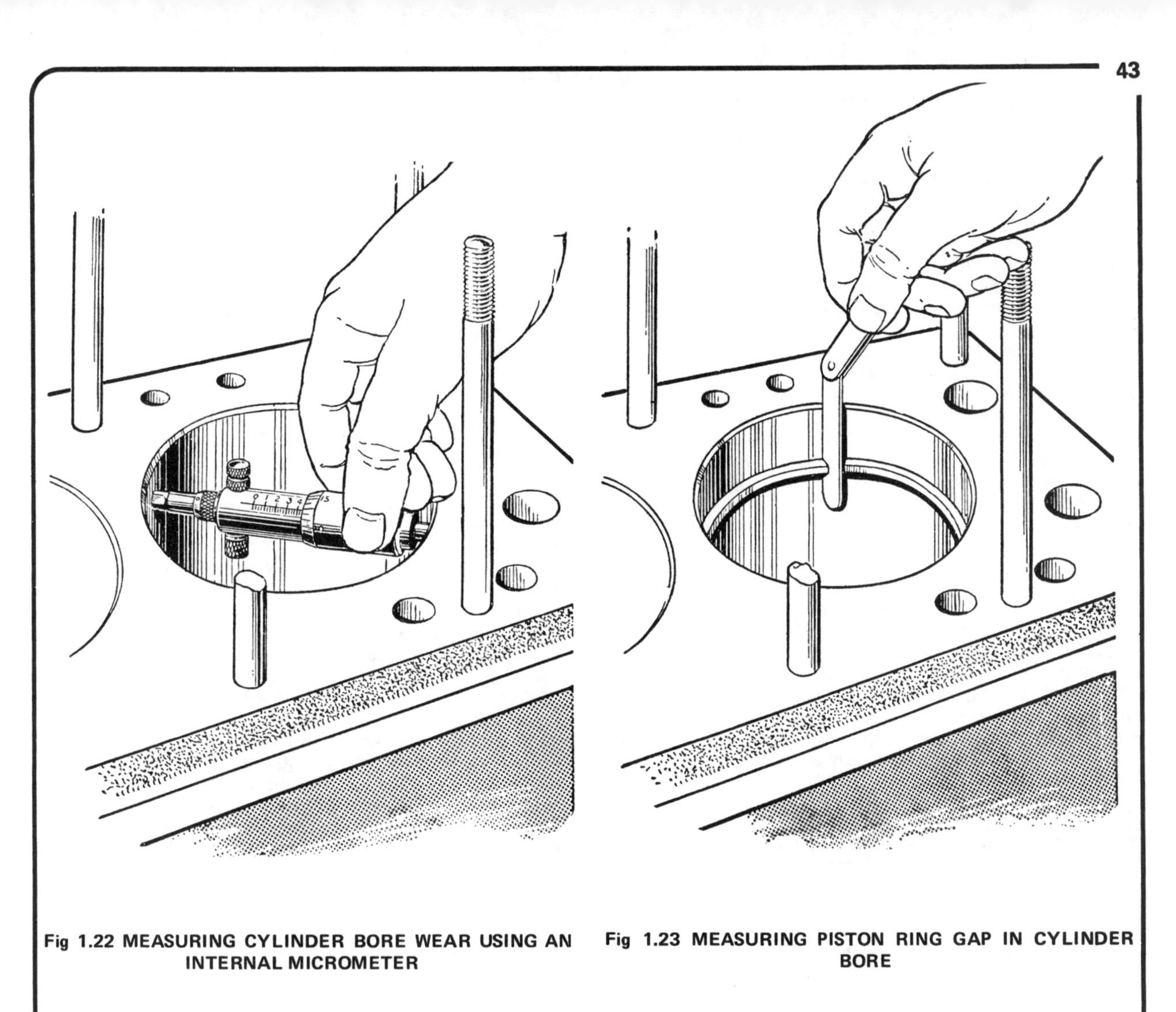

Fig 1.22 MEASURING CYLINDER BORE WEAR USING AN INTERNAL MICROMETER

Fig 1.23 MEASURING PISTON RING GAP IN CYLINDER BORE

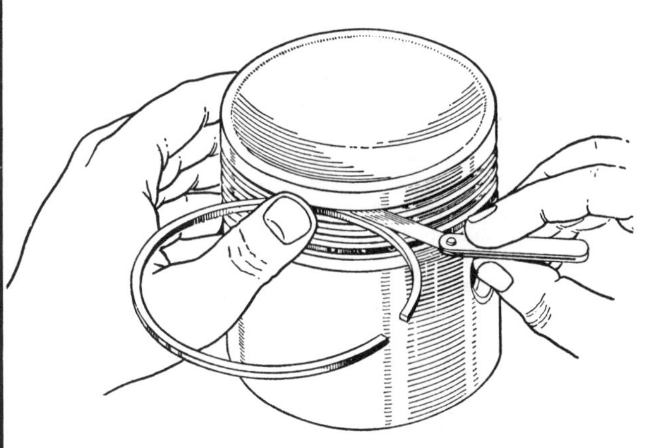

Fig 1.24 MEASURING PISTON RING GROOVE CLEARANCE

Fig 1.25 USING A SUCTION VALVE GRINDING TOOL

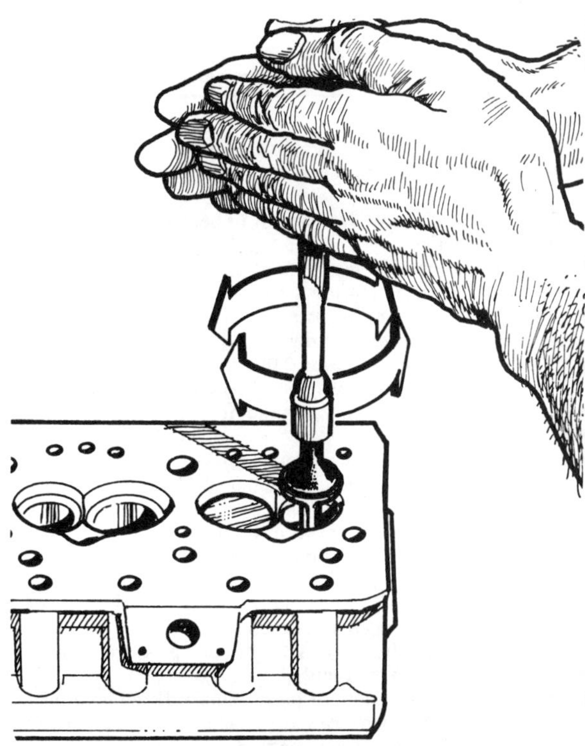

2 Check the rocker arms for wear of the rocker bushes, for wear at the rocker arm face which bears on the valve stem, and for wear of the adjusting ball ended screws. Wear in the rocker arm bush can be checked by gripping the rocker arm tip and holding the rocker arm in place on the shaft, noting if there is any lateral rocker arm shake. If shake is present, and the arm is very loose on the shaft, remedial action must be taken. Forged rocker arms which have worn bushes may be taken to your local BLMC agent or engineering works to have the old bush drawn out and a new bush fitted.
3 Check the tip of the rocker arm where it bears on the valve head for cracking or serious wear of the case hardening which will be indicated by flats. If none is present re-use the rocker arm. Check the lower half of the ball on the end of the rocker arm adjusting screw. On Healey engines wear on the ball and top of the pushrod is easily noted by the unworn 'pip' which fits in the small central oil hole in the ball. The larger this 'pip' the more wear has taken place to both the ball and the pushrod. Check the pushrods for straightness by rolling them on the bench. Renew any which are bent.

37 Tappets - examination and renovation

Examine the bearing surface of the tappets which lie on the camshaft. Any indentation in this surface or any cracks indicate serious wear and the tappets should be renewed. Thoroughly clean them out, removing all traces of sludge. It is most unlikely that the sides of the tappets will be worn, but if they are a very loose fit in their bores and can readily be rocked they should be exchanged for new units.

38 Flywheel starter ring - examination and renovation

1 If the teeth on the flywheel starter ring are badly worn, or if some are missing, it will be necessary to remove the ring. This is achieved by splitting the ring with a cold chisel. The greatest care should be taken not to damage the flywheel during this process.
2 Refitment must be carried out by a specialist engineer for the new ring requires heating before tapping on. A special oven or oxy-acetylene is needed.

39 Oil pump - examination and renovation

Early type

Thoroughly clean all the component parts in petrol and check the pump in the following manner:

1 Check if there is any slackness of fit between the inner rotor and the pump body.
2 Position the rotors in the pump and place the straight edge of a steel ruler across the joint face of the pump. Measure the gap between the bottom of the straight edge and the top of the rotors with a feeler gauge as shown in Fig 1.26.
3 If the measurement exceeds 0.005 inch then check the lobe clearances as described in the following paragraphs. If the lobe clearances are correct, then lap the joint face on a sheet of plate glass.
4 Measure with feeler gauges the gap between the inner and outer rotors. It should not be more than 0.010 inch as shown in Fig 1.27.
5 Then measure the gap between the outer rotor and the side of the pump body which should not exceed 0.008 inch.
6 It is essential to renew the pump if the measurements are outside these figures. It can be safely assumed that at any major reconditioning the pump will need renewal.
7 As will be seen from Fig 1.18 the pump is driven by the shaft from the gear on the camshaft.
8 If the case of total loss of oil pressure is being investigated this can usually be attributed to a failure of the shaft drive. Reassembly is the reverse sequence to dismantling.

Later type

Thoroughly clean all the component parts in petrol and check the pump in the following manner:
1 Check the end float which must not exceed 0.002 inch by placing a straight edge across the end faces of the gears and the oil pump body. Using feeler gauges measure the clearance between the end face of each gear and the underside of the straight edge.
2 The radial clearance between the gears and the oil pump body should not exceed 0.0025 inch and may be determined using feeler gauges.
3 Reassembly of the oil pump is the reverse sequence to dismantling. It should be noted that when refitting the driving gear it should be pressed onto the oil pump shaft until the lower end face of the shaft is 0.312 inch below the end face of the gear.

40 Cylinder head - decarbonisation

1 This can be carried out with the engine either in or out of the car. With the cylinder head off carefully remove with a wire brush and blunt scraper, all traces of carbon deposits from the combustion chamber and the ports. The valve head stems and valve guides should also be freed from any carbon deposits. Wash the combustion chamber and ports down with petrol and scrape the cylinder head surface free of any foreign matter with the side of a steel rule, or a similar article.
2 Clean the pistons and top of the cylinder bores. If the pistons are still in the block it is essential that great care is taken to ensure that no carbon gets into the cylinder bores as this could scratch the cylinder walls or cause damage to the pistons and rings. To ensure this does not happen, first turn the crankshaft so that two of the pistons are at the top of their bores. Stuff rag into the other four bores or seal them off with paper and masking tape. The waterways should also be covered with small pieces of masking tape to prevent particles of carbon entering the cooling system and damaging the water pump.
3 There are two schools of thought as to how much carbon should be removed from the piston crown. One school recommends that a ring of carbon should be left around the edge of the piston and on the cylinder bore wall as an aid to low oil consumption. Although this is probably true for early engines with worn bores, on later engines the thought of the second school may be applied; which is that for effective decarbonisation all traces of carbon should be removed.
4 If all traces of carbon are to be removed, press a little grease into the gap between the cylinder walls and the two pistons which are to be worked on. With a blunt scraper carefully scrape away the carbon from the piston crown, taking great care not to scratch the aluminium. Also scrape away the carbon from the surrounding lip of the cylinder wall. When all carbon has been removed, scrape away the grease which will now be contaminated with carbon particles, taking care not to press any into the bores. To assist prevention of carbon build-up the piston crown can be polished with a metal polish such as Brasso. Remove the rags or masking tape from the other four cylinders and turn the crankshaft so that two more pistons are not at the top. Place rag or masking tape in the cylinders which have been decarbonised and proceed as just described. Repeat once more to complete the cycle.
5 If a ring of carbon is going to be left round the piston then this can be helped by inserting an old piston ring into the top of the bore to rest on the piston and ensure that carbon is not accidentally removed. Check that there are no particles of carbon in the cylinder bores. Decarbonising is now complete.

41 Valve guides - examination and renovation

Examine the valve guides internally for wear. If the valves are a very loose fit in the guides and there is the slightest suspicion of lateral rocking, then new guides will have to be

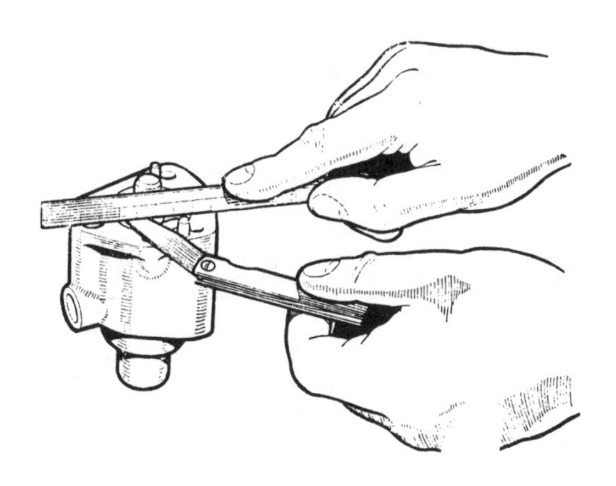

Fig 1.26 MEASURING THE OIL PUMP ROTOR END FLOAT WITH A STRAIGHT EDGE AND FEELER GAUGE. NO MORE THAN 0.005 INCH SHOULD BE PERMITTED

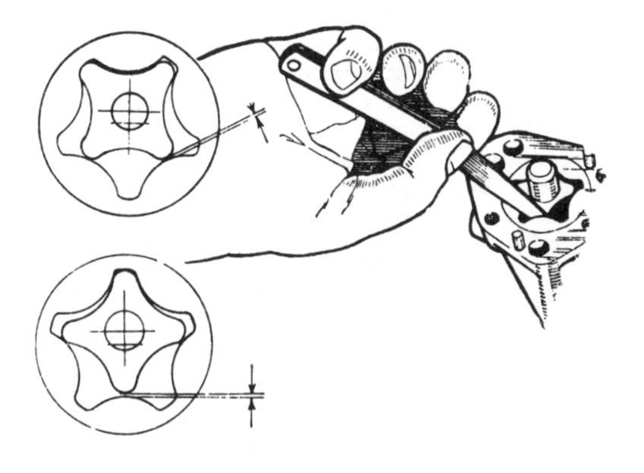

Fig 1.27 MEASURE THE LOBE CLEARANCES WITH A FEELER GAUGE IN THE POSITIONS INDICATED. THE GAPS SHOULD NOT EXCEED 0.010 INCH

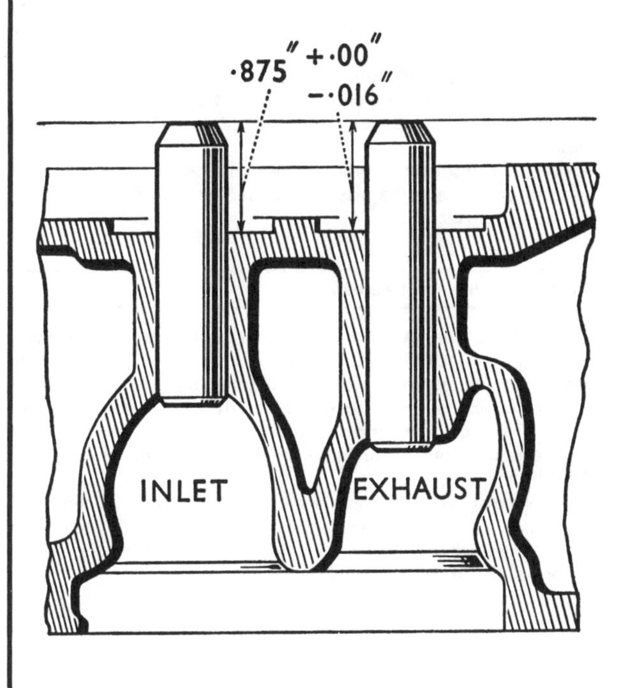

Fig 1.28 CORRECT POSITIONING OF VALVE GUIDES IN CYLINDER HEAD

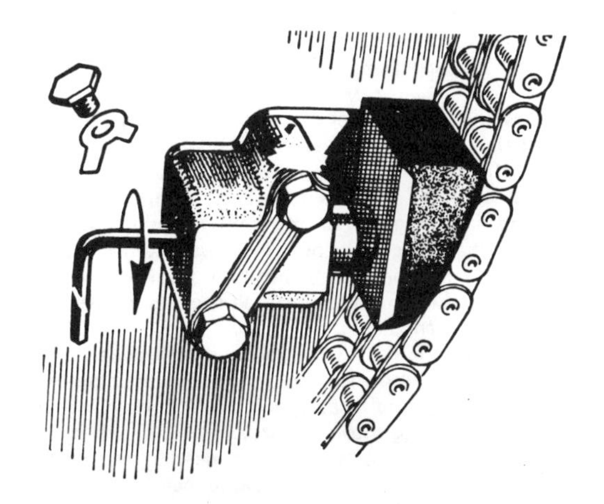

Fig 1.29 SETTING TIMING CHAIN TENSIONER WITH ALLEN KEY

refitted. If the valve guides have been removed compare them internally by visual inspection with a new guide as well as testing them for rocking with the valves.

42 Sump - examination and renovation

Thoroughly wash out the sump with petrol and then inspect the cork packing in the semi-circular crankshaft seal housings. If they are flattened new ones should be fitted. Remove the old packing pieces, carefully pushing them right down into the grooves. Should the packing material stand out more than 1/16 inch above the sump flange it must be cut back to this figure.

43 Engine reassembly - general

To ensure maximum life and reliability from a rebuilt engine, not only must everything be correctly assembled, but everything must be spotlessly clean, all the oilways must be clear, locking washers and spring washers must be fitted where indicated and all bearing and other working surfaces must be thoroughly lubricated during assembly. Before assembly begins renew any bolts or studs with damaged threads and whenever possible use new spring washers. Apart from your normal tools, a supply of clean rag, an oil can filled with engine oil (an empty plastic detergent bottle thoroughly cleaned and washed out will do just as well) a new supply of assorted spring washers, a set of new gaskets and preferably a torque spanner should be to hand.

44 Crankshaft - replacement

Ensure that the crankcase is thoroughly clean and that all oilways are clear. A thin-twist drill is useful for cleaning them out. If possible, blow them out with compressed air. Treat the crankshaft in the same fashion, and then inject engine oil into the crankshaft oilways.

Commence work on rebuilding the engine by replacing the crankshaft and main bearings. Carefully clean away all the protective grease with which new bearings are coated, with petrol.

1 Fit the three upper halves of the main bearing shells to their location in the crankcase, after wiping the locations clean.

2 Note that the back of each bearing is a tab which engages in locating grooves in either the crankcase of the main bearing cap housings.

3 With the three upper bearing shells securely in place, wipe the lower bearing cap housings and fit the three lower shell bearings to their caps ensuring that the right shell goes into the right cap if the old bearings are being refitted.

4 Wipe the recesses either side of the centre main bearing which locate the upper halves of the thrust washers.

5 Generously lubricate the crankshaft journals and the upper and lower main bearing shells and carefully place the crankshaft in position.

6 Introduce the upper halves of the thrust washers (the halves without tabs) into their grooves either side of the centre main bearing, rotating the crankshaft in the direction towards the main bearing tabs (so that the main bearing shells do not slide out). At the same time feed the thrust washers into their locations with their oil grooves outwards away from the bearing.

7 Fit the main bearing caps in position ensuring they locate properly. The mating surfaces must be spotlessly clean or the caps will not seat correctly.

8 When replacing the centre main bearing cap ensure the thrust washers, generously lubricated, are fitted with their oil grooves facing outwards, and the locating tab of each washer is in the slot in the bearing cap as shown (photo).

9 Replace the washers over the main bearing cap bolts and replace the main bearing cap nuts, screwing them up finger tight.

10 Test the crankshaft for freedom of rotation. Should it be very stiff to turn or possess high spots a most careful inspection must be made, preferably by a qualified mechanic with a micrometer to get to the root of the trouble. It is very seldom that any trouble of this nature will be experienced when fitting the crankshaft.

11 Tighten the main bearing nuts to a torque wrench setting of 75 lb ft and recheck the crankshaft for freedom of rotation (photo).

45 Piston and connecting rod - reassembly

If the same pistons are being used, then they must be mated to the same connecting rod with the same gudgeon pin. If new pistons are being fitted, it does not matter which connecting rod they are being used with but the gudgeon pins should be fitted on the basis of selective assembly.

This involves trying each of the pins in each of the pistons in turn and fitting them to the ones they fit best as is detailed below.

To avoid any damage to the piston it is best to heat it in boiling water and the gudgeon pin will slide in easily.

Lay the correct piston adjacent to each connecting rod and remember that the same rod and piston must go back into the same bore. If new pistons are being used it is only necessary to ensure that the right connecting rod is placed in each bore.

To assemble the pistons to the connecting rods, proceed as follows:

1 Locate the small end of the connecting rod in the piston, with the marking 'FRONT' on the piston crown towards the front of the engine.

2 Lubricate the gudgeon pin and slide it in through the piston and connecting rod little end bush until it is located centrally.

3 On early produced engines refit the pinch bolt and spring washer.

4 On later produced engines fit new circlips to either end of the gudgeon pin bore in the piston and double check that the circlips are seating correctly.

46 Piston ring - replacement

1 Check that the piston ring grooves and oilways are thoroughly clean and unblocked. Piston rings must always be fitted over the head of the piston and never from the bottom.

2 The easiest method to use when fitting rings is to wrap a 0.020 feeler gauge round the top of the piston and place the rings one at a time, starting with the bottom oil control ring, over the feeler gauge.

3 The feeler gauge, complete with ring, can then be slid down the piston over the other piston ring grooves until the correct groove is reached. The piston ring is then slid gently off the feeler gauge into the groove (photo).

4 An alternative method is to fit the rings by holding them slightly open with the thumbs and both of your index fingers. This method requires a steady hand and great care as it is easy to open the ring too much and break it.

47 Piston replacement

The pistons, complete with connecting rods, can be fitted to the cylinder bores in the following sequence:

1 With a wad of clean rag wipe the cylinder bores clean.

2 The pistons, complete with connecting rods, are fitted to their bores from above.

3 As each piston is inserted into its bore ensure that it is the correct piston/connecting rod assembly for that bore and that the connecting rod is the right way round and that the front of the piston is towards the front of the bore, ie: towards the front of the engine.

4 The piston will only slide into the bore as far as the oil control ring. It is then necessary to compress the piston rings

44.8

44.11

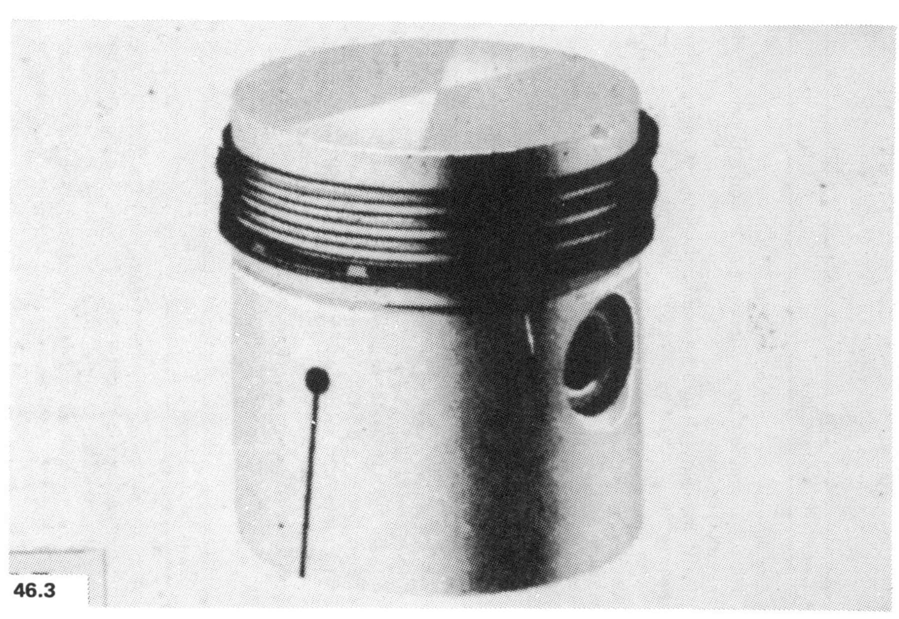
46.3

48.3

into a clamp and to gently tap the piston into the cylinder bore with a wooden or plastic hammer. If a proper piston ring clamp is not available then a suitable jubilee clip does the job very well.

48 Connecting rod to crankshaft - reassembly

As the big ends on the connecting rods are offset it will be obvious if they have been inserted the wrong way round as they will not fit over the crankpins. The centre two connecting rods should be fitted with the offset part of the rods adjacent, and the connecting rods at each extremity of the engine should have the offset part of the rods facing outwards.

1 Wipe the connecting rod half of the big end bearing cap and the underside of the shell bearing clean, and fit the shell bearing in position with its locating tongue engaged with the corresponding groove in the connecting rod.

2 If the old bearings are nearly new and are being refitted then ensure they are replaced in their correct locations on the correct rods.

3 Generously lubricate the crankpin journals with engine oil and turn the crankshaft so that the crankpin is in the most advantageous position for the connecting rod to be drawn onto it (photo).

4 Wipe the connecting rod bearing cap and back of the shell bearing clean and fit the shell bearing in position ensuring that the locating tongue at the back of the bearing engages with the locating groove in the connecting rod cap.

5 Generously lubricate the shell bearing and offer up the connecting rod bearing cap to the connecting rod.

6 Fit new connecting rod self-locking nuts if originally fitted or bolts, with one piece lock tab washer, and tighten the nuts/bolts to a torque wrench setting of 50 lb ft. With a cold chisel or pair of pliers tap or bend the locking washer tabs against the bolt head (photo).

7 When all the connecting rods have been fitted, rotate the crankshaft to check that everything is free, and there are no high spots causing binding.

49 Camshaft - replacement

1 Lay the cylinder block on its side.

2 Wipe the camshaft bearing journals clean and lubricate them generously with engine oil.

3 Insert the camshaft into the crankcase gently, taking care not to damage the camshaft bearings with the cams.

4 With the camshaft inserted into the block as far as it will go, rotate it to make sure that it is free in its bearings without any signs of binding.

5 Replace the camshaft locating plate and tighten down the retaining bolts and washers.

50 Oil pump and drive shaft - replacement

1 Fit the thrust washer to the oil pump driveshaft to be located as shown in Fig 1.19.

2 Fit a new joint washer to the driveshaft boss of the oil pump and then refit the driveshaft making sure that the rectangular drive end is located correctly in the end of the driving gear shaft. Place a new joint washer on the two oil pump mounting studs and offer the oil pump up into the cylinder block. Secure in position with the nuts and spring washers (photos).

51 Sump - replacement

1 After the sump has been thoroughly cleaned, scrape all traces of the old sump gasket from the sump flange, and fit new main bearing cap oil seals if required.

2 Wipe clean the inside of the crankcase, including the camshaft bearing surfaces. Thoroughly clean and scrape the crankshaft to sump flange.

3 With a new sump gasket held lightly in position offer up the sump to the crankcase (photo).

4 Bolt the sump in position with the large flat washer next to the sump flange and the starred or spring washer under the bolt head (photo).

5 Take care not to overtighten the sump bolts as they strip their threads very easily. Their correct tightening torque is 6 lb ft.

52 Timing gear, chain tensioner, cover - replacement

1 Before reassembly begins check that the packing washers are in place on the crankshaft nose. If new gearwheels are being fitted it may be necessary to fit additional washers (see paragraph 7, this Section). These washers ensure that the crankshaft gearwheel lines up correctly with the camshaft gearwheel.

2 Replace the Woodruff keys in their respective slots in the crankshaft and camshaft and ensure that they are fully seated. If their edges are burred they must be cleaned with a fine file.

3 Rotate the crankshaft and camshaft so that the keyways are approximately at the TDC position when seen from the front.

4 Double the timing chain, bringing both bright links together so giving a short and long portion of the timing chain on either side of the bright links.

5 With the shorter part of the timing chain on the right, ie: the bright links facing the operator, and the longer part on the left, engage the camshaft sprocket tooth marked with a 'dimple' on the top right link, and the camshaft sprocket with the tooth marked with a 'dimple' coinciding with the other bright link.

6 Place the sprockets in their respective positions on the camshaft and crankshaft complete with the chain and push the assembly home. Take care to ensure that the sprockets are kept in line with each other at all times so as to avoid straining the chain.

7 Note: If new gearwheels are being fitted they should be checked for alignment before being finally fitted to the engine. Place the gearwheels in position without the timing chain and place a straight edge of a steel ruler from the side of the camshaft gear teeth to the side of the crankshaft gearwheel and measure the gap between the steel rule and the gearwheel, If a gap exists a suitable number of packing washers must be placed on the crankshaft nose to bring the crankshaft gearwheel onto the same plane as the camshaft gearwheel.

8 Next assemble the timing chain tensioner by inserting one end of the spring into the plunger and fitting the other end of the spring into the cylinder.

9 Compress the spring until the cylinder enters the plunger bore and ensure the peg in the plunger engages the helical slot. Insert and turn an Allen key clockwise until the end of the cylinder is below the peg and the spring is held compressed.

10 Fit the backplate and secure the assembly to the cylinder block with the two bolts. Turn up the tabs of the lockwasher. On later produced engines refit the timing chain vibration damper, locating it on its dowel, and secure in position with the setscrew and spring washer.

11 With the timing chain in position, the tensioner can now be relaxed. Insert the 1/8 inch Allen key and turn it in a clockwise direction so that the slipper head moves forward under spring pressure against the chain. Do not under any circumstances turn the key anticlockwise or force the slipper head into the chain.

12 Refit the oil thrower with the concave side facing forwards, onto the end of the crankshaft. Fit the locking washer to the camshaft gearwheel with its locating tab in the gearwheel keyway.

13 Screw on the camshaft gearwheel retaining nut and tighten securely.

14 Bend up the locking tab of the locking washer to securely hold the camshaft retaining nut.

15 Generously lubricate the chain and gearwheel.

16 Ensure that the interior of the timing cover and timing cover flange is clean and generously lubricate the oil seal in the timing cover. Then, with a new gasket in position, fit the timing cover to the cylinder block.

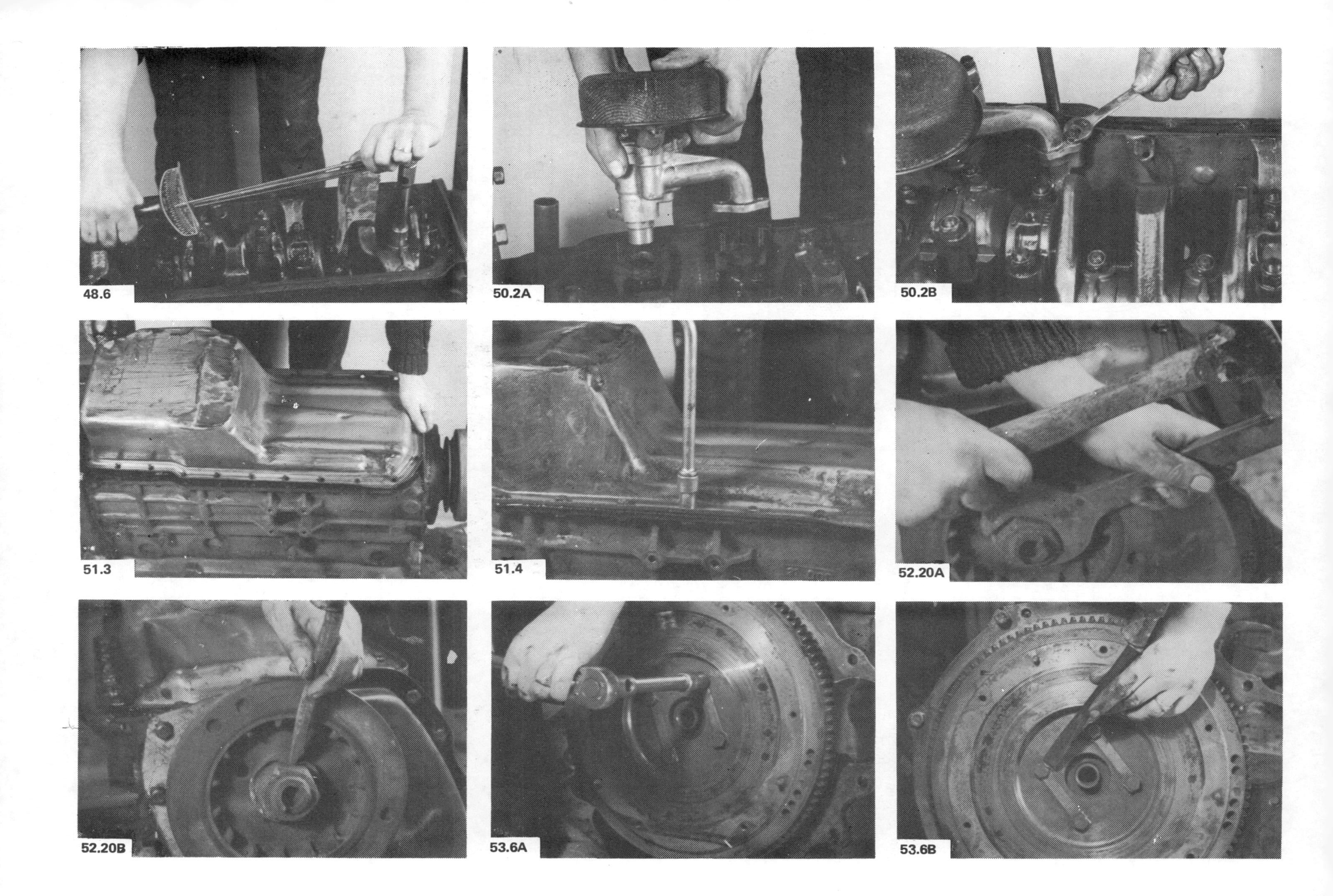

48.6

50.2A

50.2B

51.3

51.4

52.20A

52.20B

53.6A

53.6B

17 Screw in the timing cover retaining bolts with the flat washers next to the cover flange and under the spring washer. Do not fully tighten the bolts yet.
18 Fit the crankshaft pulley to the nose of the crankshaft ensuring that the keyway engages with the second Woodruff key. This action will centralise the oil seal in the timing cover with the pulley.
19 Tighten the timing cover bolts in a diagonal manner. The ¼ inch bolts should be tightened to a torque wrench setting of 6 lb ft and the 5/16 inch bolts to a torque wrench setting of 14 lb ft. It may be necessary to draw the pulley from the end of the crankshaft to gain access to the lower bolts but providing that the top ones are tight the cover will not move.
20 Fit the crankshaft retaining bolt locking washer in position and screw on the crankshaft pulley retaining dog. Tighten this to a torque wrench setting of 70 lb ft (photos).

53 Backplate and flywheel - replacement

1 To replace the engine backplate first fit a new paper gasket in place on the rear of the cylinder block and hold it in place with jointing compound.
2 Remember to fit into its groove the square section cork which will fit into the bottom of the rear main bearing cap. If this is not done the plate will have to be removed later on.
3 Replace the end plate, insert the securing bolts with locking tab washers and tighten in a diagonal manner. Pull up the tabs on the lockwashers.
4 When replacing the flywheel ensure that the TDC notch on the crankshaft pulley is set in line with the pointer on the timing cover as shown in Fig 1.31.
5 Wipe the mating faces of the crankshaft flange and flywheel clean and then fit the flywheel so that the Figure 1 stamped on the periphery is set vertical or alternatively with the timing rod on the periphery of the flywheel in line with and on the same side as the first and sixth throws of the crankshaft (Fig 1.30).
6 Replace the four high tensile steel bolts and lockplates and tighten the bolts to a torque wrench setting of 50 lb ft (photos).
7 Replace the clutch as described in Chapter 5, Section 6.

Fig 1.30 FLYWHEEL ALIGNMENT MARK

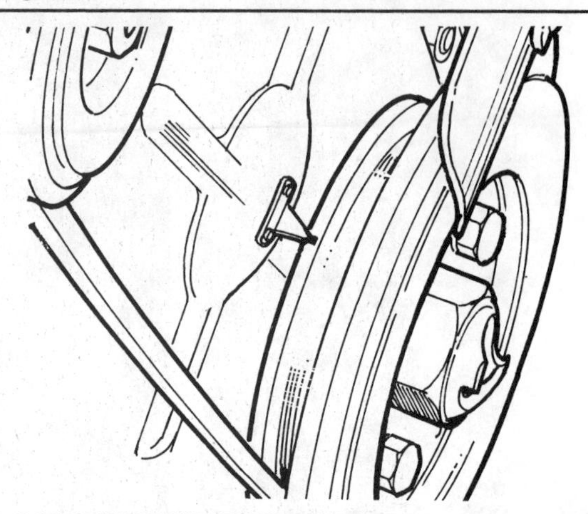

Fig 1.31 ALIGNMENT OF NOTCH IN PULLEY WITH POINTER ON TIMING COVER

54 Valve and valve spring assembly

To refit the valves and valve springs to the cylinder head, proceed as follows:
1 Rest the cylinder head on its side, or if the manifold studs are still fitted, with the gasket surface downwards.
2 Fit each valve and valve spring in turn, wiping down and lubricating each valve stem as it is inserted into the same valve guide from which it was removed.
3 As each valve is inserted slip the oil control rubber ring into place just under the bottom of the cotter groove (use a new rubber ring if possible).
4 Move the cylinder head towards the edge of the workbench if it is facing downwards and slide it partially over the edge of the bench so as to fit the bottom half of the valve spring compressor to the valve head.
5 Slip the valve spring, shroud and cap over the valve stem.
6 With the base of the valve compressor on the valve head, compress the valve spring until the cotters can be slipped into place in the cotter grooves. Gently release the compressor and fit the circlip in position in the grooves in the cotters.
7 Repeat this procedure until all twelve valves and valve springs are fitted.

55 Rocker shaft assembly

1 To reassemble the rocker shaft fit the split pin, flat washer and spring washer at the rear end of the shaft and then slide on the rocker arms, rocker shaft pedestals, and spacing springs in the same order in which they were removed.
2 With the front pedestal in position, screw in the rocker shaft locating screw and slip the locating plate into place. Finally, fit to the front of the shaft the spring washer, plain washer and split pin, in that order.

56 Tappet and pushrod replacement

1 Generously lubricate the tappets internally and externally and insert them in the bores from which they were removed through the tappet chest (photo).
2 With the cylinder head in position fit the pushrods in the same order in which they were removed. Ensure that they locate properly in the stems of the tappets and lubricate the pushrod ends before fitment.

57 Cylinder head - replacement

After checking that both the cylinder block and cylinder head mating faces are perfectly clean, generously lubricate each

56.1

57.1

57.4

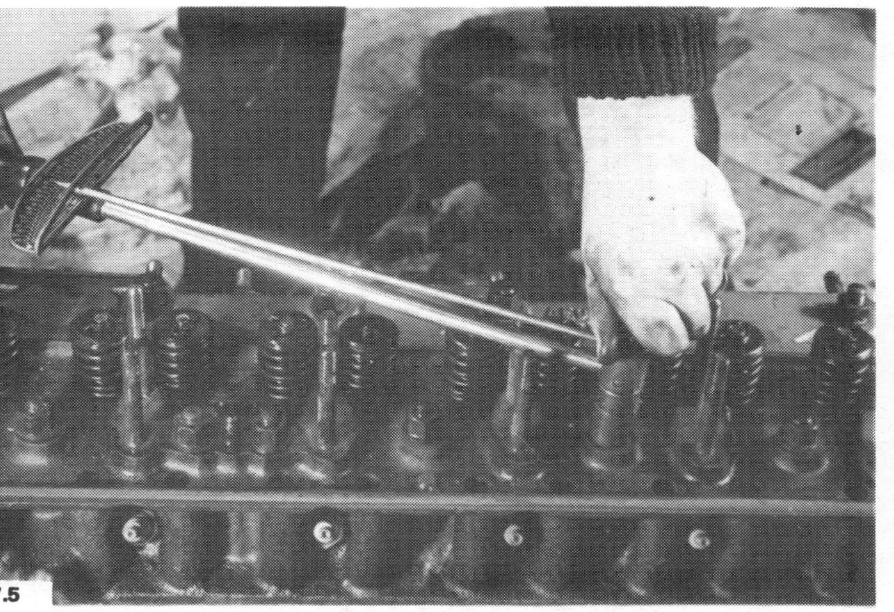
57.5

cylinder with engine oil.

1 Always use a new cylinder head gasket as the old gasket will be compressed and not capable of giving a good seal (photo).

2 Never smear grease on either side of the gasket as when the engine heats up the grease will melt and may allow compression leaks to develop. Gasket cement should be unnecessary on this application.

3 The cylinder head gasket is marked 'FRONT' and 'TOP' and should be fitted in position according to the markings.

4 With the gasket in position carefully lower the cylinder head onto the block (photo).

5 With the cylinder head in position fit the sixteen cylinder head nuts and washers and tighten in the manner shown in Fig 1.7 to a torque wrench setting of 75 lb ft (photo).

6 Fit the pushrods as detailed in the previous section.

7 The rocker shaft assembly can now be lowered over its twelve locating studs. Take care that the rocker arms are the right way round. Lubricate the ball joints, and insert the rocker arm ball joints in the pushrod cups (photo). Note: Failure to place the ball joints in the cups can result in the ball joints seating on the edge of a pushrod or outside it when the rocker assembly is pulled down tight.

8 Fit the rocker pedestal securing nuts and washers and tighten them down evenly in a diagonal manner to a torque wrench setting of 25 lb ft (photo).

9 Replace the rocker shaft feed pipe (photo).

58 Rocker arm/valve adjustment

1 The valve adjustments should be made with the engine cold. The importance of correct rocker arm/valve stem clearances cannot be overstressed as they vitally affect the performance of the engine.

2 If the clearances are set too open, the efficiency of the engine is reduced as the valves open late and close earlier than was intended. If, on the other hand, the clearances are set too close there is a danger that the stems will expand upon heating and not allow the valves to close properly which will cause burning of the valve head and seat and possible warping.

3 If the engine is in the car remove the two holding down studs from the rocker cover and lift the rocker cover and gasket away.

4 It is important that the clearance is set when the tappet of the valve being adjusted is on the heel of the cam, ie: opposite the peak. This can be done by carrying out the adjustment in the following order, which also avoids turning the crankshaft more than necessary:

Valve fully open	Check and adjust
Valve No 12	Valve No 1
6	7
4	9
11	2
8	5
3	10
1	12
7	6
9	4
2	11
5	8
10	3

5 The correct valve clearance is 0.012 inch but for competition work Mk II and Mk III models the clearance should be increased to 0.015 inch.

59 Distributor and distributor drive - replacement

It is important to set the distributor drive correctly as otherwise the ignition timing will be totally incorrect. It is easy to set the distributor drive in apparently the right position but exactly 180° out by failing to select the correct cylinder which must not only be at TDC but must also be on its firing stroke with both valves closed. The distributor drive should therefore not be fitted until the cylinder head is in position and the valves can be observed.

1 Rotate the crankshaft until No 6 piston is at TDC on its compression stroke. When the valves on Number 1 cylinder are 'rocking' ie: where the exhaust valve is just closing and the inlet valve is just opening. Number 6 piston is at the top of its compression stroke.

2 The engine may be set so that the groove in the crankshaft pulley is in line with the pointer on the timing chain cover, or the 'dimple' marks and bright links of the timing chain and gears are in line, the piston is exactly at TDC.

3 Screw a 5/16 inch UNF bolt into the threaded end of the distributor drive gear and insert it with the large segment away from the engine and the slot approximately in the 4 o'clock position (photo).

59.3

4 As the gear on the driveshaft engages with the camshaft the slot will turn in an anticlockwise direction until it just passes the horizontal position as shown in Fig 1.33.

5 Unscrew and remove the bolt from the drive gear and insert the drive housing. Secure it in position with the hexagon setscrew and shakeproof washer (photo).

59.5

57.7

Fig 1.32 CORRECT METHOD OF ADJUSTING ROCKER ARM/VALVE CLEARANCE

1 Feeler gauge
2 Rocker
3 Locknut
4 Adjusting screw

57.8

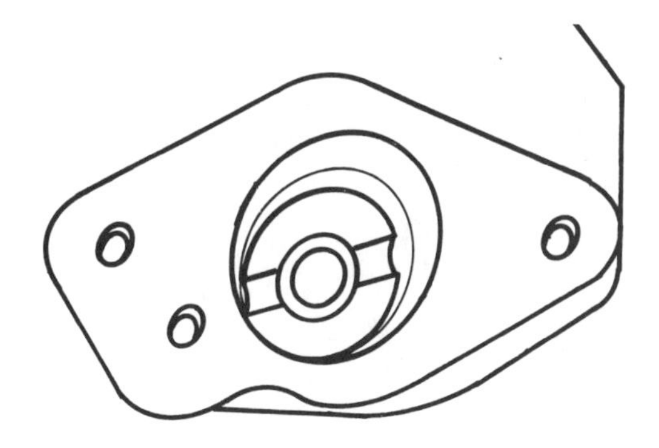

Fig 1.33 DISTRIBUTOR DRIVING GEAR WITH SLOT IN CORRECT POSITION FOR REPLACING DISTRIBUTOR

57.9

6 The distributor may now be refitted and the two securing bolts and spring washers which hold the distributor clamping plate to the distributor housing tightened.

7 If the clamp bolt on the clamping plate was not previously loosened and the distributor body was not tuned in the clamping plate, then the ignition timing will be as previously set. If the clamping bolt has been loosened then it will be necessary to retime the ignition and full information will be found in Chapter 4, Section 10.

60 Engine - final assembly

1 The stage at which ancillary equipment is fitted will depend on the method of refitting the engine which is controlled by the equipment available.

2 Whether the engine be in or out of the car the following items must be refitted, the procedure for which will be, in the main, the reverse sequence to removal: Engine side cover, exhaust manifolds, starter motor, water thermostat, oil filter, distributor, dynamo, carburettors, rocker cover and air cleaner.

3 Make sure that the oil filter is full of engine oil otherwise there will be a delay in the oil reaching the bearings while the oil filter refills.

4 The rocker cover should be refitted using a new cork gasket, do not overtighten the two shaped nuts otherwise the rocker cover will distort, causing oil leaks.

5 The engine side covers should be refitted using new cork gaskets and the securing bolts tightened to a torque wrench setting of 2 lb ft. Do not overtighten.

61 Engine - replacement

This operation is a direct reversal of the removal procedure but the following items should be checked if applicable to the method of refitting.

1 The engine front mounting clearance should be checked between the rebound buffer and mounting rubber as shown in Figs 1.34 and 1.35 and any adjustment made by inserting shims between the rebound rubber and its mounting bracket.

2 Check all electric cables, pipes and hoses to make sure that all have been refitted correctly. If necessary refer to the wiring diagrams in Chapter 10.

3 Refill the gearbox and overdrive as described in Chapter 6.

4 Check the carburettor adjustment as described in Chapter 3.

5 Check the ignition timing as described in Chapter 4.

6 Refill the cooling system as described in Chapter 2.

7 Refill the engine with oil as described in Section 2 of this Chapter.

62 Engine - initial start-up after overhaul or major repair

1 Make sure that the battery is fully charged and that all lubricants, coolants and fuel are replenished.

2 If the fuel system has been dismantled it will require several revolutions of the engine on the starter motor to get the petrol up to the carburettor. An initial 'prime' of about 1/6 of a cupful of petrol poured down each carburettor intake will help the engine to fire quickly. Do not overdo this as flooding may result.

3 As soon as the engine fires and runs keep it going at a fast tickover only (no faster) and bring it up to normal working temperature.

4 As the engine warms up there will be odd smells and some smoke from parts getting hot and burning off oil. Look for leaking oil and water which should be obvious if serious. Check also the clamp connections of the exhaust pipes to the manifolds as these do not always 'find' their gas tight position until warm and settled and it is almost certain that they will need tightening further. Do this with the engine switched off.

5 When running temperature has been reached adjust the idling speed as described in Chapter 3.

6 Stop the engine and wait a few minutes to see if any lubricant or coolant is dripping out. Rectify if necessary.

7 Road test the car to check that the timing is correct and giving the necessary smoothness and power. Do not race the engine - if new bearings and/or pistons and rings have been fitted it should be treated as a new engine and run it at reduced revolutions for 500 miles.

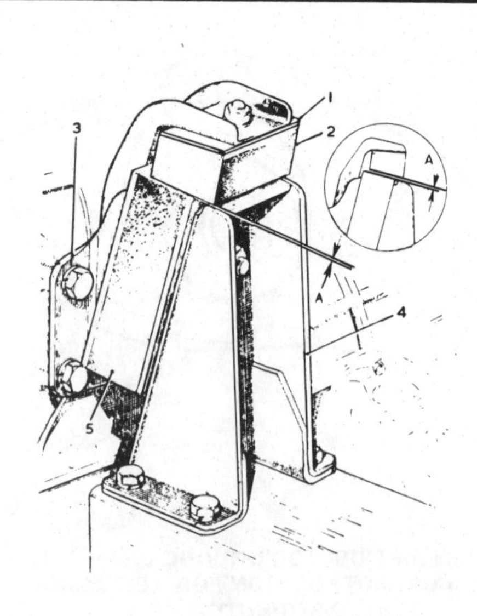

Fig 1.34 CORRECT SETTING FOR THE FRONT ENGINE MOUNTING (EARLY TYPE)

1 Shims
2 Rebound buffer
3 Mounting bracket on crank-case
4 Front mounting bracket
5 Mounting rubber
A 1/22 inch (0.8 mm) clearance

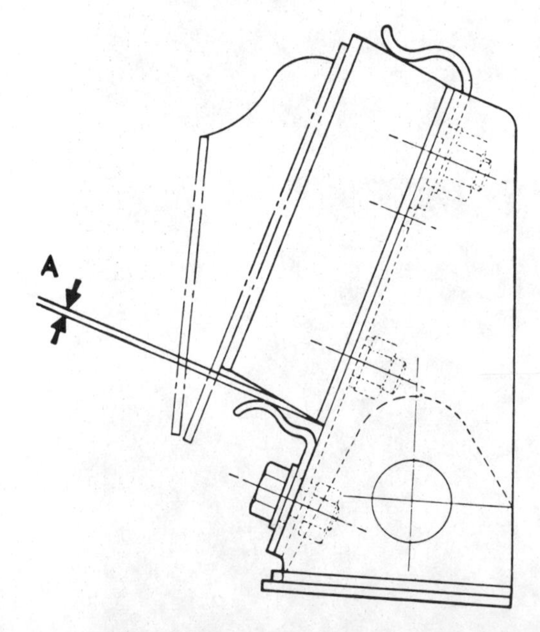

Fig 1.35 ENGINE FRONT MOUNTING (LATER TYPE) THE CLEARANCE 'A' MUST BE SET AT THE LOWER CHECK BRACKET BEFORE INSTALLING THE ASSEMBLY IN THE CAR

1A

1B

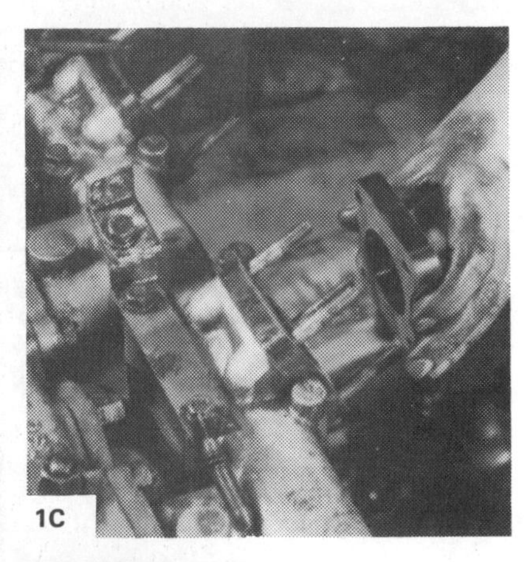
1C

1D

1E

REFITTING THE INLET MANIFOLD AND CARBURETTORS

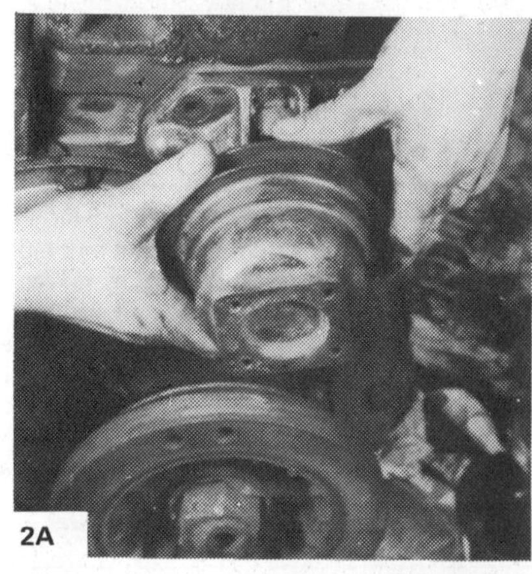
2A

2B

REFITTING THE WATER PUMP

3A

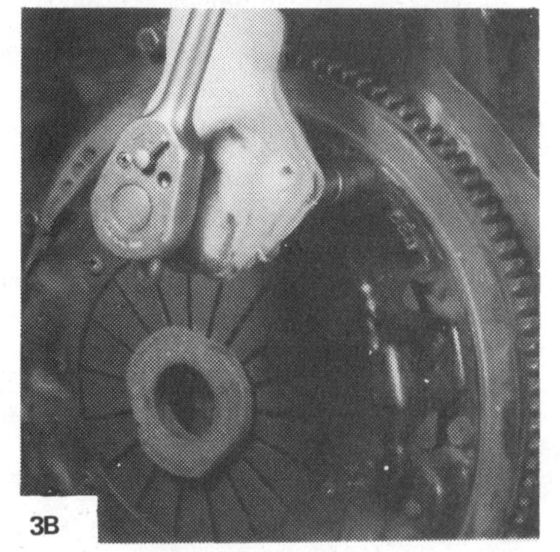
3B

REFITTING THE CLUTCH

63 Fault diagnosis

Cause	Trouble	Remedy
SYMPTOM: ENGINE FAILS TO TURN OVER WHEN STARTER BUTTON OPERATED		
No current at starter motor	Flat or defective battery	Charge or replace battery. Push-start car (manual transmission only).
	Loose battery leads	Tighten both terminals and earth ends of earth lead.
	Defective starter or solenoid switch or broken wiring	Run a wire direct from the battery to the starter motor or by-pass the solenoid.
	Engine earth strap disconnected	Check and retighten strap.
Current at starter motor	Jammed starter motor drive pinion	Place car in gear and rock from side to side (manual transmission only). Alternatively free exposed square end of shaft with spanner.
	Defective starter motor	Remove and recondition.
SYMPTOM: ENGINE TURNS OVER BUT WILL NOT START		
No spark at spark plug	Ignition damp or wet	Wipe dry the distributor cap and ignition leads.
	Ignition leads to spark plugs loose	Check and tighten at both spark plug and distributor cap ends.
	Shorted or disconnected low tension leads	Check the wiring on the CB and SW terminals of the coil and to the distributor.
	Dirty, incorrectly set, or pitted contact breaker points	Clean, file smooth, and adjust.
	Faulty condenser	Check contact breaker points for arcing, remove and fit new.
	Defective ignition switch	Bypass switch with wire.
	Ignition leads connected wrong way round	Remove and replace leads to spark plugs in correct order.
	Faulty coil	Remove and fit new coil.
	Contact breaker point spring earthed or broken	Check spring is not touching metal part of distributor. Check insulator washers are correctly placed. Renew points if the spring is broken.
No fuel at carburettor float chamber or at jets	No petrol in petrol tank	Refill tank!
	Vapour lock in fuel line (in hot conditions or at high altitude)	Blow into petrol tank, allow engine to cool, or apply a cold wet rag to the fuel line.
	Blocked float chamber needle valve	Remove, clean and replace.
	Fuel pump filter blocked	Remove, clean and replace.
	Choked or blocked carburettor jets	Dismantle and clean.
	Faulty fuel pump	Remove, overhaul and replace. Check contact breaker points on SU pumps.
Excess of petrol in cylinder or carburettor flooding	Too much choke allowing too rich a mixture to wet plugs	Remove and dry spark plugs or with wide open throttle, push-start the car.
	Float damaged or leaking or needle not seating	Remove, examine, clean and replace float and needle valve as necessary.
	Float lever incorrectly adjusted	Remove and adjust correctly.
SYMPTOM: ENGINE STALLS AND WILL NOT START		
No spark at spark plug	Ignition failure - sudden	Check over low and high tension circuits for breaks in wiring.
	Ignition failure - misfiring precludes total stoppage	Check contact breaker points, clean and adjust. Renew condenser if faulty.
	Ignition failure - in severe rain or after traversing water splash	Dry out ignition leads and distributor cap.
No fuel at jets	No petrol in petrol tank	Refill tank.
	Petrol tank breather pipe choked	Remove petrol cap and clean out breather hole or pipe.
	Sudden obstruction in carburettors	Check jets, filter, and needle valve in float chamber for blockage.
	Water in fuel system	Drain tank and blow out fuel lines.
SYMPTOM: ENGINE MISFIRES OR IDLES UNEVENLY		
Intermittent sparking at spark plug	Ignition leads loose	Check and tighten as necessary at spark plug and distributor cap ends.

Cause	Trouble	Remedy
SYMPTOM: ENGINE MISFIRES OR IDLES UNEVENLY		
Intermittent sparking at spark plug	Battery leads loose on terminals	Check and tighten terminal leads.
	Battery earth strap loose on body attachment point	Check and tighten earth lead to body attachment point.
	Engine earth lead loose	Tighten lead.
	Low tension leads to SW and CB terminals on coil loose	Check and tighten leads if found loose.
	Low tension lead from CB terminal side to distributor loose	Check and tighten if found loose.
	Dirty, or incorrectly gapped plugs	Remove, clean and regap.
	Dirty, incorrectly set, or pitted contact breaker points	Clean, file smooth and adjust.
	Tracking across inside of distributor cover	Remove and fit new cover.
	Ignition too retarded	Check and adjust ignition timing.
	Faulty coil	Remove and fit new coil.
Fuel shortage at engine	Mixture too weak	Check jets, float chamber needle valve, and filters for obstruction. Clean as necessary. Carburettors incorrectly adjusted.
	Air leak in carburettors	Remove and overhaul carburettor.
	Air leak at inlet manifold to cylinder head, or inlet manifold to carburettor	Test by pouring oil along joints. Bubbles indicate leak. Renew manifold gasket as appropriate.
Mechanical wear	Incorrect valve clearances	Adjust rocker arms to take up wear.
	Burnt out exhaust valves	Remove cylinder head and renew defective valves.
	Sticking or leaking valves	Remove cylinder head, clean, check and renew valves as necessary.
	Weak or broken valve springs	Check and renew as necessary.
	Worn valve guides or stems	Renew valve guides and valves.
	Worn pistons and piston rings	Dismantle engine, renew pistons and rings.
SYMPTOM: LACK OF POWER AND POOR COMPRESSION		
Fuel/air mixture leaking from cylinder	Burnt out exhaust valves	Remove cylinder head, renew defective valves.
	Sticking or leaking valves	Remove cylinder head, clean, check and renew valves as necessary.
	Worn valve guides and stems	Remove cylinder head and renew valves and valve guides.
	Weak or broken valve springs	Remove cylinder head, renew defective springs.
	Blown cylinder head gasket (accompanied by increase in noise)	Remove cylinder head and fit new gasket.
	Worn pistons and piston rings	Dismantle engine, renew pistons and rings.
	Worn or scored cylinder bores	Dismantle engine, rebore, renew pistons and rings.
Incorrect adjustments	Ignition timing wrongly set. Too advanced or retarded.	Check and reset ignition timing.
	Contact breaker points incorrectly gapped	Check and reset contact breaker points.
	Incorrect valve clearances	Check and reset rocker arm to valve stem gap.
	Incorrectly set spark plugs	Remove, clean and regap.
	Carburation too rich or too weak	Tune carburettors for optimum performance.
Carburation and ignition faults	Dirty contact breaker points	Remove, clean and replace.
	Fuel filters blocked causing top end fuel starvation	Dismantle, inspect, clean and replace all fuel filters.
	Distributor automatic balance weights or vacuum advance and retard mechanisms not functioning correctly	Overhaul distributor.
	Faulty fuel pump giving top end fuel starvation	Remove, overhaul or fit exchange reconditioned fuel pump.
SYMPTOM: EXCESSIVE OIL CONSUMPTION		
Oil being burnt by engine	Badly worn, perished or missing valve stem oil seals	Remove, fit new oil seals to valve stems.
	Excessively worn valve stems and valve guides	Remove cylinder head and fit new valve and valve guides.
	Worn piston rings	Fit oil control rings to existing pistons or

Cause	Trouble	Remedy
		purchase new pistons.
	Worn pistons and cylinder bores	Fit new pistons and rings, rebore cylinders.
	Excessive piston ring gap allowing blow-up	Fit new piston rings and set gap correctly.
	Piston oil return holes choked	Decarbonise engine and pistons.
Oil being lost due to leaks	Leaking oil filter gasket	Inspect and fit new gasket as necessary.
	Leaking rocker cover gasket	" " " " "
	Leaking tappet chest gasket	" " " " "
	Leaking timing case gasket	" " " " "
	Leaking sump gasket	" " " " "
	Loose sump plug	Tighten, fit new gasket if necessary.
SYMPTOM: UNUSUAL NOISES FROM ENGINE		
Excessive clearances due to mechanical wear	Worn valve gear (noisy tapping from rocker box)	Inspect and renew rocker shaft, rocker arms and ball pins as necessary.
	Worn big end bearing (regular heavy knocking)	Drop sump, if bearings broken up clean out oil pump and oilways, fit new bearings. If bearings not broken but worn fit bearing shells.
	Worn timing chain and gears (rattling from front of engine)	Remove timing cover, fit new timing wheels and timing chain.
	Worn main bearings (rumbling and vibration)	Drop sump, remove crankshaft, if bearings worn but not broken up, renew. If broken up strip oil pump and clean out oilways.
	Worn crankshaft (knocking, rumbling and vibration)	Regrind crankshaft, fit new main and big end bearings.

Chapter 2 Cooling system

Contents

Specifications

Type	Pressurised. Pump impeller and fan assisted. Centrifugal water pump.
Water pump drive	V belt from crankshaft pulley
Tension of fan belt	½ inch movement between water pump pulley and crankshaft pulley
Pressure cap opens	10 lb/sq in
Thermostat opening temperature:	
Mk I	70°C
Mk II and III (from engine number 29F/2592	83°C
Capacity (without heater)	18 pints

1 General description

The engine cooling water is circulated by a thermosyphon water pump assisted system and the coolant is pressurised. This is to prevent both the loss of water down the overflow pipe with the radiator cap in position and premature boiling in adverse motoring conditions.

The radiator is pressurised and increases the boiling point to approximately 225°F. If the water temperature exceeds this figure and the water boils, the pressure in the system forces the internal part of the cap off its seat, thus exposing the overflow pipe down which steam from the water escapes thus relieving the pressure.

It is, therefore, important to check that the radiator cap is in good condition and that the spring behind the sealing washer has not weakened. Most garages have special equipment on which radiator caps can be tested.

The cooling system comprises the radiator, top and bottom water hoses, heater hoses, impeller, water pump, the thermostat, the two drain taps and a fan mounted on the water pump spindle.

The system functions in the following manner. Cold water in the bottom of the radiator circulates up the lower hoses to the water pump where it is pushed round the water passages in the cylinder block, helping to keep the cylinder bores and pistons cool.

The water then travels up into the cylinder head and circulates round the combustion chambers and valve seats absorbing more heat and then, when the engine is at its correct operating temperature, travels out of the cylinder head, past the open thermostat into the upper radiator hose and so into the radiator header tank.

The water travels down the radiator where it is rapidly cooled by the inrush of cold air, through the radiator matrix, which is created by both the fan and the motion of the car. The water, now cooler, reaches the bottom of the radiator, when the cycle is repeated.

When the engine is cold the thermostat (which is a valve, able to open and close, depending on the temperature of the water) maintains the circulation of the same water in the engine.

Only when the correct minimum operating temperature has been reached as shown in the Specifications, does the thermostat begin to open, allowing water to return to the radiator.

2 Routine maintenance

1 Check the level of the water in the radiator once a week or more frequently if necessary, and top up with a soft water (rain water is excellent) as required.
2 Once every 6000 miles check the fan belt for wear and correct tension and renew or adjust the belt as necessary. (See Sections 11 and 12 for further details.)
3 Once every 12,000 miles unscrew the plug from the top of the water pump and press in a little grease using finger pressure only. Do not overgrease or the carbon seal may be rendered inoperative. Replace the plug and screw down. This plug is shown in Fig 2.2.

3 Cooling system - draining

1 With the car on level ground drain the system as follows:
2 If the engine is cold remove the filler cap from the radiator by turning the cap anticlockwise. If the engine is hot, having just been run, then turn the filler cap very slightly until the pressure in the system has had time to disperse. Use a rag over the cap to protect your hand from escaping steam. If, with the engine very hot, the cap is suddenly released, the drop in pressure can result in the water boiling. With the pressure released the cap can be removed.
3 If antifreeze is in the radiator drain it into a clean bowl having a capacity of at least 18 pints (19 pints with heater) for re-use.
4 Open the two drain taps. When viewed from the front the radiator drain tap is on the bottom right hand side of the radiator and the engine drain tap is halfway down the right hand side of the cylinder block under the exhaust manifold. These

taps are shown in Figs 2.3A and 2.3B. A short length of rubber tubing over the radiator drain tap nozzle will assist draining the coolant into a container without splashing.
5 When the water has finished running, probe the drain tap orifices with a short piece of wire to dislodge any particles of rust or sediment which may be blocking the taps and preventing all the coolant draining out.

4 Cooling system - flushing

1 With time the cooling system will gradually lose its efficiency as the radiator becomes choked with rust scales, deposits from the water and other sediment. To clean the system out, remove the radiator cap and drain tap and leave a hose running in the radiator cap orifice for ten to fifteen minutes.
2 In very bad cases the radiator should be reverse flushed. This can be done with the radiator in position. The cylinder block tap is closed and a hose placed over the open radiator drain tap. Water, under pressure is then forced up through the radiator and header tank filler orifice.
3 The hose is then removed and placed in the filler orifice and the radiator washed out in the usual fashion.

5 Cooling system - filling

1 Close the two drain taps.
2 Fill the system slowly to ensure that no air locks develop. If a heater is fitted, check that the valve to the heater is open, otherwise an air lock may form in the heater. The best type of water to use in the cooling system is rain water so use this whenever possible.
3 Do not fill the system higher than within ½ inch of the filler orifice. Overfilling will merely result in wastage, which is especially to be avoided when antifreeze solution is in use.
4 Only use antifreeze with a glycerine or ethylene glycol base.
5 Replace the filler cap and turn it firmly clockwise to lock it in position.

6 Radiator - removal, inspection, cleaning and replacement

1 To remove the radiator first drain the cooling system as described in Section 3 of this Chapter.
2 Undo the clip that holds the top water hose to the cylinder head thermostat pipe outlet.
3 Pull the top hose off the thermostat outlet, if necessary using a screwdriver to break the seal.
4 Undo the clips securing the lower radiator hose to the water pump, heater outlet pipe and bottom radiator tanks.
5 Carefully remove the lower radiator hose.
6 Disconnect the thermometer element from the radiator header tank.
7 Undo and remove the six nuts and spring washers that secure the radiator to the mounting flanges (there are three nuts and spring washers each side).
8 Carefully lift the radiator up and out of the engine compartment together with the top hose and overflow pipe. The fan blades must not touch the fragile matrix.
9 With the radiator out of the car any leaks can be soldered up or repaired with a substance such as Cataloy. Clean out the inside of the radiator by flushing out as detailed in Section 4. When the radiator is out of the car it is advantageous to turn it upside down for reverse flushing. Clean the exterior of the radiator by hosing down the radiator matrix with a strong jet of water to clean away road dirt, dead flies etc.
10 Inspect the radiator hoses for cracks, internal and external perishing, and damage caused by overtightening of the securing clips. Renew any hoses as necessary. Examine the radiator hose securing clips and renew them if they are rusted or distorted. The drain tap should be renewed if showing signs of leaking.
11 Refitting the radiator is the reverse sequence to removal.

7 Thermostat - removal, testing and replacement

1 To remove the thermostat, partially drain the cooling system (8 pints is enough), loosen the upper radiator hose at the thermostat elbow end and pull it off the elbow.
2 Unscrew and remove the two nuts and spring washers from the thermostat housing (Fig 2.5) and lift the housing and paper gasket away from the top of the cylinder head. If the housing is difficult to move, gently tap with a soft faced hammer so as to release any corrosion holding it in position.
3 Lift out the thermostat if necessary lifting the lip with a screwdriver should it be tight.
4 Test the thermostat for correct functioning by immersing it in a saucepan of cold water together with a thermometer. Suspend the two items with string and do not allow them to touch the side or bottom.
5 Heat the water and note when the thermostat begins to open. Refer to the Specifications at the beginning of this Chapter for the correct opening temperature.
6 Discard the thermostat if it opens too early. Continue heating the water until the thermostat is fully open. Then let it cool down naturally. If the thermostat will not open fully on boiling or does not close down as the water cools, then it must be exchanged for a new one.
7 If the thermostat is stuck open when cold this will be apparent when removing it from the housing.
8 Replacement of the thermostat is the reverse sequence to removal. Remember to use a new paper gasket between the thermostat housing elbow and the thermostat. If the thermostat housing is badly eaten away it should be renewed.
9 It should be noted that on later produced engines instead of the bellows type thermostat being fitted a new wax element type thermostat was used. When refitting this type of thermostat, which is interchangeable with the earlier type, make sure that the threaded stem of the unit faces upwards.

8 Water pump - removal and replacement

1 The water pump is of the centrifugal impeller design and is attached on a common spindle to the fan blades. It operates in a casing which is bolted to the front end of the cylinder block.
2 The watertight seal is by means of a spring loaded hard faced seal bearing on a seating on the impeller hub.
3 To remove the water pump, first partially drain the cooling system as described in Section 3 of this Chapter, usually 8 pints is enough.
4 Refer to Section 6 and remove the radiator from the engine compartment.
5 Remove the fan belt as described in Section 12 of this Chapter.
6 Undo and remove the four bolts that secure the fan blades to the pulley and lift away the fan blades.
7 Slacken the hose clip securing the hose to the water pump body and ease the hose from the elbow.
8 Undo the four nuts that secure the water pump to the front of the cylinder block and lift away the nuts and spring washers followed by the water pump and its gasket.
9 Refitting is a straightforward reversal of the removal sequence. Make sure that the mating faces are clean and always use a new gasket. Note that the fan belt tension must be correct when all is reassembled. If the belt is too tight undue strain will be placed on the water pump and dynamo bearings, and if the belt is too loose it will slip and wear rapidly as well as giving rise to low electrical output from the dynamo.

9 Water pump - dismantling and reassembly

It should be noted that later engines are fitted with a water pump having a one-piece bearing and spindle assembly. The fan pulley is an interference press fit onto the spindle instead of

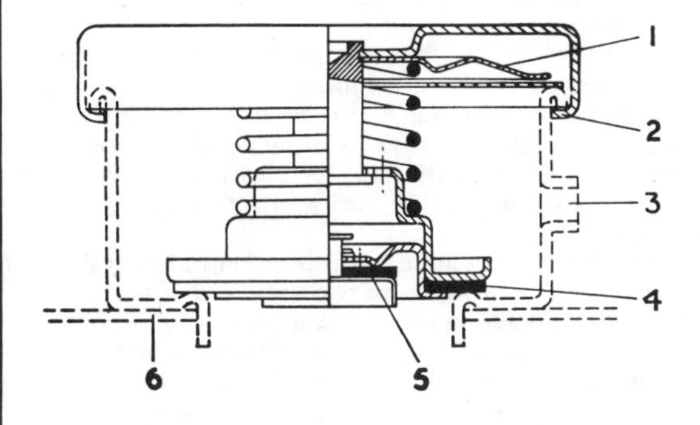

Fig 2.1 RADIATOR CAP COMPONENT PARTS

1 Spring friction plate
2 Retaining lugs
3 Pressure release pipe
4 Pressure valve seal
5 Vacuum valve seal
6 Header tank

Fig 2.2 LUBRICATION PLUG ON WATER PUMP BODY

1 Thermostat cover securing nuts
2 Thermostat cover
3 Lubrication plug

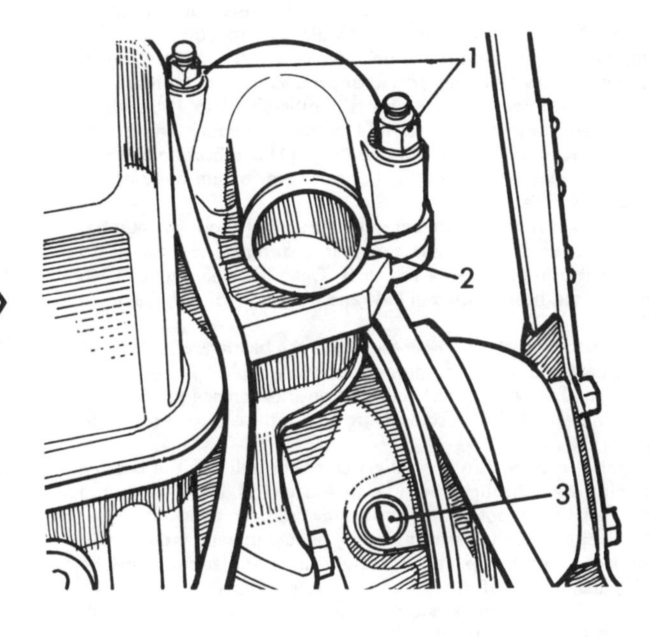

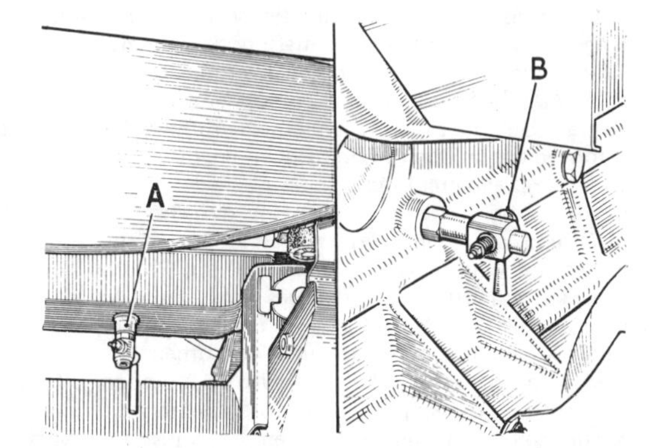

Fig 2.3 COOLING SYSTEM DRAIN TAPS

A Radiator tap
B Cylinder block tap
Both are turned to horizontal position for draining

Fig 2.4 CROSS SECTION THROUGH LATER TYPE WATER PUMP

When assembled, the hole in the bearing (A) must coincide with the lubricating hole in the pump body and there must be a clearance of .020 to .030 inch between the vane and the pump body. The recessed face of the hub (B) must be flush with the end of the spindle.

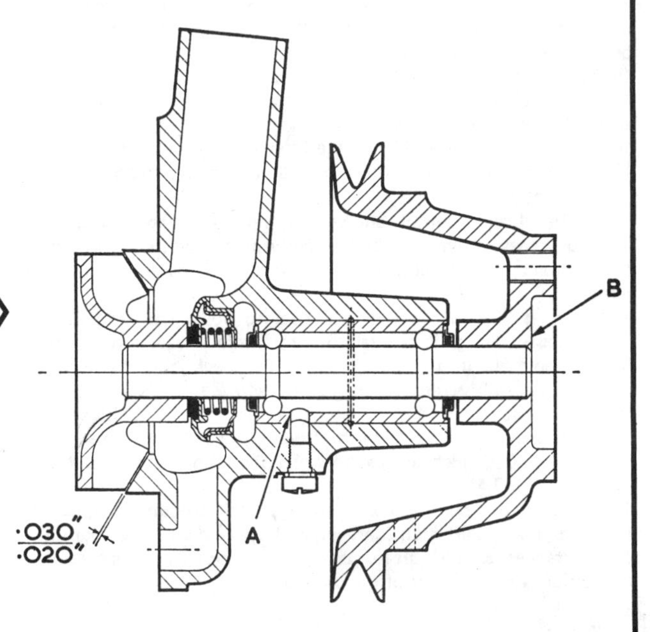

being keyed and secured by a nut and washers as used on the earlier type. Information on this later type pump will be found in paragraphs 8 to 13 (dismantling) and from 16 to 19 (reassembly).

1 Early type. Undo and remove the large nut, spring washer and plain washer from the front end of the spindle. The fan pulley may now be removed.

2 Ease off the Woodruff key from the spindle with a screwdriver and with a fine file remove any burrs from the keyway side.

3 With a pair of pliers release and withdraw the circlip and then grease the retaining washer.

4 Using a soft faced hammer and holding the pump body in the hand, gently tap out the spindle towards the rear. The spindle will bring with it the impeller vane and water seal which if required, may be either drifted or pressed from the spindle.

5 With reference to Fig 2.5 parts numbered 13 to 16 inclusive and 18 are removed and assembled from the front end of the pump body, whereas the front bearing (11) and bearing distance piece (12) may be drifted out from the rear of the body with a suitable soft metal drift.

6 Although it is recommended that a special tool, part number 18G61 be used to remove the rear bearing, it is possible to remove it using a suitable size drift. Once this bearing has been removed the bearing oil seal (14) and housing (15) may be taken off.

7 Inspect all parts for wear especially the bearings and the seal. If any parts are suspect fit new as necessary.

8 Later type. As the fan pulley is an interference press fit on the spindle it will be necessary to use a three legged puller to remove the pulley.

9 With a pair of thin nose pliers carefully pull out the bearing retaining wire through the hole in the top of the water pump body, which is exposed once the pulley has been removed.

10 The spindle and bearing assembly are combined (and are only supplied on exchange as a complete unit) and should now be gently tapped out of the rear of the water pump.

11 The oil seal assembly and the impeller will also come out with the spindle and bearing assembly.

12 The impeller vane is removed from the spindle by judicious tapping and levering, or preferably, to ensure no damage and for ease of operation, with an extractor. The oil seal assembly can then be slipped off.

13 Check the spindle and bearing assembly for signs of wear and if one or both parts are worn a new assembly must be obtained. Also check the water seal if the carbon face shows signs of wear.

14 Early type. Reassembly of the early type is a direct reversal of the dismantling procedure. It will be necessary to repack the bearings before refitting them.

15 Take great care that when drifting the front and rear bearings into position they are fitted squarely. Also line up the spacer between the bearings.

16 Later type. Reassembly of the later type is a direct reversal of the dismantling procedure but there are three additional points to be noted.

17 If the oil seal assembly showed any sign of damage or wear it must be renewed and the gasket between the water pump and the cylinder block should be renewed every time the pump is removed.

18 There is a small hole in the bearing body (A) (Fig 2.4). When reassembling, it is vital that this hole lines up with the lubrication hole in the pump body. To check that this is so, prior to reassembly, remove the greasing screw and check visually that the hole is in the correct position directly below the greasing aperture. Check with feeler gauges that a gap of 0.020 to 0.030 inch exists between the impeller and pump body (see Fig 2.4).

19 Regrease the bearing by pushing a small amount of grease into the greaser and then screwing in the grease screw. Under no circumstances should grease be applied under pressure as it could ruin the efficiency of the oil seal.

10 Antifreeze mixture

1 In circumstances where it is likely that the temperature will drop to below freezing point, it is essential that some water is drained and an adequate amount of ethylene glycol antifreeze (Castrol Antifreeze) to BS 3151 or BS 3152 is added to the cooling system. Never use an antifreeze with an alcohol base as evaporation is too high.

2 Castrol Antifreeze with an anti-corrosion additive can be left in the cooling system for up to two years, but after the first six months it is advisable to have the specific gravity of the coolant checked at your local garage, and thereafter every three months.

3 By referring to the right hand column in the table below the amounts of antifreeze required for various degrees of frost protection will be found:

Anti-freeze	Commence to freeze		Frozen solid		Amount of antifreeze
%	^{o}C	^{o}F	^{o}C	^{o}F	Pints
25	-13	9	-26	-15	5¼
$33^{1}/3$	-19	-2	-36	-33	7
50	-36	-33	-48	-53	10½

4 Always mix a pint or so of additional antifreeze solution which is to be used for topping up as necessary.

11 Fan belt - adjustment

1 It is important to keep the fan belt correctly adjusted and it should be checked regularly.

2 If the belt is too loose it will slip, wear rapidly, and cause the dynamo and water pump to malfunction. If the belt is too tight the dynamo and water pump bearings will wear rapidly causing premature failure of these components.

3 The fan belt tension is correct when there is ½ inch of lateral movement at the midpoint position of the belt between the water pump and crankshaft pulleys.

4 To adjust the fan belt, slacken the dynamo securing bolts (Fig 2.6) and move the dynamo either in or out until the correct tension is obtained. It is easier if the dynamo bolts are only slackened a little so it requires some force to move the dynamo. In this way the tension of the belt can be arrived at more quickly than by making frequent adjustment.

5 If, with the dynamo bolts only slightly loosened, difficulty is experienced in moving the dynamo away from the engine, a long spanner placed behind the dynamo and resting against the block serves as a very good lever and can be held in this position while the dynamo bolts are finally tightened.

12 Fan belt - removal and replacement

1 To remove the fan belt refer to Fig 2.6 and slacken all four dynamo mountings. Move the dynamo towards the cylinder block so as to release the tension on the belt.

2 Lift the belt from the dynamo pulley and then from the crankshaft pulley. It may then be lifted over the fan blades.

3 To refit the fan belt pass it over the fan blades and then onto the fan hub pulley. Next place the belt over the dynamo pulley.

4 Place the belt in the right hand side of the crankshaft pulley and with a large spanner or the starting handle rotate the pulley, so drawing the belt onto the groove of the pulley.

5 Adjust the fan belt tension as detailed in Section 11 of this Chapter.

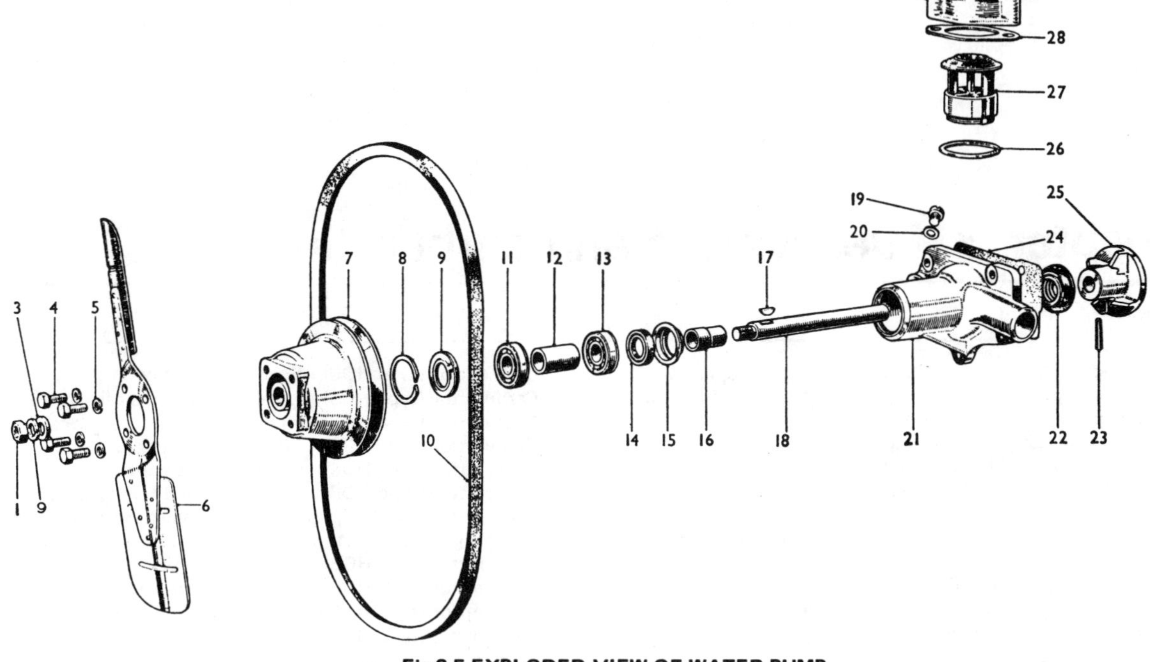

Fig 2.5 EXPLODED VIEW OF WATER PUMP

1 Spindle nut
2 Spring washer
3 Plain washer
4 Fan blade setpins
5 Spring washer
6 Fan. Some early models 2 blades
7 Fan pulley
8 Split ring
9 Grease retainer
10 Fan belt
11 Ball race
12 Distance piece
13 Ball race
14 Rubber seal
15 Seal housing
16 Distance piece
17 Key
18 Spindle
19 Oiling plug
20 Fibre washer
21 Pump body
22 Water seal
23 Locking pin
24 Joint washer
25 Impeller
26 Joint washer
27 Thermostat
28 Joint washer
29 Thermostat cover

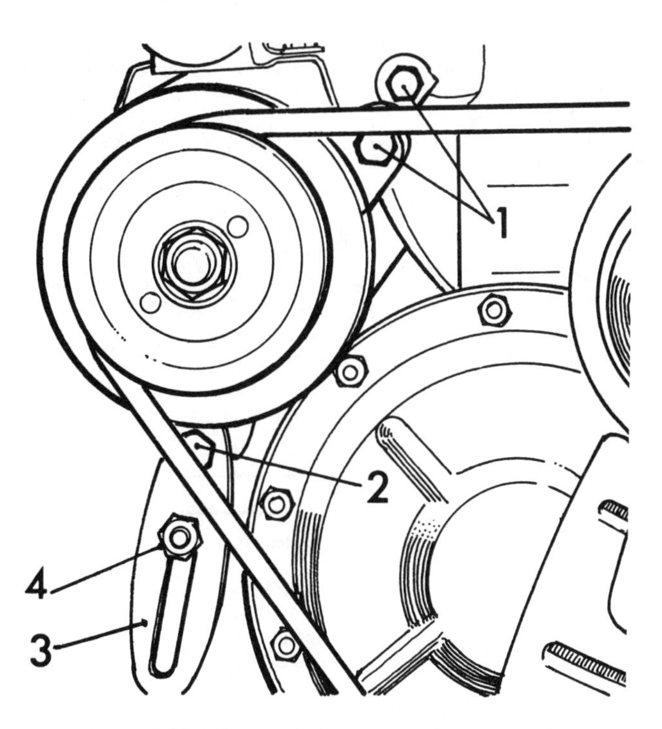

Fig 2.6 FAN BELT ADJUSTMENT POINTS

1 & 2 Dynamo securing bolts
3 Swinging link
4 Locknut

Chapter 3 Fuel system and carburation

Contents

Specifications

Fuel pump ... SU electric

Types:

BN4 up to engine number 60412	Single HP
Mk II and early Mk III	Single LCS
Later Mk III	Single AUF 300
Later Mk III convertible	Single AUF 301

Test data:

	HP	LCS	AUF 300/301
Maximum flow (gallons per hour)	12½	12½	15
Suction head (inches)	30	33	18
Delivery head (feet)	4	4	4
Cut off pressure (lb/sq in)	3.8	3.8	3.8

Carburettors

Model	Series	Carburettor Number	Type	Throttle Diameter	Spring Colour	Rich	Standard	Weak
100-Six	BN4	2	H4	1½ in	Red	4	AJ	MI
100-Six	BN4 6 port head	2	HD6	1¾ in	Yellow	RD	CV	SQ
100-Six	BN6	2	HD6	1¾ in	Yellow	RD	CV	SQ
3000 Mk I	BN7/BT7	2	HD6 or HD6 thermo	1¾ in	Green	RD	CV	SQ
3000 Mk II	BN7/BT7	3	HS4	1½ in	Red	DK	DJ	DH
3000 Mk II	BJ7	2	HS6	1¾ in	Red	RD	BC	TZ
3000 Mk III	BJ8	2	HD8	2 in	Red and Green	UN	UH	UL

Spring identification:

Colour	Load at length
Red	4½ oz at 2.625 in
Yellow	8 oz at 2.750 in
Green	12 oz at 3.000 in

Jet size:
- 100-Six and 3000 Mk I ... ... 0.100 inch
- 3000 Mk II ... ... 0.090 inch
- 3000 Mk II series BJ7 ... ... 0.100 inch
- 3000 Mk III ... ... 0.125 inch

Float setting
- All models except below ... ... 7/16 inch
- Mk I and II (series BN7 and BT7) * ... ... 5/16 inch
- Mk II and III (series BJ7 and BJ8) * ... ... 5/16 inch

* Nylon float only

Air cleaners
- 2639 cc engine ... ... Twin 'Pancake' type
- 2912 cc engine ... ... Oil wetted

Fuel tank capacity ... ... 12 gallons

Main fuel pipe outside diameter ... ... 0.375 inch

Torque wrench setting
- Carburettor stud nuts ... ... 2 lb ft (0.3 kg m)

1 General description

The fuel system fitted to models covered by this manual comprise the fuel tank having a capacity of 12 gallons and mounted at the rear of the car, an electric fuel pump and twin SU carburettors except for 3000 Mk II Series BN7/BT7 models which were fitted with three SU carburettors. On some carburettor installations an electrically operated automatic choke was fitted instead of the conventional SU method of richening the petrol/air charge.

2 Air cleaners

1 Every 3000 miles the air cleaners should be removed and cleaned as described below.
2 Undo and remove the two set bolts and spring washer securing each air cleaner to the carburettor air intake flange and carefully lift away each air cleaner and paper gasket. Note which way round the air cleaners and paper gaskets are fitted with reference to the two air drillings above the set bolt holes in the carburettor air intake flanges. Always fit new paper gaskets if the air cleaners are removed.
3 Thoroughly wash the air cleaner gauze in paraffin and when clean allow the paraffin to drain off.
4 Apply a little Castrol GTX engine oil to the gauze and refit the air cleaner, noting the comments made in the latter half of paragraph 2.

3 Fuel pump - general description

Four different types of SU fuel pump have been fitted during the production life of the car. Each pump is dealt with individually where a considerable difference arises so that no confusion can arise whilst servicing.

The operation of each type of pump is basically identical so the following description can be applied to a particular type.

The SU 12 volt electric fuel pump consists of a long outer body casing housing and diaphragm, armature and solenoid assembly, with at one end the contact breaker assembly protected by a bakelite cover, and at the other end a short casting containing the inlet and outlet ports, filter, valves and pumping chamber. The joint between the bakelite cover and the body casing is protected with a rubber sheath.

The pump operates in the following manner. When the ignition is switched on current travels from the terminal on the outside of the bakelite cover through the coil located round the solenoid core which becomes energised and acting like a magnet draws the armature towards it. The current then passes through the points to earth.

When the armature is drawn forward it brings the diaphragm with it against the pressure of the diaphragm spring. This creates sufficient vacuum in the pump chamber to draw in fuel from the tank through the fuel filter and non-return inlet valve.

As the armature nears the end of its travel a 'throw-over' mechanism operates which separates the points so breaking the circuit.

The diaphragm return spring then pushes the diaphragm and armature forwards into the pumping chamber so forcing the fuel in the chamber out to the carburettor through the non-return valve. When the armature is nearly fully forward the throw-over mechanism again functions, this time closing the points and re-energising the solenoid, so repeating the cycle.

5 Fuel pump - removal and refitting

The fuel pump on all models except the BN7, BT7, BJ7 and BJ8 (early models) series is fitted under the left hand seat pan. The BN7 series has the fuel pump fitted on the left hand side of the car and access to it is gained through the hinged portion of the spare wheel floor. The fuel pump on later BN7 models is accessible through the hinged portion of the spare wheel floor but now located on the right hand side.

1 To remove the fuel pump from all models except the BN7, BT7, BJ7 and BJ8 (early models) series first disconnect the positive terminal from the battery. Lift away the left hand seat pan. (Later BN7, BT7, BJ7 and BJ8 models, right hand seat pan).

2 Detach the electrical cable connection from the contact breaker end of the pump.

3 Unscrew the inlet and outler pump connections from the unions on the pump. On the LCS type pump slacken the clips securing the pump inlet and outlet hoses to the pump unions and carefully ease the hoses from the unions. Make a note of which are the inlet and outlet hoses so that they may be refitted in their original positions.

4 Undo and remove the nuts, bolts and washers securing the pump mounting bracket to the body and lift away the pump and bracket.

5 On early BN7 series cars, partially withdraw the spare wheel in the boot and raise the hinged lid located in the spare wheel floor.

6 Disconnect the positive terminal from the battery.

7 Unscrew the inlet and outlet pipe connections from the unions on the pump. Make a note of which are the inlet and outlet hoses so that they may be refitted in their original positions.

8 Undo and remove the nuts, bolts and washers securing the pump mounting bracket to the body and lift away the pump and bracket.

9 From car number 17547 (BN7) and 17352 (BT7) the fuel pump and fuel lines were transferred from the left hand side to the right hand side of the car so isolating them from the area of the exhaust system to diminish any possibility of fuel vaporisation.

10 On the BT7, BJ7 and BJ8 models access to the fuel pump is gained once the right hand seat pan has been removed.

5 Fuel pump (single HP) - dismantling

1 Unscrew the filter plug (14) (Fig 3.1) and remove the plug, washer (13) and filter (12).

2 Unscrew the inlet union (1) and remove the union and washer (2). Note that this washer is coloured orange and is thicker than that used under the valve cage (5).

3 Lift out the valve cage (5), valve cage fibre washer (6), suction valve disc (7) and spring (31).

4 Remove the circlip (3) that retains the delivery valve disc (4) in the valve cage (5) and lift away the valve disc.

5 Mark the flanges adjacent to each other and separate the magnet housing (27) from the pump body (8) by unscrewing the six screws holding both halves of the pump together. Take great care not to tear or damage the diaphragm (9) as it may stick to either of the flanges as they are separated.

6 The armature spindle which is attached to the armature head and diaphragm (9) is unscrewed anticlockwise from the trunnion at the contact breaker end of the magnet housing. Lift out the armature, spindle, diaphragm (9) and diaphragm spring (28). Recover the rollers (10) which are fitted behind the diaphragm (9).

7 An impact washer may be fitted under the head of the armature so as to quieten the noise of the armature head hitting the solenoid core (17), lift away this washer.

8 Slide off the protective rubber sheath and unscrew the terminal nut, connector (where fitted), and washer (22) from the terminal screw (20). Remove the bakelite contact breaker cover (21).

9 Unscrew the five BA screws which hold the contact spring blade (24) in position and remove it together with the blade and washer.

10 Remove the cover retaining nut on the terminal screw (20), and cut through the lead washer under the nut on the terminal screw with a pocket knife and push the terminal down a short way so that the tag on the coil end is free on the terminal.

11 Remove the two bakelite pedestal retaining screws complete with spring washers which hold the pedestal to the solenoid housing, remove the braided copper earth lead, and the coil lead, from the terminal screw.

12 Remove the pin on which the rollers pivot by pushing it out sideways and remove the rocker assembly. The pump is now fully dismantled. It is not possible to remove the solenoid core and coil and the rocker assembly (25, 26) must not be broken down as it is only supplied in exchange as a complete assembly.

6 Fuel pump (single HP) - inspection, servicing and reassembly

1 Remove the filter as has already been detailed and thoroughly clean. At the same time clean the points by gently drawing a piece of thin card between them. Do this very carefully so as not to disturb the tension of the spring blade. If the points are burnt or pitted they must be renewed and a new blade and rocker assembly fitted.

2 On the pump fuel starvation combined with rapid operation is indicative of an air leak on the suction side. To check whether this is so, undo the fuel line at the top of the float chamber, and immerse the end of the pipe in a jam jar half filled with petrol. With the ignition on and the pump functioning, should a regular stream of air bubbles emerge from the end of the pipe, air is leaking in on the suction side.

3 If the filter is coated with gum-like substance very like varnish, serious trouble can develop in the future unless all traces of this gum (formed by deposits from the fuel) are removed.

4 To do this boil all steel and brass parts in a 20 per cent solution of caustic soda, then dip them in nitric acid and clean them in boiling water. Alloy parts can be cleaned with a clean rag after they have been left to soak for a few hours in methylated spirits.

5 With the pump stripped right down, wash and clean all the parts thoroughly in paraffin and renew any that are worn, damaged, fractured or cracked. Pay particular attention to the gaskets and diaphragm.

6 To reassemble, first ensure that all parts are really clean and then fit the delivery valve disc (4) with its smooth side downwards in the valve cage (5).

7 Refit the shaped spring clip (3) to the valve cage (5) and the red fibre washer (6) to the underside of the cage.

8 Place the suction valve disc (7) in the pump body (8) followed by the spring (31) and the previously assembled valve cage. Insert the thick orange fibre washer (2) and screw down the outlet union (1).

9 Insert the filter (12) into the pump body (8) and secure in position with the filter plug (14) and thick orange fibre washer (13).

10 Assemble the contact breaker onto its pedestal so that the rockers are free in their mounting without appreciable side play.

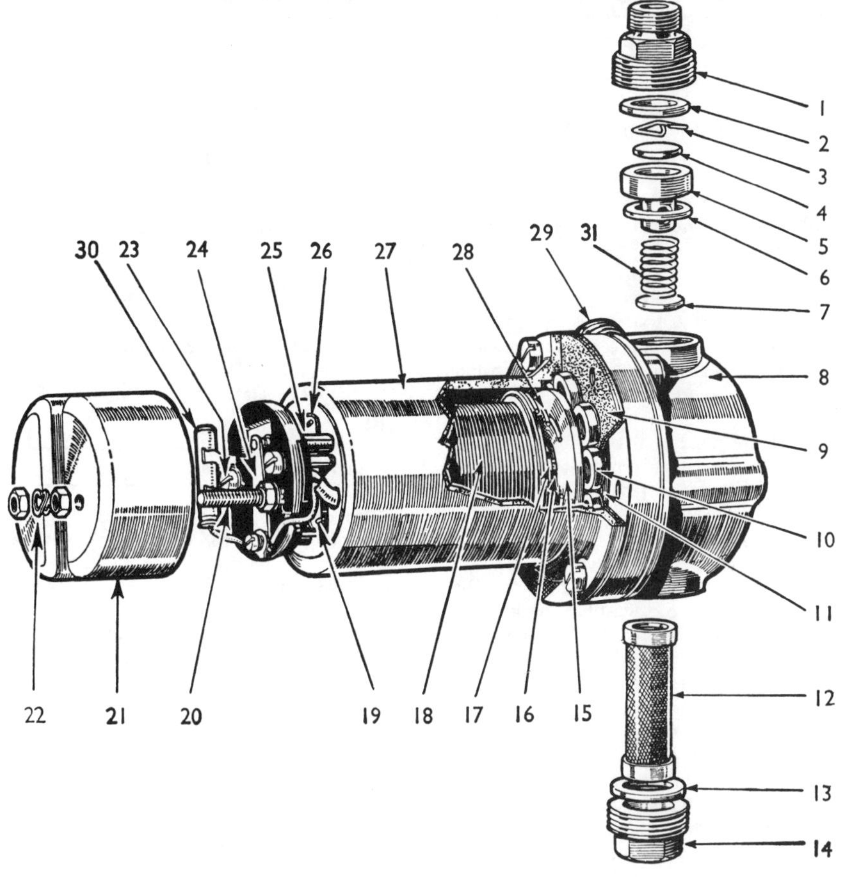

Fig 3.1 FUEL PUMP COMPONENTS (TYPE HP)

1 Outlet union
2 Fibre washer (thick,orange)
3 Spring clip
4 Delivery valve disc
5 Valve cage
6 Fibre washer
7 Suction valve disc
8 Pump body
9 Diaphragm assembly
10 Armature guide rollers
11 Retaining plate
12 Filter
13 Fibre washer (thick,orange)
14 Filter plug
15 Steel armature
16 Push rod
17 Magnet iron core
18 Magnet coil
19 Rocker hinge pin
20 Terminal screw
21 Cover
22 Cover and terminal nuts
23 Earth terminal screw
24 Spring blade
25 Inner rocker
26 Outer rocker
27 Magnet housing
28 Volute spring
29 Inlet union
30 Condenser
31 Suction valve spring

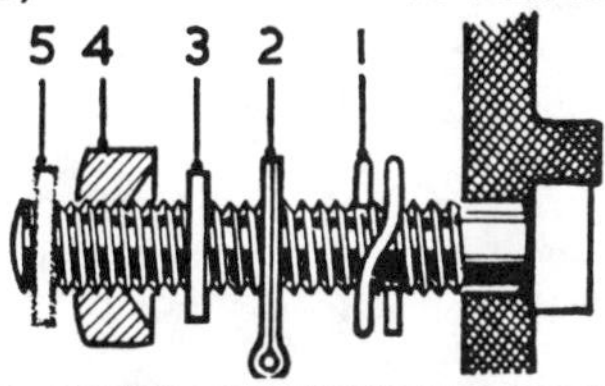

Fig 3.2 THE COMPONENTS OF THE TERMINAL SCREW ASSEMBLED IN CORRECT ORDER

1 Spring washer
2 Wiring tag
3 Lead washer
4 Recessed nut
5 Seal

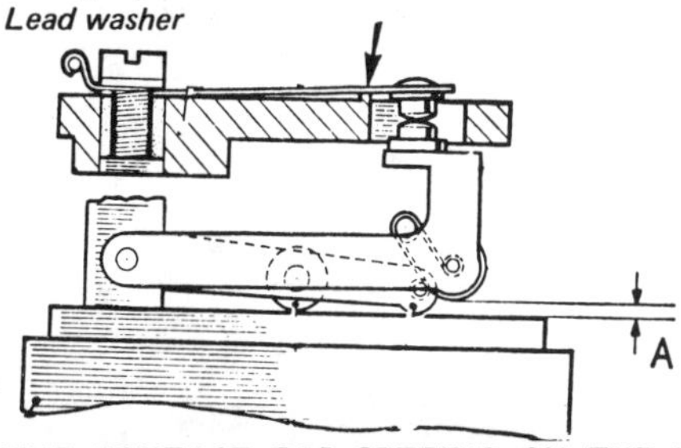

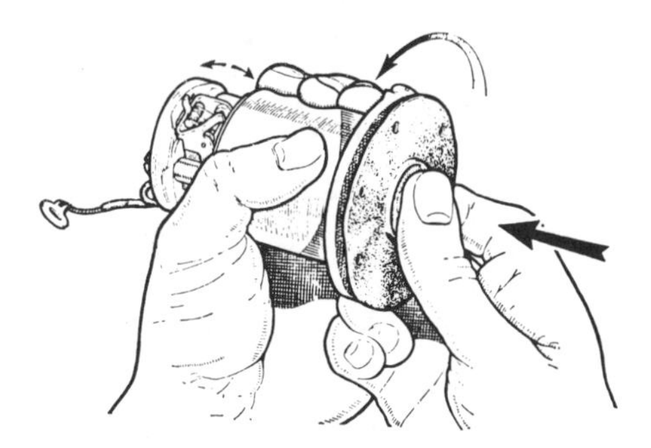

Fig 3.4 SCREW IN THE DIAPHRAGM UNTIL THE ROCKER THROW OVER STOPS

Fig 3.3 THE CONTACT GAP SETTING ON THE ROCKER ASSEMBLY

A = 0.030 inch (8 mm)

Any excessive side play on the outer rocker will allow the points to be out of line, whilst excessive tightness will interfere with the action of the pump through sluggish contact breaker operation. Should tightness be evident it may be rectified by using a pair of thin nosed pliers to square up the ends of the outer rocker.

11 Slide in the rocker pivot pin. The pin is case hardened and wire or any other substitute should never be used if the pin is lost.

12 Attach the copper earth wire from the outer rocker immediately under the head of the nearest pedestal securing screw and fit the pedestal to the solenoid housing with the two pedestal securing screws and lockwashers. It is unusual to fit an earth wire immediately under the screw head but in this case the spring washer has been found not to be a particularly good conductor.

13 Fit the lockwasher under the head of the spring blade (24) contact securing screws, then the last lead from the coil and then the spring blade so that there is nothing between it and the bakelite pedestal. It is important that this order of assembly is adhered to. Tighten the screw lightly.

14 The static position of the pump when it is not in use is with the contact points making firm contact and this forces the spring blade to be bent slightly back. Move the outer rocker arm up and down and position the spring blade so that the contacts on the rocker or blade wipe over the centre line of the other points. When open the blade should rest against the small edge on the bakelite pedestal just below the points. The points should come into contact with each other when the rocker is halfway forward. To check that this is correct, press the middle of the blade gently so that it rests against the ridge with the points just having come into contact. It should now be possible to slide a 0.030 inch feeler gauge between the rocker rollers and the solenoid housing. If the clearance is not correct bend the tip of the blade very carefully until it is.

15 Tighten down the blade retaining screw, and check that a considerable gap exists between the underside of the spring blade and the pedestal ledge, with the rocker contact bearing against the blade contact and the rocker fully forward in the normal static position. With the rocker arm down, ensure that the underside of the blade rests on the ledge of the pedestal. If not, remove the blade and very slightly bend it until it does.

16 Place the impact washer on the underside of the armature head, fit the diaphragm return spring with the wider portion of the coil against the solenoid body, place the brass rollers in position under the diaphragm and insert the armature spindle through the centre of the solenoid core, and screw the spindle into the rocker trunnion.

17 It will be appreciated that the amount the spindle is screwed into the rocker trunnion will vitally affect the function of the pump. To set the diaphragm correctly, turn the steel blade to one side, and screw the armature spindle into the trunnion until if the spindle was screwed in a further sixth of a turn, the throw-over rocker would not operate the points closed to points open position. Now screw out the armature spindle four holes (2/3 of a turn) to ensure that wear in the points will not cause the pump to stop working. Turn the blade back into its normal position.

18 Reassembly of the valves, filter and nozzles into the pumping chamber is a reversal of the dismantling process. Use new washers and gaskets throughout.

19 Fit the bakelite cover (13) and replace the shakeproof washer, Lucar connector (if fitted), cover nut and terminal knob (22) to the terminal screw (20). Then replace the terminal lead and cover nut so locking the lead between the cover nut and the terminal nut. Reassembly is now complete.

7 Fuel pump (single LCS) - dismantling

An exploded view of the LCS type pump is shown in Fig 3.5 and upon comparison with Fig 3.1 it will be seen that the magnet housing and contacts are similar, the main differences being in the body. To dismantle, proceed as follows:

1 Unscrew the six 2BA screws which secure the top cover plate (6) to the body (17). Lift away the top cover plate.

2 Mark the lower cover plate (18) and body (17) with a file or scriber so that it may be refitted correctly and undo the six 2BA screws which secure the lower cover plate (17) to the body. Lift away the cover plate.

3 Remove the filter (19) from the body (17).

4 Unscrew the outlet valve cage (5) and lift out the inlet valve disc (3).

5 To dismantle the outlet valve cage (5) extract the spring clip and lift out the outlet valve (4).

6 Unscrew the inlet and outlet unions (1, 15) and lift away the union together with the rubber sealing rings (2, 16).

7 Dismantling is now similar to that for the HP type pump as described in Section 5, paragraphs 5 onwards.

8 Fuel pump (single LCS) - inspection, servicing and reassembly

Inspection and servicing for this type of pump is basically identical to that for the HP type pump and full information will be found in Section 6, paragraphs 1 to 7 inclusive. To reassemble proceed as follows:

1 First make sure that all parts are really clean and then fit the outlet valve disc (4) (Fig 3.5) smooth side downwards, into the valve cage (5). Refit the shaped spring clip to the valve cage.

2 Insert the inlet valve disc (3) smooth side downwards into the pump body and screw in the valve cage.

3 Replace the top cover plate (6) and secure in position with the six 2BA screws.

4 Invert the pump body (17) and replace the filter (19). Refit the lower cover plate (18) making sure that it engages correctly with the filter and secure in position with the six 2BA screws.

5 Refit the inlet and outlet unions (1, 15) not forgetting to use new rubber rings (2, 16) and lightly tighten.

6 Reassembly is now basically identical to that for the HP type pump and full information will be found in Section 6, paragraph 12 onwards.

9 Fuel pump (single AUF 300 and 301) - dismantling

1 Undo the one 2BA screw (48) (Fig 3.6) which retains the inlet air bottle cover (45). Remove the screw (48), spring washer (47), dished washer (46), air bottle cover (45) and its cover joint (44) from the pump body.

2 Undo the four screws (52) securing the cover (51) to the pump body (1) and lift away the cover (51), spring and spring cap (if fitted), rubber O ring (53), plastic diaphragm (54) and sealing washer (55). On some models a plastic diaphragm barrier and rubber diaphragm may also be fitted and these should also be removed.

3 Mark the pump body and coil housing mating flanges so that the parts may be refitted in their original position and unscrew and remove the six screws securing the two halves together. Separate the two parts taking great care not to tear or damage the diaphragm as it may stick to either of the flanges as they are separated.

4 To remove the valves from the pump body (1) undo the two screws (35) securing the clamp plate (34) to the pump body and lift away the clamp plate. Referring to Fig 3.6 and noting the order of fitting, remove the two valve caps (36), inlet valve (37), outlet valve (38), two sealing washers (39), filter (40) and a further sealing washer.

5 The sequence for dismantling is now similar to that for the HP type pump and full information will be found in Section 5, paragraph 6 onwards.

10 Fuel pump (single AUF 300 and 301) - inspection, servicing and reassembly

Inspection and servicing for these types of pumps is basically

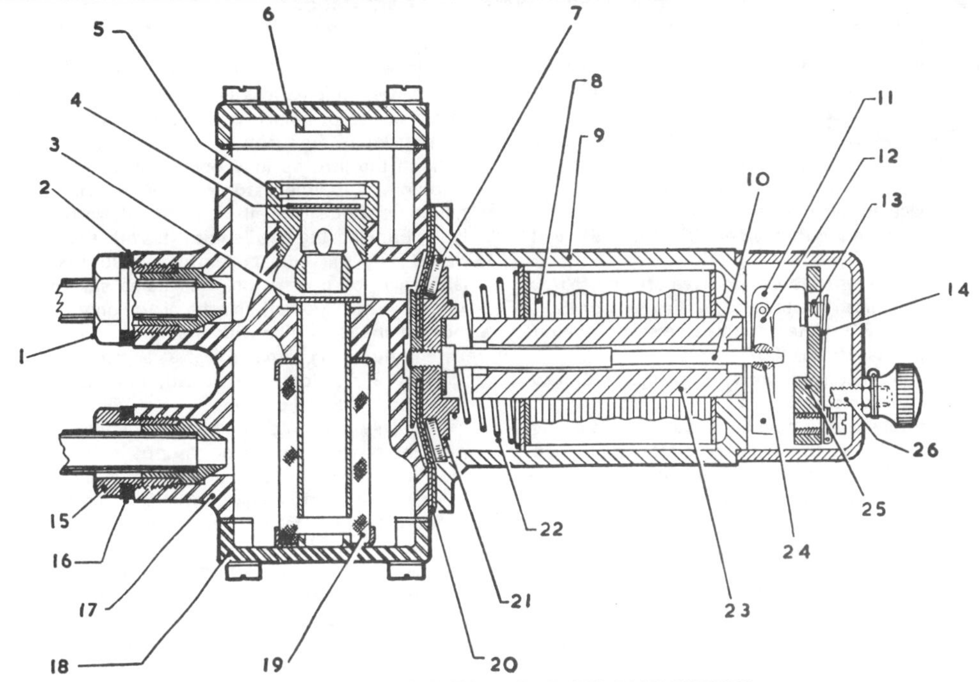

Fig 3.5 LCS TYPE FUEL PUMP COMPONENTS

1 Outlet union
2 Rubber ring
3 Inlet valve
4 Outlet valve
5 Outlet valve cage
6 Top cover plate
7 Spherical rollers
8 Magnet coil
9 Iron coil housing
10 Bronze rod
11 Outer rocker
12 Inner rocker
13 Tungsten points
14 Spring blade
15 Inlet union
16 Rubber ring
17 Body
18 Lower cover plate
19 Filter
20 Diaphragm
21 Armature
22 Armature spring
23 Magnet core
24 Trunnion
25 Bakelite moulding
26 Terminal screw

Note: For clarity reasons the inlet and outlet union connections are shown 90° out of position

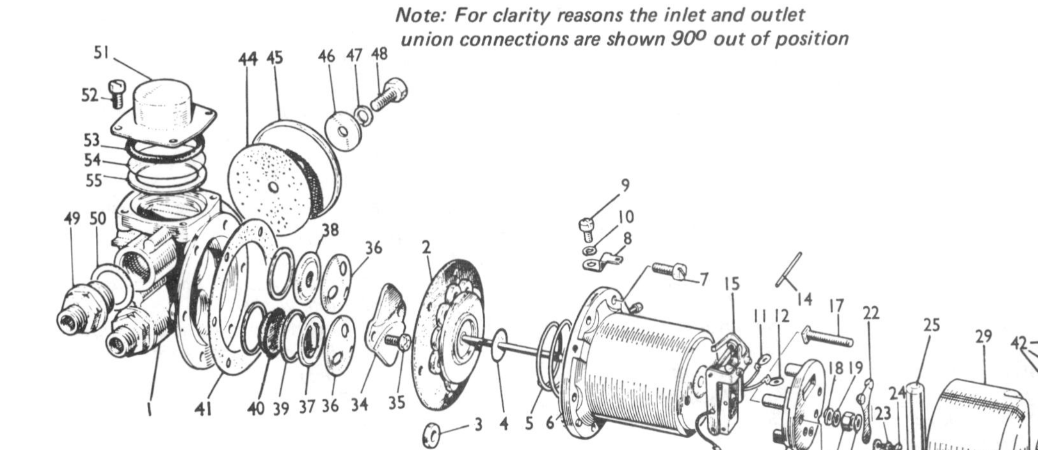

Fig 3.6 AUF 300 AND 301 FUEL PUMP COMPONENTS

1 Pump body
2 Diaphragm and spindle assembly
3 Armature centralising roller
4 Impact washer
5 Armature spring
6 Coil housing
7 Setscrew
8 Earth connector
9 Setscrew
10 Spring washer
11 Terminal tag
12 Terminal tag
13 Earth tag
14 Rocker pivot pin
15 Rocker mechanism
16 Pedestal
17 Terminal stud
18 Spring washer
19 Lead washer
20 Terminal nut
21 End cover seal washer
22 Contact blade
23 Washer
24 Contact blade screw
25 Condenser
26 Condenser clip
27 Spring washer
28 Screw
29 End cover
30 Shakeproof washer
31 Connector
32 Nut
33 Insulating sleeve
34 Clamp plate
35 Screw
36 Valve cap
37 Inlet valve
38 Outlet valve
39 Sealing washer
40 Filter
41 Diaphragm gasket
42 Vent valve
43 Sealing band
44 Inlet air bottle cover joint
45 Inlet air bottle cover
46 Dished washer
47 Spring washer
48 Cover securing screw
49 Outlet connection
50 Sealing washer
51 Delivery air bottle
52 Cover securing screw
53 Rubber O ring
54 Plastic diaphragm
55 Sealing washer

identical to that for the HP type pump and full information will be found in Section 6, paragraphs 1 to 7. The plastic and rubber diaphragms (if fitted) should be carefully inspected for perishing or distortion, which, if evident, new parts should be obtained. To reassemble proceed as follows:

1 First make sure that all parts are really clean and then fit the valves to the body. Place one sealing washer into the outlet valve bore and follow this with the outlet valve (38) Fig 3.6) and valve cap (36).

2 Insert one sealing washer into the inlet valve bore and follow this with the filter (40) conical face inwards, sealing washer (39), inlet valve and valve cap (36). Retain the valve assemblies with the clamp plate (34) and two screws (35).

3 Refit the inlet and outlet connections (49) together with new sealing washer (50) to the pump body.

4 The air bottle assembly should now be refitted. Insert a new sealing washer (55) and follow this with the plastic diaphragm (54) and rubber O ring (53). If a rubber diaphragm and plastic barrier were fitted these should also be refitted in their locations as noted in removal.

5 Refit the delivery air bottle cover (1) and secure with the four screws (52).

6 Fit the inlet air bottle cover joint (44) followed by the cover (45), conical washer (46), spring washer (47) and securing screw.

7 Reassembly is now similar to that for the HP type pump and full information will be found in Section 6, paragraph 12 onwards.

11 SU carburettor - description

The variable choke SU carburettor is a relatively simple instrument and is basically the same irrespective of its size and type. It differs from most other carburettors in that instead of having a number of various sized fixed jets for different conditions, only one variable jet is fitted to deal with all possible conditions.

Air passing rapidly through the carburettor choke draws petrol from the jet so forming the petrol/air mixture. The amount of petrol drawn from the jet depends on the position of the tapered carburettor needle, which moves up and down the jet orifice according to engine load and throttle opening, thus effectively altering the size of the jet so that exactly the right amount of fuel is metered for the prevailing road conditions.

The position of the tapered needle in the jet is determined by engine vacuum. The shank of the needle is held at its top end in a piston which slides up and down the dashpot in response to the degree of manifold vacuum. This is directly controlled by the position of the throttle.

With the throttle fully open, the full effect of inlet manifold vacuum is felt by the piston which has an air bleed into the choke tube on the outside of the throttle. This causes the piston to rise fully, bringing the needle with it. With the accelerator partially closed only slight inlet manifold vacuum is felt by the piston (although, of course, on the engine side of the throttle the vacuum is now greater) and the piston only rises a little, blocking most of the jet orifice with the metering needle. To prevent the piston fluttering, and to give a richer mixture when the accelerator is suddenly depressed, an oil damper and light spring are fitted inside the dashpot.

The only portion of the piston assembly to come into contact with the piston chamber or dashpot is the actual central piston rod. All the other parts of the piston assembly including the lower choke portion, have sufficient clearances to prevent any direct metal to metal contact which is essential if the carburettor is to work properly.

The correct level of the petrol in the carburettor is determined by the level of the float in the float chamber. When the level is correct the float rises and by means of a lever resting on top of it closes the needle valve in the cover of the float chamber. This closes off the supply of fuel from the pump. When the level in the float chamber drops as fuel is used in the carburettor the float sinks. As it does, the float needle comes away from its seat so allowing more fuel to enter the float chamber and restore the correct level.

12 Twin carburettor installation - removal and refitting

1 Turn the battery master switch to the off position or disconnect the battery positive terminal.

2 Disconnect the petrol feed pipe from the union on the front carburettor.

3 Undo and remove the air cleaner securing bolts and spring washers from the rear of the air intake flange and lift away the two air cleaners and joint washers. Note the location of the air holes above the two bolt holes so that on reassembly they are not blocked by fitting the air cleaner or gasket the wrong way round.

4 Undo and remove the two float chamber overflow pipe union

Fig 3.7 COMPONENT PARTS OF SU H4 CARBURETTOR

1 Body - bare - front carburettor
1a Body - bare - rear carburettor
2 Adaptor - ignition union
2 Ignition union
4 Suction chamber and piston
5 Piston damper unit
7 Skid washer - piston spring
8 Fibre washer - damper
9 Piston spring (red colour identification)
10 Suction chamber screw
12 Jet needle
13 Locking screw - jet needle
15 Jet with head
16 Jet locking screw
17 Jet adjusting nut
19 Lock spring - adjusting nut
20 Jet sealing ring (brass)
21 Jet sealing ring (cork)
22 Copper washer (bottom half)
23 Jet bearing (bottom half)
24 Jet gland (cork)
25 Jet gland (brass)
26 Jet gland spring
27 Jet bearing (top half)
28 Copper washer (top half)
29 Jet return spring
34 Jet lever
35 Jet link - front carburettor
35a Jet link - rear carburettor
36 2 BA bolt
37 2 BA nut
38 2 BA spring washer
39 Jet link pin
40 Fork pivot pin
41 Swivel pin - jet lever
42 1/16 in split pin
43 Fork joint
44 2 BA connecting rod
45 Tension rod
46 Float chamber (bare)
47 Float chamber lid
52 Float
53 Seating and needle
55 Float hinged lever
56 Pivot pin - hinged lever
60 Filter
61 Banjo bolt
61a Inlet nozzle
63 Fibre washer - banjo bolt
64 Cap nut
65 Cover cap
66 Holding up bolt
67a Brass skid washer - holding-up bolt
67B Fibre washer - holding-up bolt
68 Fibre washer
81 Throttle spindle - rear carburettor
81a Throttle spindle - front carburettor
82 Throttle disc
83 Throttle disc screw
86 Return spring - throttle
87 Retaining clip assembly - return spring
88 Anchor plate return spring
91 ¼ in dia. taper pin
92 Stop lever - rear carburettor
92a Stop lever - front carburettor
93 Throttle adjusting screw
93a Fast-idle adjusting screw
94 Lock spring throttle adjusting screw
94a Lock spring fast idle adjusting screw
95 Cam - fast idle
96 Pivot bolt - cam
96a Spring washer - cam
97 Plain washer - cam
99 Flexible coupling
100 5/16 in dia. coupling rod
101 4 BA bolt
102 4 BA nut
103 4 BA washer

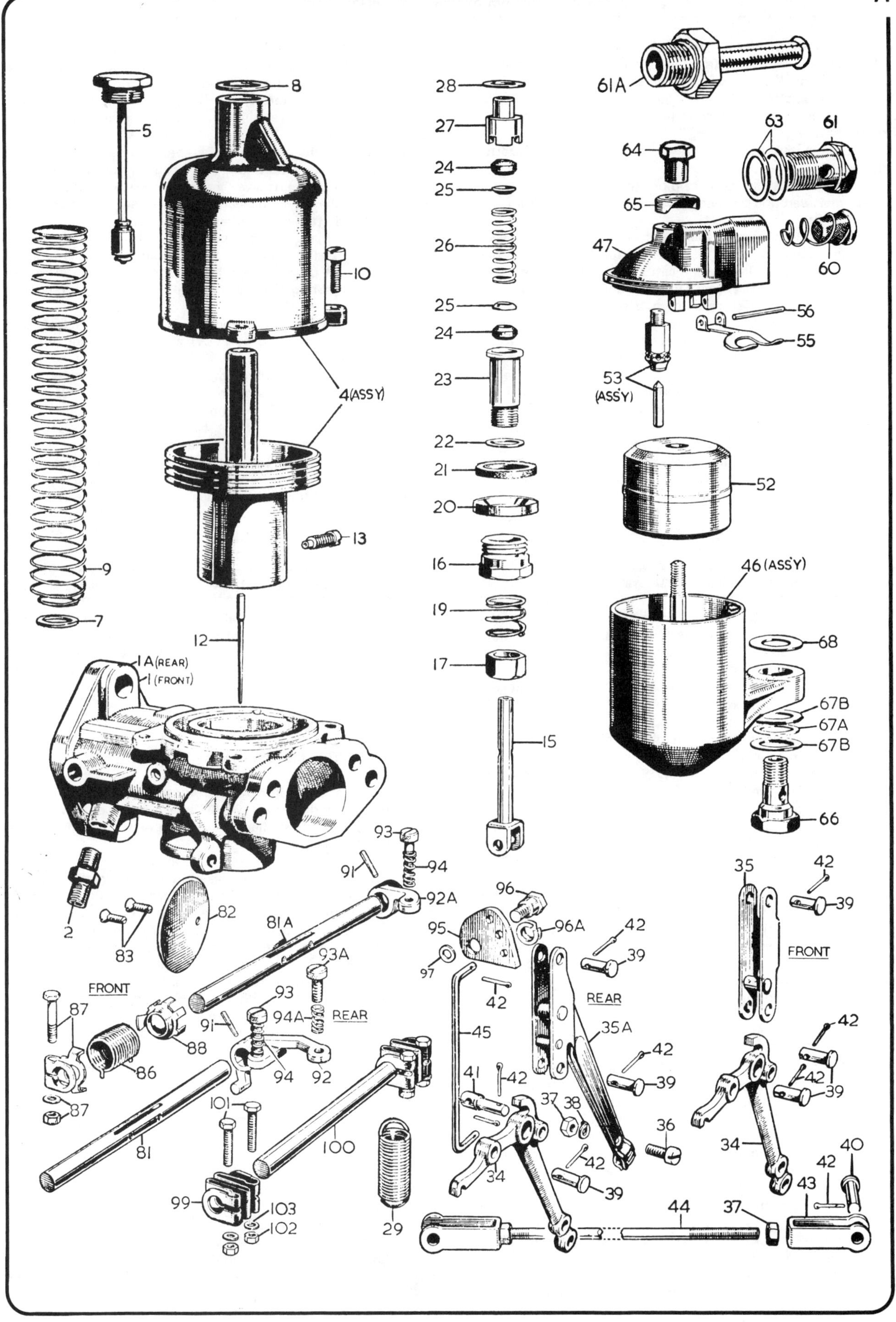

61A
63
61
64
65
47
60
56
55
53
(ASS'Y)
52
46 (ASS'Y)
68
67B
67A
67B
66
28
27
24
25
26
25
24
23
22
21
20
16
19
17
15
8
5
10
4(ASSY)
13
9
7
12
1A(REAR)
1 (FRONT)
2
83
82
81A
93
91
94
92A
96
95
96A
97
42
39
35
42
39
FRONT
93A
FRONT
87
93
91
94A
REAR
88
86
94
92
87
45
REAR
35A
42
42
39
42
42
39
41
42
101
81
100
37
38
36
34
34
42
39
40
42
43
44
37
99
103
102
29

bolts, fibre washers and overflow pipes.
5 Detach the choke control inner wire from the jet hand control lever on the rear carburettor and the outer cable from the clamping bracket.
6 Detach the throttle valve rod from the lever on the carburettor throttle shaft.
7 Disconnect the accelerator link rod from the carburettor control shaft.
8 Unscrew the vacuum advance pipe from its union on the top of the rear carburettor body. Separate the pipe from the carburettor body.
9 Undo and remove the four nuts and washers from each carburettor flange securing the carburettor to the mounting studs. Remove the bracket which locates the rear extension of the throttle shaft from the engine bulkhead and carefully pull the carburettors from the studs in such a manner that they are removed together.
10 It should be noted that if only one carburettor is to be removed the interconnecting petrol pipe must be disconnected. It is also necessary to separate the throttle shaft at the centre connecting W clip. Also the connecting rod for the jet hand control lever must be released by taking out the clevis pins from the yokes.
11 Lift away the insulating packing and gaskets.
12 Refitting the carburettor/s is the reverse sequence to removal. It is recommended that new gaskets are fitted at the carburettor and air cleaner mounting flanges.
13 Refer to Section 23 for full information on resetting the linkage and tuning the carburettors.

13 Triple carburettor installation - removal and refitting

1 The procedure for removal of the triple carburettor is basically identical to that for the twin carburettor installation.
2 Refer to Section 12 and follow the instructions given in paragraphs 1 to 8 inclusive.
3 Remove the top of the auxiliary enrichment device solenoid and detach the cables from the two terminals. Refit the solenoid top securely.
4 Slacken the unions that secure the external feed pipe and remove the union completely from the inlet manifold.
5 Refer to Section 12 and follow the instructions given in paragraphs 9, 11, 12 and 13.

14 Carburettor (type H4) - dismantling and reassembly

1 All reference numbers refer to Fig 3.7. Unscrew the piston damper (5) and lift away from the chamber and piston assembly (4). Using a screwdriver or small file scratch identification marks on the suction chamber and carburettor body (1) so that they may be fitted together in their original position. Remove the four screws (10) and lift the suction chamber from the carburettor body (1) leaving the piston in situ.
2 Lift the piston spring (9) from the piston noting which way round it is fitted and remove the piston. Invert it and allow the oil in the damper bore to drain out. Recover the piston spring skid washer (7). Place the piston in a safe place so that the needle will not be touched or the piston roll onto the floor. It is recommended that the piston be placed on the neck of a narrow jam jar with the needle inside so acting as a stand.
3 Mark the position of the float chamber lid (47) relative to the body (46) and unscrew the cap nut (64). Lift away the cap nut cover (65). Remove the lid (47) and withdraw the pin (56) thereby releasing the float lever (55). Using a spanner or socket remove the needle valve assembly (53).
4 Invert the float chamber and recover the float (52) noting which way round it is fitted.
5 With a screwdriver or file mark the position of the float chamber (46) relative to the body (1) to ensure correct position on reassembly and unscrew the holding up bolt (66). Remove the bolt (66) together with the brass skid washer (67A) and two fibre washers (67B) from the bolt head side of the float chamber and the fibre washer (68) from the carburettor body mounting side of the float chamber mounting boss.
6 Release the jet return spring (29), extract the split pin (42) and withdraw the swivel pin (41) so releasing the jet (15) from the jet lever (34).
7 Extract the split pin (42) and withdraw the jet link pin (39). Undo the cam pivot bolt (96) and lift away the spring washer (96A) and cam (95). Lift away the jet link assembly (35).
8 Withdraw the jet with head (15). Unscrew the jet adjusting screw (17) and lift away together with the spring (19).
9 Unscrew the jet locking screw (16) and lift away followed by the brass jet sealing ring (20), cork sealing ring (21), copper washer (22), jet bearing (lower) (23), cork jet gland (24), brass jet gland (25), spring (26), jet bearing (upper) (27), and copper washer (28).
10 To remove the throttle and actuating spindle, release the retainer clip locknut and bolt (87) and slide off the retainer, spring (86) and anchor plate (88).
11 Bend straight the ends of the two throttle disc screws (83) and remove the screws securing the throttle disc (82) to the throttle spindle (81). Make a note of the tapered edges of the throttle disc (82) and slide it out of the spindle (81). The spindle may now be drawn out of the carburettor body (1).
12 Reassembly of the carburettor is the reverse sequence to removal. It will, however, be necessary to centre the jet and adjust the carburettor settings as described later in this Chapter.

15 SU carburettor (type H4) - examination and repair

The SU carburettor generally speaking is most reliable, but even so it may develop one of several faults which may not be readily apparent unless a careful inspection is carried out. The common faults are:

1 Piston sticking
2 Float needle sticking
3 Float chamber flooding
4 Water or dirt in the carburettor

In addition the following parts are susceptible to wear after long mileages and as they vitally affect the economy of the engine they should be checked and renewed where necessary, every 24,000 miles:

a) The carburettor needle. If this has been incorrectly fitted at some time so that it is not centrally located in the jet orifice, then the metering needle will have a tiny ridge worn on it. If a ridge can be seen then the needle must be renewed. SU carburettor needles are made to very fine tolerances and should a ridge be apparent no attempt should be made to rub the needle down with fine emery paper. If it is wished to clean the needle it can be polished lightly with metal polish.
b) The carburettor jet. If the needle is worn it is likely that the rim of the jet will be damaged where the needle has been striking it. It should be renewed as otherwise fuel consumption will suffer. The jet can also be badly worn or ridged on the outside from where it has been sliding up and down between the jet bearing every time the choke has been pulled out. Removal and renewal is the only answer as well.
c) Check the edges of the throttle and the choke tube for wear. Renew if worn.
d) The washers fitted to the base of the jet and under the float chamber lid may leak after a time and can cause a great deal of fuel wastage. It is wisest to renew them automatically when the carburettor is stripped down.
e) After high mileages the float chamber needle and seat are bound to be ridged. They are not an expensive item to replace and must be renewed as a set. They should never be renewed separately.

16 SU carburettor (type H4) - piston sticking

1 The hardened piston rod which slides in the centre guide tube in the middle of the dashpot is the only part of the piston assembly (which comprises the jet needle, suction disc and piston choke) which should make contact with the dashpot. The piston rim and the choke periphery are machined to very fine tolerances so that they will not touch the dashpot or the choke tube walls.
2 After high mileages wear in the centre guide tube may allow the piston to touch the dashpot wall. This condition is known as sticking.
3 If piston sticking is suspected or it is wished to test for this condition, rotate the piston about the centre guide tube at the same time as sliding it up and down inside the dashpot. If any portion of the piston makes contact with the dashpot then that portion of the wall must be polished with a metal polish until clearance exists. In extreme cases fine emery cloth can be used. The greatest care should be taken to remove only the minimum amount of metal to provide the clearance as too large a gap will cause air leakage and will upset the functioning of the carburettor. Clean down the walls of the dashpot and the piston rim and ensure that there is no oil on them. A trace of oil may be judiciously applied to the piston rod.
4 If the piston is sticking, under no circumstances try to clear it by trying to alter the tension of the light return spring.

17 SU carburettor (type H4) - float needle sticking

1 If the float needle sticks, the carburettor will soon run dry and the engine will stop despite there being fuel in the tank.
2 The easiest way to check a suspected sticking float needle is to remove the inlet pipe at the carburettor and, where a mechanical fuel pump is fitted, turn the engine over on the starter motor by pressing the solenoid rubber button. Where an electrical pump is fitted turn on the ignition but do not start the engine. If fuel spurts from the end of the pipe (direct it towards the ground or into a wad of cloth or jar) then the fault is almost certain to be a sticking float needle.
3 Remove the float chamber, dismantle the valve and clean the housing and float chamber out thoroughly.

18 SU carburettor (type H4) - float chamber flooding

If fuel emerges from the small breather hole in the cover of the float chamber this is known as flooding. It is caused by the float chamber needle not seating properly in its housing; normally this is because a piece of dirt or foreign matter is jammed between the needle and the needle housing. Alternatively the float may have developed a leak or be maladjusted so that it is holding open the float chamber needle valve even though the chamber is full of petrol. Remove the float chamber cover, clean the needle assembly, check the setting of the float as detailed later in this Chapter and shake the float to verify if any has leaked into it.

19 SU carburettor (type H4) - water or dirt in the carburettor

1 Because of the size of the jet orifice, water or dirt in the carburettor is normally easily cleaned. If dirt in the carburettor is suspected, lift the piston assembly and flood the float chamber. The normal level of the fuel should be about 1/16 inch below the top of the jet, so that on flooding the carburettor the fuel should flow out of the jet hole.
2 If little or no petrol appears, start the engine (the jet is never completely blocked) and with the throttle fully open, blank off the air intake. This will cause a partial vacuum in the choke tube and help suck out any foreign matter from the jet tube. Release the throttle as soon as the engine speed alters considerably. Repeat this procedure several times, stop the engine and then check the carburettor as detailed in the first paragraph of this Section.
3 If this failed to do the trick then there is no alternative but to remove and blow out the jet.

20 SU carburettor (type H4) - jet centering

1 This operation is always necessary if the carburettor has been dismantled; but to check if this is necessary on a carburettor in service, first screw up the jet adjusting nut as far as it will go without forcing it, and lift the piston and then let it fall under its own weight. It should fall onto the bridge making a soft metallic click. Now repeat the above procedure but this time with the adjusting nut screwed right down. If the soft metallic click is not audible in either of the two tests proceed as follows.
2 Disconnect the jet link (34) (Fig 3.7) from the jet head (15) by extracting the split pin (42) and withdrawing the jet link pin (39). Gently slide the jet from the underside of the carburettor body (1). Next unscrew the jet adjusting nut (17) and lift away the nut and the locking spring (19). Refit the adjusting nut without the locking spring and screw it up as far as possible without forcing. Replace the jet.
3 Slacken the jet locking screw (16) so that it may be rotated with the fingers only. Unscrew the piston damper (5) and lift away the damper. Gently press the piston down onto the bridge and tighten the jet locking screw (16). Lift the piston and check that it is able to fall freely under its own weight. Now lower the jet locking screw (16) and check once again, and, if this time there is a difference in the two metallic clicks repeat the centering procedure until the sound is the same for both tests.
4 Gently remove the jet (15) and unscrew the adjusting nut (17). Refit the locking spring and jet adjusting nut. Top up the damper with oil if necessary and replace the damper. Reconnect the jet link to the jet head.

21 SU carburettor (type H4) - float chamber fuel level adjustment

1 It is essential that the fuel level in the float chamber is always correct as otherwise excessive fuel consumption may occur. On reassembly of the float chamber check the level before replacing the float chamber cover in the following manner.
2 Invert the float chamber cover so that the needle valve is closed. It should just be possible to place a 7/16 inch bar parallel to the float chamber cover without fouling the float hinged lever as shown in Fig 3.8, or if the lever stands proud of the bar then it is necessary to bend the float lever slightly until the clearance is correct.

22 SU carburettor (type H4) - needle replacement

1 Should it be found necessary to fit a new needle, first remove the piston and suction chamber assembly, marking the chamber for correct reassembly in its original position.
2 Slacken the needle clamping screw and withdraw the needle from the piston.
3 Upon refitting a new needle it is important that the shoulder on the shank is flush with the underside of the piston. Use a straight edge such as a metal rule for the adjustment. Refit the piston and suction chamber and check for freedom of piston movement.

23 SU carburettor (type H4) - adjustment and tuning

1 To adjust and tune the SU carburettor proceed in the following manner. Check the colour of the exhaust at idling speed with the choke fully in. If the exhaust tends to be black and the tailpipe interior is also black it is a fair indication that the mixture is too rich. If the exhaust is colourless and the deposit in

the exhaust pipe is very light grey it is likely that the mixture is too weak. This condition may also be accompanied by intermittent misfiring, while too rich a mixture will be associated with 'hunting'. Ideally the exhaust should be colourless with a medium grey pipe deposit.
2 The exhaust pipe deposit should only be checked after a good run of at least 20 miles. Idling in city traffic and stop/start motoring is bound to produce excessively dark exhaust pipe deposits.
3 Once the engine has reached its normal operating temperature, detach the carburettor air intake cleaners.
4 Only two adjustments are provided on the SU carburettor. Idling speed is governed by the throttle adjusting screw (2) (Fig 3.9) and the mixture strength by the jet adjusting nut (1). The SU carburettor is correctly adjusted for the whole of its engine revolution range when the idling mixture strength is correct.
5 Idling speed adjustment is effected by the idling adjusting screw (3). To adjust the mixture set the engine to run at about 1000 rpm by screwing in the idling screw.
6 Check the mixture strength by lifting the piston of the carburettor approximately 1/32 inch with the piston lifting pin so as to disturb the air flow as little as possible. If:

a) the speed of the engine increases appreciably the mixture is too rich
b) the engine speed immediately decreases the mixture is too weak
c) the engine speed increases very slightly the mixture is correct

To enrich the mixture rotate the adjusting nut, which is at the bottom of the underside of the carburettor, in an anticlockwise direction, ie: upwards. Only turn the adjusting nut a flat at a time and check the mixture strength between each turn. It is likely that there will be a slight increase or decrease in rpm after the mixture adjustment has been made so that the throttle idling adjusting screw should now be turned so that the engine idles at between 550 - 600 rpm.

24 Twin carburettor installation (type H4) - synchronisation

1 First ensure that the mixture is correct in each carburettor by disconnecting each carburettor as described previously in this Chapter. With a twin SU carburettor installation, not only have the carburettors to be individually set to ensure correct mixture, but also the idling suction must be equal on both. It is best to use a vacuum synchronising device such as that produced by Crypton. If this is not available it is possible to obtain fairly accurate synchronisation by listening to the hiss made by the air flow into the intake throat of each carburettor. A rubber tube held to the ear is useful for this adjustment.
2 The aim is to adjust the throttle butterfly disc so that an equal amount of air enters each carburettor. Slacken the throttle shaft levers on the throttle shaft which connects the two throttle discs together. Listen to the hiss from each carburettor intake and if a difference in intensity is noticed between them, then unscrew the throttle adjusting screw on the other carburettor until the hiss from both the carburettors is the same.
3 With the vacuum synchronising device all that is necessary to do is to place the instrument over the intake of each carburettor in turn and adjust the adjusting screws until the reading on the gauge is identical for both carburettors.
4 Tighten the levers on the interconnecting linkage to connect the two throttle discs of the two carburettors together, at the same time holding down the throttle adjusting screws against their idling stops. Synchronisation of the two carburettors is now complete.

25 SU carburettor (type HD6) - dismantling and reassembly

1 All reference numbers refer to Fig 3.10. Unscrew the piston damper (3) and lift away from the chamber and piston assembly (1, 5). Using a screwdriver or small file scratch identification marks on the suction chamber (1) and carburettor body (11) so that they may be fitted together in their original position. Remove the three screws (4) and lift the suction chamber (1) from the carburettor body (11) leaving the piston

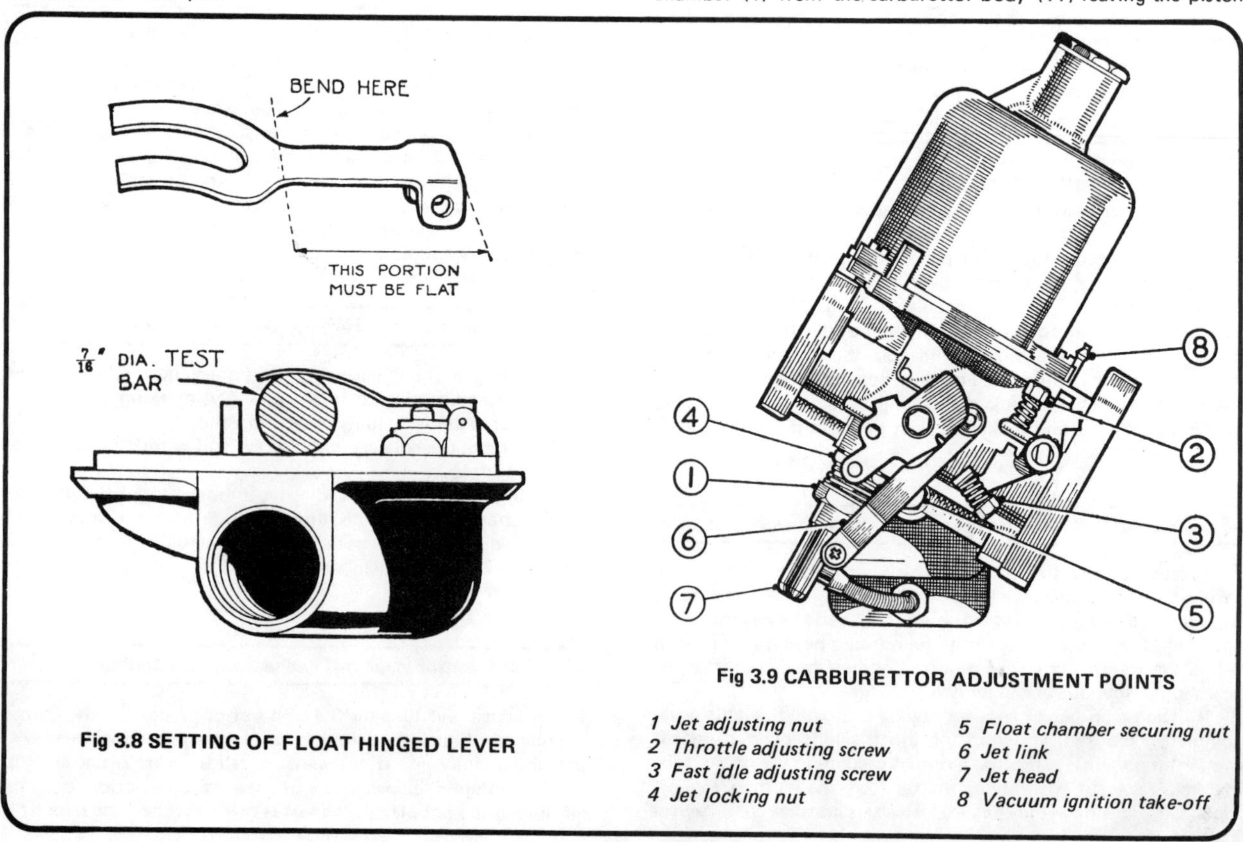

Fig 3.8 SETTING OF FLOAT HINGED LEVER

Fig 3.9 CARBURETTOR ADJUSTMENT POINTS

1 Jet adjusting nut
2 Throttle adjusting screw
3 Fast idle adjusting screw
4 Jet locking nut
5 Float chamber securing nut
6 Jet link
7 Jet head
8 Vacuum ignition take-off

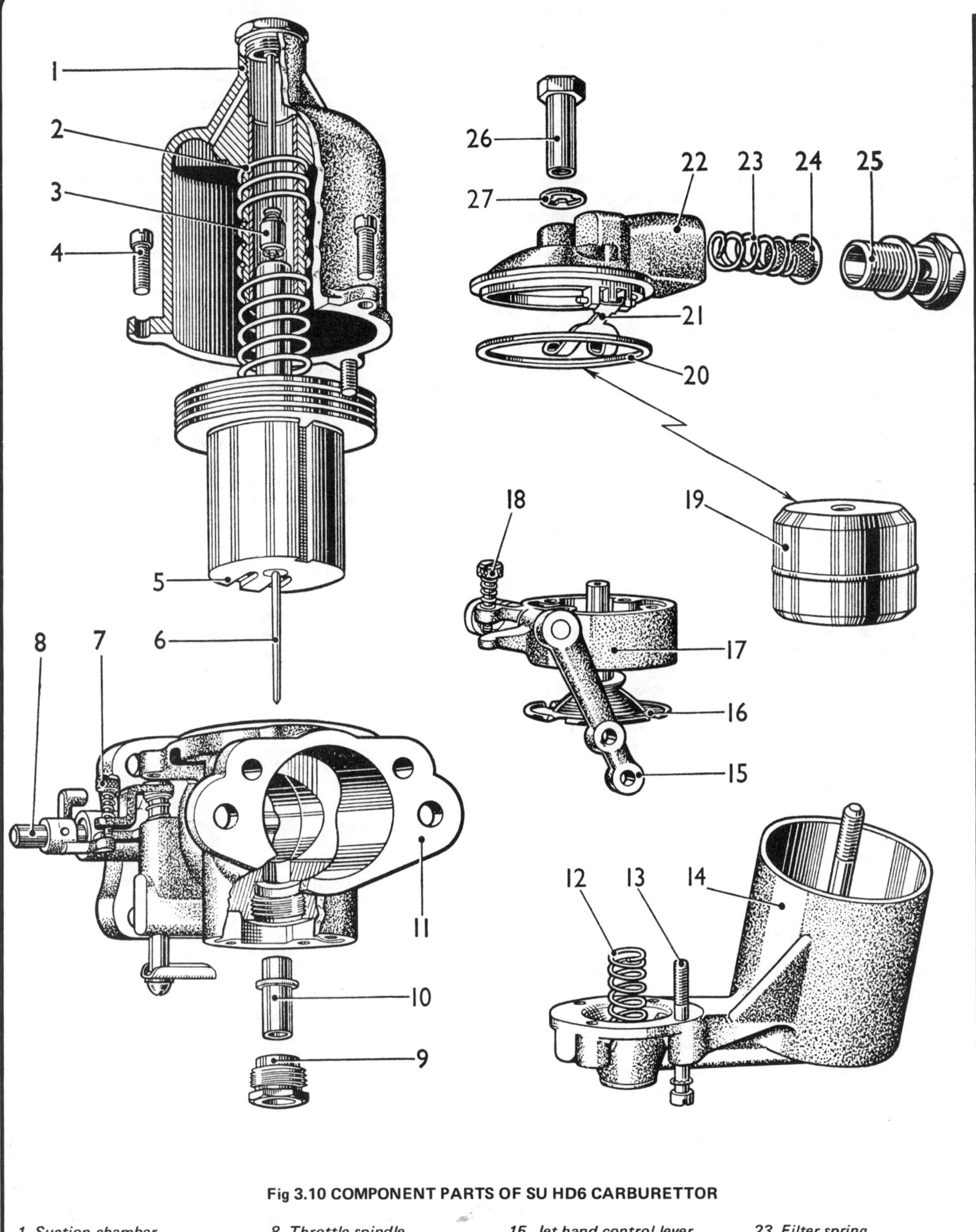

Fig 3.10 COMPONENT PARTS OF SU HD6 CARBURETTOR

1 Suction chamber
2 Piston spring
3 Hydraulic damper
4 Suction chamber screw
5 Piston
6 Needle
7 Throttle stop lever adjusting screw
8 Throttle spindle
9 Jet nut
10 Jet bearing
11 Carburettor body
12 Jet return spring
13 Float chamber securing screw
14 Float chamber
15 Jet hand control lever
16 Jet and diaphragm
17 Diaphragm casing
18 Jet adjusting screw
19 Float
20 Cover joint washer
21 Float lever
22 Float chamber cover
23 Filter spring
24 Filter
25 Inlet union
26 Float chamber cover screw
27 Fibre washer

in position.
2 Lift the piston spring (2) from the piston noting which way round it is fitted and remove the piston (5). Place the piston spring in a safe place so that the needle will not be touched or the piston roll onto the floor. It is recommended that the piston be placed on the neck of a narrow jam jar with the needle inside so acting as a stand.
3 Mark the position of the float chamber lid (22) relative to the body (14) and unscrew the cover screw (26). Lift away the cover screw (26) and shaped fibre washer (27). Remove the lid (22) and joint washer (20) and withdraw the pin thereby releasing the float lever (21). Using a spanner or socket remove the needle valve assembly.
4 Invert the float chamber and recover the float (19) noting which way round it is fitted.
5 With a screwdriver or file mark the position of the float chamber (14) relative to the body (11) to ensure correct position on reassembly and unscrew the four float chamber securing screws (13).
6 Lift away the float chamber (14). Lift out the jet spring (12). Mark the jet diaphragm (16) opposite one of the screw holes in the jet housing and withdraw the jet assembly. Lift off the jet housing (17).
7 Using a ring spanner slacken and remove the jet locking nut together with the jet bearing.
8 Close the throttle and mark the relative positions of the throttle disc and the carburettor flange.
9 Straighten the ends of the two disc retaining screws and remove the two screws. Withdraw the disc from its slot in the throttle spindle. As the disc is oval it will easily jam so take care when removing it.
10 Slide out the spindle from its bearings. The spindle sealing glands should not be removed as they do not require any servicing.
11 Unscrew and remove the slow running valve complete with spring, seal and brass washer. These parts are shown in Fig 3.11.
12 Undo and remove the two screws and shakeproof washers retaining the vacuum ignition take off plate and union. Lift off the plate and gasket.
13 The piston lifting pin may be removed only if it requires renewal, by extracting the circlip from its groove with the pin pressed upwards. Withdraw the pin downwards.
14 Reassembly of the carburettor is the reverse sequence to dismantling. It will however be necessary to centre the jet and adjust the carburettor settings as described in this Chapter.

26 SU carburettor (type HD6) - examination and repair

For full information refer to Section 15 of this Chapter.

27 SU carburettor (type HD6) - piston sticking

For full information refer to Section 16 of this Chapter.

28 SU carburettor (type HD6) - float needle sticking

For full information refer to Section 17 of this Chapter.

29 SU carburettor (type HD6) - float chamber flooding

For full information refer to Section 18 of this Chapter.

30 SU carburettor (type HD6) - water or dirt in the carburettor

For full information refer to Section 19 of this Chapter.

31 SU carburettor (type HD6) - jet centering

The piston should fall freely onto the carburettor bridge with a click when the lifting pin is released with the jet in the fully up position. If it will only do this with the jet lowered then the jet unit requires centering. To do this proceed as follows:
1 Mark the position of the jet housing and float chamber in relation to the carburettor body for reassembly.
2 Remove the plate retaining screw and withdraw the cam rod assembly.
3 Unscrew and remove the float chamber securing screws (13) (Fig 3.10).
4 Remove the float chamber and the jet housing and release the jet assembly.
5 Slacken the jet locking nut (9), using a ring spanner, until the jet bearing (10) is just free to move.
6 Remove the piston damper, hold the jet in the 'fully up' position and apply light pressure on the top of the piston rod. Tighten the jet locking nut (9).
7 Check again as in the first paragraph of this Section and ensure that the jet moves down the bearing freely.
8 Reassemble, ensuring that the jet and diaphragm are kept to the same angular position and that the beaded edge of the diaphragm is located in the housing groove.
9 Refill the piston damper with oil.

32 SU carburettor (type HD6) - fuel level adjustment

For full information refer to Section 21 of this Chapter.

33 SU carburettor (type HD6) - needle replacement

For full information refer to Section 22 of this Chapter.

34 SU carburettor (type HD6) - adjustment and tuning

1 To adjust and tune the SU carburettor proceed in the following manner. Check the colour of the exhaust at idling speed with the choke fully in. If the exhaust tends to be black and the tailpipe interior is also black it is a fair indication that the mixture is too rich. If the exhaust is colourless and the deposit on the exhaust pipe is very light grey it is likely that the mixture is too weak. This condition may also be accompanied by intermittent misfiring, while too rich a mixture will be associated with 'hunting'. Ideally the exhaust should be colourless with a medium grey pipe deposit.
2 The exhaust pipe deposit should only be checked after a good run of at least 20 miles. Idling in city traffic and stop/start motoring is bound to produce excessively dark exhaust pipe deposits.
3 Once the engine has reached its normal operating temperature, detach the carburettor air intake cleaners.
4 Unscrew the fast idle adjusting screw (4) (Fig 3.11) to clear the throttle stop with the throttle closed.
5 Screw down the slow running valve (1) onto its seating and then unscrew it 3½ turns.
6 Remove the piston/suction chamber unit as described in Section 25 paragraphs 1 and 2.
7 Turn the jet adjusting screw (3) until the jet is flush with the bridge on the carburettor.
8 Refit the piston/suction chamber unit.
9 Check that the piston falls freely onto the bridge when the lifting pin is released. If not refer to Section 31 and centre the jet.
10 Lower the jet by turning the jet adjusting screw (3) down 2½ turns.
11 Restart the engine and adjust the slow running valve to give the desired idling speed of approximately 1000 rpm. The ignition warning light should just be glowing.

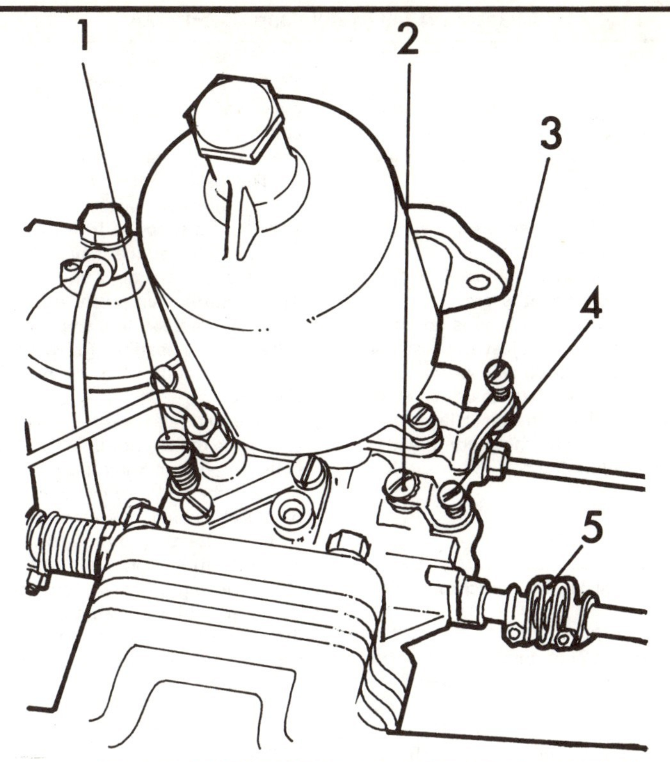

Fig 3.11 CARBURETTOR ADJUSTMENT POINTS (HD6)

1 Slow-run valve
2 Top plate securing screw
3 Jet adjusting screw
4 Throttle stop screw
5 Throttle shaft interconnection clip

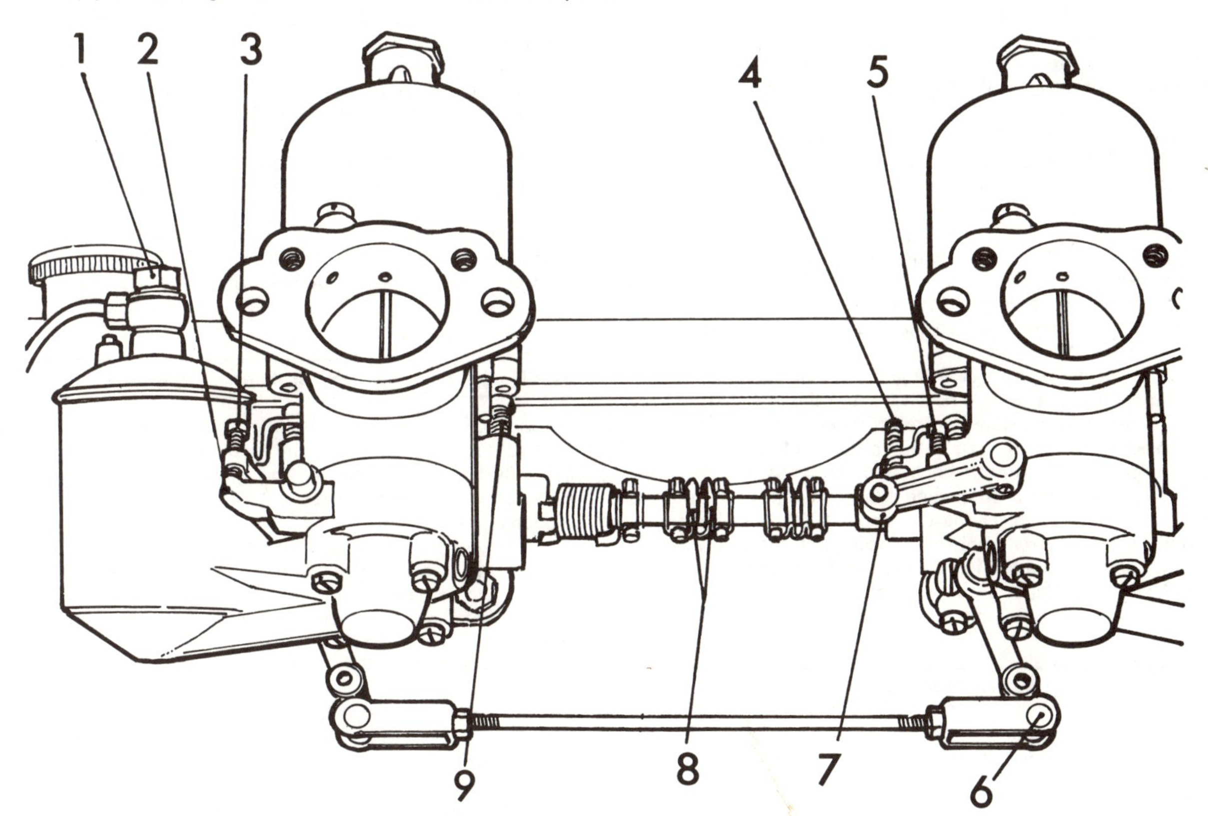

Fig 3.12 TWIN HD6 CARBURETTOR INSTALLATION

1 Petrol overflow pipe union
2 Jet adjusting screw stop
3 Jet adjusting screw
4 Throttle stop screw
5 Jet adjusting screw
6 Jet lever connecting yoke
7 Jet lever
8 Throttle shaft interconnection clip
9 Slow run valve

12 Turn the jet adjusting screw (3) up to weaken or down to enrich until the fastest idling speed consistent with even running is obtained.
13 Re-adjust the slow running valve (1), if necessary, to give correct idling.
14 Check for correct mixture by gently pushing the lifting pin up about 1/32 inch after free movement has been taken up.
15 Re-adjust the mixture strength as necessary.
16 Reconnect the mixture control wire with about 1/16 inch free movement before it starts to pull on the jet lever.
17 Pull the mixture control knob until the linkage is about to move the carburettor jet operating arm and adjust the fast idle screw (2) to give an engine speed of about 1000 rpm when hot.
18 Return the control knob and check that there is some clearance between the fast idle screw (2) and the throttle stop.
19 Finally top up the piston damper with the recommended engine oil until the level is ½ inch below the top of the hollow piston rod.

35 SU carburettor installation (type HD6) - synchronisation and linkage adjustment

1 Slacken a clamping bolt (8) (Fig 3.12) on one of the throttle spindle interconnection couplings between the carburettors.
2 Disconnect the jet control interconnecting rod at the forked end (6).
3 Restart the engine and turn the slow running valve (9), or throttle adjusting screws, an equal amount on each carburettor to give the desired idling speed.
4 Compare the intensity of the intake hiss on all carburettors and alter the slow running valves (9), or throttle adjusting screws, until the hiss is the same.
5 Turn the jet adjusting screw (3) an equal amount on all carburettors, up to weaken or down to enrich, until the fastest idling speed consistent with even running is obtained.
6 Re-adjust the slow running valves (9) if necessary.
7 Check the mixture by raising the lifting pin of the front carburettor 1/32 inch after free movement has been taken up.
8 Repeat the operation on the other carburettor(s) and after adjustment recheck as the carburettors are interdependent.
9 Tighten the clamp bolt (8) of the throttle spindle interconnections with the pin of the link pin lever resting against the edge of the pick-up lever hole. This provides the correct delay in opening the front carburettor throttle. When forked levers are fitted, set the cranked levers so that the pin is 0.006 inch from the lower edge of the fork.
10 Reconnect the jet control linkage (6) so that the jet operating arms move simultaneously; if necessary turn the fork end(s).
11 Reconnect the mixture control wire with about 1/16 inch free movement before it starts to pull on the jet levers.
12 Pull the mixture control knob until the linkage is about to move the carburettor jet operating arms, and adjust the fast idle screws (4) to give an engine speed of about 1000 rpm when hot.
13 Return the control knob and check that there is a small clearance between the fast idle screws and the throttle stops.
14 Refit the air cleaners and recheck for correct mixture as described in paragraphs 7 and 8.

36 SU carburettor (type HD6) thermo - description

1 The auxiliary carburettor is used on certain installations to provide automatically differing degrees of mixture enrichment at:

a) starting
b) idling and light cruising conditions
c) full throttle conditions

It may be used with single or multi-carburettor installations and is shown in Fig 3.13.

2 The auxiliary carburettor is a separate unit attached to the main carburettor. When fitted to 'H' type carburettors the construction of the main carburettor jet assembly differs from the normal in the method of mixture adjustment.
3 The device consists of a solenoid operated valve and a fuel metering needle which draws its fuel from the base of the auxiliary jet supplied from the main carburettor.
4 When the device is operated, air is drawn in through the air intake into a chamber and is mixed with fuel as it passes the jet. The mixture then passes upwards past the shank of the needle, through a passage, and so past the aperture provided between the valve and its seating. From here it passes directly to the main induction manifold through the external feed pipe as shown in Fig 3.13.
5 Solenoid and valve. This unit is brought into action by energising the solenoid. The iron is thus raised carrying with it the ball-jointed disc valve against the load of the conical spring so opening the aperture between the valve and seating.
6 Valve seating. A cup washer is fitted against the solenoid face to centralise the conical spring. Any leakage between the valve and its seating would allow the device to operate and affect the idling setting of the main carburettor(s). If the solenoid is energised while the engine is idling the valve will not normally lift owing to the high manifold depression; the act of opening the throttle will reduce manifold depression and allow the device to operate.
7 Fuel level. The fuel level in the auxiliary carburettor is controlled by the main carburettor float chamber. It can be seen in Fig 3.13 that this results in a reservoir of fuel remaining in the well of the auxiliary carburettor.
8 Fuel well. When starting with the device in operation, this fuel is drawn into the induction manifold to provide the rich mixture necessary for instant cold starting.
9 Needle and disc. When the valve has lifted, the needle disc chamber is in direct communication with the inlet manifold and the depression, dependent on throttle opening, varies the position of the needle by exerting a downward force upon the suction disc and needle assembly. Thus:

a) At idling the relatively high depression will draw the needle into the jet until the needle head abuts against the adjustable stop.
b) At larger throttle opening a reduced depression is communicated to the needle disc chamber and the spring will tend to overcome the downward movement of the needle, thus increasing mixture strength.

37 SU carburettor (type HD6) thermo - adjustment

As both the main and auxiliary carburettors operate when starting from cold, the main carburettor(s) must be tuned correctly before attempting any adjustment to the auxiliary carburettor. Reference should be made to the appropriate sections.

Tuning of the auxiliary carburettor is confined to adjustment of the stop nut which limits the downward movement of the needle, and is carried out with the engine running at normal temperature and the main carburettor(s) tuned. Proceed as follows:
1 Switch on the auxiliary carburettor:

a) where the thermostat has automatically broken the circuit, energise the solenoid by short circuiting the thermostatic switch to earth, or if this is inaccessible, earth the appropriate terminal of the auxiliary carburettor with a separate wire.
b) where a manual switch is fitted, switch on.

2 Open the throttle momentarily to allow the valve to lift.
3 Adjust the stop nut.

a) Initially clockwise (to weaken) until the engine begins to run erratically.

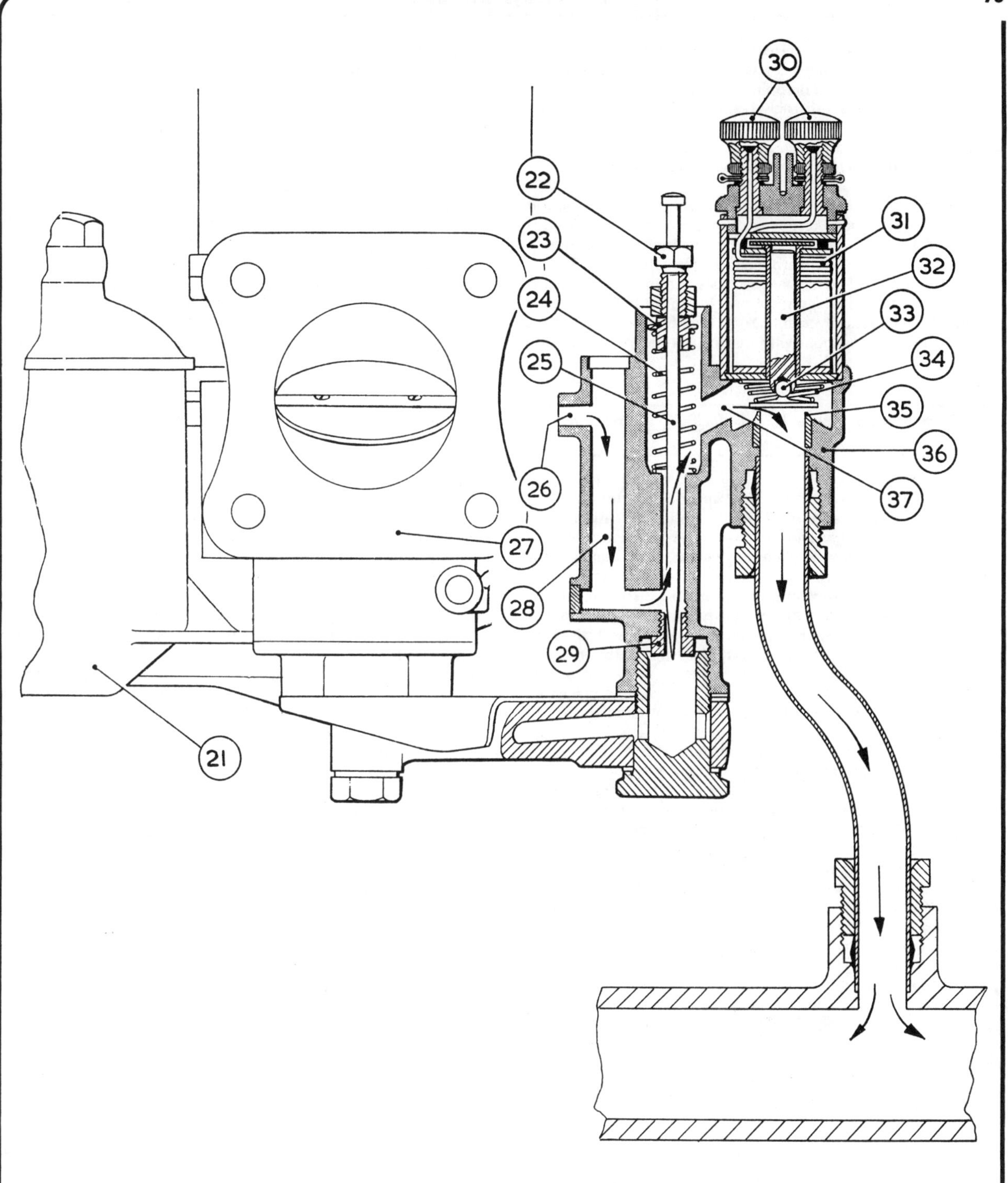

Fig 3.13 AUXILIARY ENRICHMENT CARBURETTOR (HD6)

21 Float chamber
22 Stop screw
23 Disc
24 Spring
25 Needle
26 Air intake
27 Carburettor body
28 Air passage
29 Jet
30 Terminals
31 Solenoid
32 Core
33 Valve
34 Conical spring
35 Valve seating
36 Body casting
37 Passage

b) then anticlockwise (to enrich) through the phase where the engine speed has risen markedly to the point where overrichness results in the engine speed dropping to between 800 to 1000 rpm with the exhaust gases noticeably black in colour.

38 SU carburettor (types HS4 and HS6) - description

As will be seen from the Specifications at the beginning of this Chapter three HS4 carburettors were used on Healey 3000 Mk II series BN7 and BT7. These carburettors are basically identical in construction to the HS6 carburettor information on which is included in this and subsequent sections. An illustration of the triple carburettor installation is shown in Fig 3.14. The principle of operation is basically identical to that for the carburettors described in earlier sections with particular reference to Section 11.

39 SU carburettor (types HS4 and HS6) - dismantling and reassembly

1 All reference numbers refer to Fig 3.17. Unscrew the piston damper (10) and lift away from the chamber and piston assembly (8). Using a screwdriver or small file scratch identification marks on the suction chamber and carburettor body (2) so that they may be fitted together in their original position. Remove the three suction chamber retaining screws (13) and lift the suction chamber from the carburettor body leaving the piston in situ.
2 Lift the piston spring (12) from the piston, noting which way round it is fitted, and remove the piston. Invert it and allow the oil in the damper bore to drain out. Place the piston in a safe place so that the needle will not be touched or the piston roll onto the floor. It is recommended that the piston be placed on the neck of a narrow jam jar with the needle inside so acting as a stand.
3 Mark the position of the float chamber lid relative to the body and unscrew the three screws (35) holding the float chamber lid (32) to the float chamber body (25). Remove the lid and withdraw the pin (31) thereby releasing the float and float lever (30). Using a spanner or socket remove the needle valve assembly.
4 Disconnect the jet link from the base of the jet and unscrew the nut (15) holding the flexible nylon tube into the base of the float chamber (25). Carefully withdraw the jet and nylon connection tube.
5 Unscrew the jet adjustment nut (23) and lift away together with its locking spring (22). Also unscrew the jet locknut (21) and lift away together with the brass washer (20) and jet bearing (19).
6 Remove the bolt (29) securing the float chamber to the carburettor body and separate the two parts.
7 To remove the throttle and actuating spindle release the two screws (40) holding the throttle in position in the slot in the spindle (38), make a note of the tapered edges of the throttle (39) and slide it out of the spindle from the carburettor body.
8 Reassembly is a straight reversal of the dismantling sequence.

40 SU carburettor (types HS4 and HS6) - examination and repair

For full information refer to Section 15 of this Chapter.

41 SU carburettor (types HS4 and HS6) - piston sticking

For full information see Section 16 of this Chapter.

42 SU carburettor (types HS4 and HS6) - float needle sticking

For full information see Section 17 of this Chapter.

43 SU carburettor (types HS4 and HS6) - float chamber flooding

For full information see Section 18 of this Chapter.

44 SU carburettor (types HS4 and HS6) - water or dirt in carburettor

For full information see Section 19 of this Chapter.

45 SU carburettor (types HS4 and HS6) - jet centering

1 This operation is always necessary if the carburettor has been dismantled; but to check if this is necessary on a carburettor in service, first screw up the jet adjusting nut as far as it will go without forcing it, and lift the piston and then let it fall under its own weight. It should fall onto the bridge making a soft metallic click. Now repeat the above procedure but this time with the adjusting nut screwed right down. If the soft metallic click is not audible in either of the two tests proceed as follows:
2 Disconnect the jet link (49) (see Fig 3.17) from the bottom of the jet and the nylon flexible tube from the underside of the float chamber (25). Gently slide the jet and the nylon tube from the underside of the carburettor body. Next unscrew the jet adjusting nut (23) and lift away the nut and the locking spring. Refit the adjusting nut without the locking spring and screw it up as far as possible without forcing. Replace the jet and tube but there is no need to reconnect the tube.
3 Slacken the jet locking nut (21) so that it may be rotated with the fingers only. Unscrew the piston damper (10) and lift away the damper. Gently press the piston down onto the bridge and tighten the locknut (21). Lift the piston using the lifting pin (3) and check that it is able to fall freely under its own weight. Now lower the adjusting nut (23) and check once again and if this time there is a difference in the two metallic clicks repeat the centering procedure until the sound is the same for both tests.
4 Gently remove the jet and unscrew the adjusting nut. Refit the locking spring and jet adjusting nut. Top up the damper with oil if necessary and replace the damper. Connect the nylon flexible tube to the underside of the float chamber and finally reconnect the jet link.

46 SU carburettor (types HS4 and HS6) - fuel level adjustment

Metal float

1 It is essential that the fuel level in the float chamber is always correct as otherwise excessive fuel consumption may occur. On reassembly of the float chamber check the level before replacing the float chamber cover in the following manner.
2 Invert the float chamber cover so that the needle valve is closed. It should just be possible to place a 3/16 inch bar parallel to the float chamber cover without fouling the float hinged lever as shown in Fig 3.15, or if the lever stands proud of the bar then it is necessary to bend the float lever slightly until the clearance is correct.

Nylon float

1 It is essential that the fuel level in the float chamber is always correct as otherwise excessive fuel consumption may occur. On reassembly of the float chamber check the fuel level before replacing the float chamber cover in the following manner:
2 Invert the float chamber cover so that the needle valve is closed. It should be just possible to place a 3/16 inch bar parallel to the float chamber cover without fouling the float or if the float stands proud of the bar then it is necessary to bend the float lever slightly until the clearance is correct.

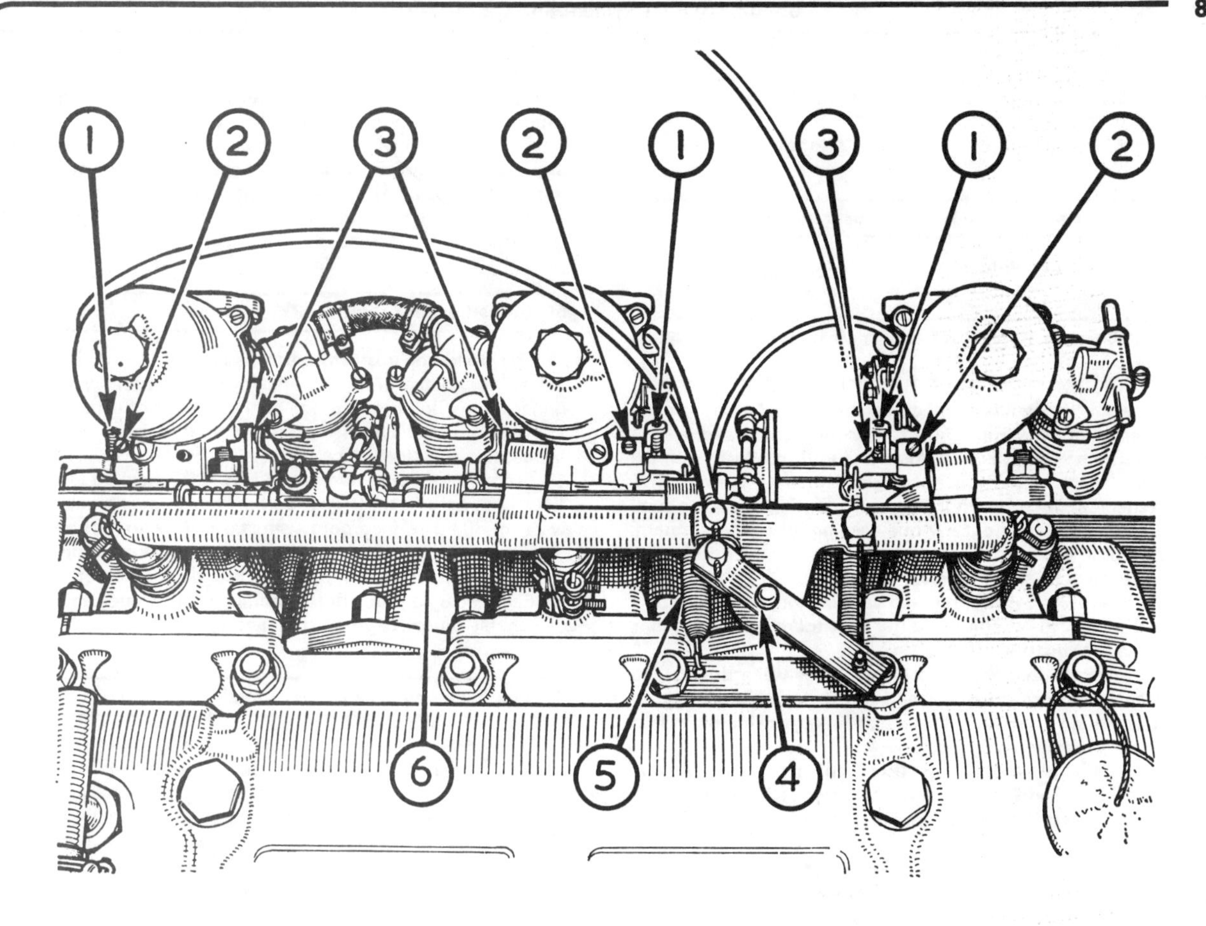

Fig 3.14 LAYOUT OF TRIPLE HS4 CARBURETTOR

1 Fast idling adjusting screws
2 Throttle adjusting screws
3 Throttle operating levers
4 Choke cable relay lever
5 Throttle return spring
6 Balance tube

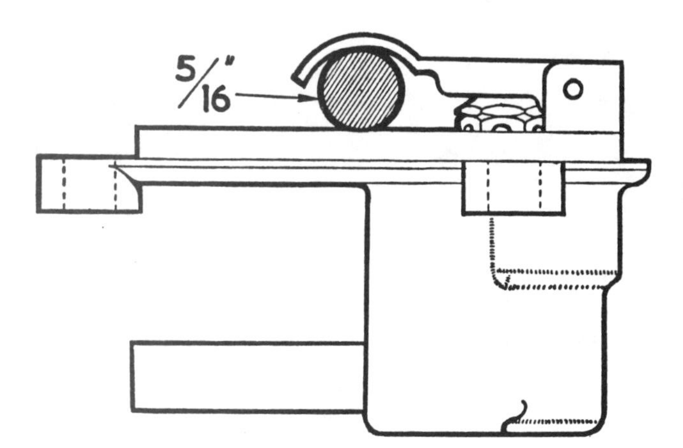

Fig 3.15 METHOD OF SETTING THE CORRECT CLEARANCE OF THE FLOAT LEVER ON EARLY CARBURETTORS

47 SU carburettor (types HS4 and HS6) - needle replacement

For full information refer to Section 22 of this Chapter.

48 SU carburettor (type HS6) - adjustment and tuning

For full information refer to Section 23 of this Chapter.

49 SU carburettor (type HS4) - adjustment, tuning and linkage adjustment

The basic procedure for adjusting a triple carburettor installation as shown in Fig 3.14 is to disconnect each carburettor linkage and then adjust each carburettor in turn. Work from the front carburettor to the middle one and then to the rear one. Then recheck each carburettor again and continue to do so until all three are checked without any one needing further adjustment. This is because each carburettor is dependent on the others. The linkage may now be set as follows:

1 The carburettor throttle on each carburettor is operated by a lever and pin, with the pin working in a forked lever which is attached to the throttle spindle.

2 It is important that there is a clearance between the pin and fork when the throttle is closed and the engine idling. This is to prevent any load from the accelerator linkage being placed on thr throttle butterfly and spindle.

3 With the throttle shaft levers (Fig 3.16) free on the throttle shaft, put a 0.012 inch feeler gauge between each throttle shaft stop at the top and the carburettor head shield.

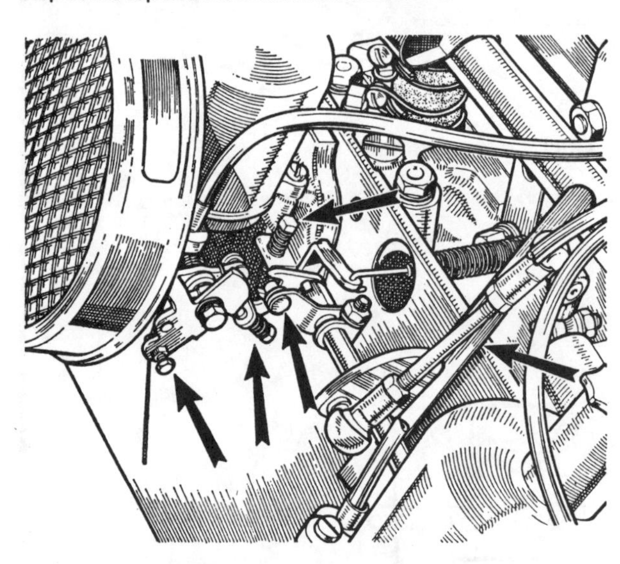

Fig 3.16 THE CARBURETTOR LINKAGE WITH A FEELER BEHIND THE THROTTLE SHAFT STOP AND THE PIN AT THE BOTTOM OF THE CLEARANCE IN THE FORKED LEVER. THE THROTTLE, FAST IDLING AND CHOKE CABLE SECURING SCREWS ARE ALSO INDICATED.

4 Slowly move each throttle shaft lever downwards in turn until the lever pin rests lightly on the lower arm of the fork in the carburettor throttle lever.

5 Tighten the clamp bolt of the throttle shaft lever when in the position set in paragraph 4.

6 When all carburettors have been set remove the feeler gauges. Check that there is a clearance between the pins on the throttle shafts and the forks.

7 Reconnect the choke cables. Make quite sure that the jet heads return fully against the lower face of the jet adjusting nuts when the choke control is pushed fully in.

8 Slowly pull out the mixture control knob on the dash panel until the linkage is about to move the carburettor jets (a minimum of ¼ inch) and adjust the fast idle cam screws to give an engine speed of about 1000 rpm at normal operating temperature.

50 SU carburettor (type HD8)

The design of this carburettor is basically identical to that of the HD6 and therefore all servicing information will be found under the section headings for the HD6 series of carburettor.

Two design features were incorporated in the carburettor installation as described below.

1 From power unit number 29E-H-1092 flexible plastic overflow pipes were fitted to each carburettor float chamber. The lids were modified to incorporate short overflow nozzles onto which the flexible pipes are a push fit. The overflow pipes may be fitted with modified lids, to earlier produced carburettors.

2 Each HD8 carburettor is attached by four studs and nuts to a detachable one-piece six port induction manifold.

51 Fuel tank - removal and refitting

1 If it is known beforehand that the fuel tank is to be removed allow the level of petrol to fall as low as possible so that the minimum amount of fuel has to be stored away from the tank.

2 Note how much fuel is in the tank, remove the hexagon drain plug and drain the fuel from the tank into a container of suitable size.

3 Working inside the luggage compartment, detach the spare wheel securing strap and lift away the spare wheel. For safety reasons disconnect the battery positive terminal.

4 Remove the carpeting from the floor of the luggage compartment.

5 Undo and remove the six Phillips screws that secure the petrol tank feed pipe cover to the body. This cover is located in the top right hand corner of the luggage compartment.

6 Disconnect the fuel feed pipe from the tank.

7 Slacken the two petrol tank filler pipe rubber union securing clips and disconnect the petrol tank filler pipe.

8 Detach the insulated lead from the petrol gauge unit terminal.

9 Working under the luggage compartment floor just in front of the rear body panel undo and remove the locknut and nut from each of the tank strap retaining studs.

10 Draw the straps through the compartment floor and hinge them back on their clevis pin anchorages.

11 The fuel tank may now be lifted away from the rear of the car.

12 Refitting the fuel tank is the reverse procedure to removal.

52 Fuel tank - cleaning

1 With time it is likely that sediment will collect in the bottom of the fuel tank. Condensation resulting in rust and other impurities will usually be found in the fuel tank of most models.

2 When the tank is removed it should be vigorously flushed out and turned upside down and, if facilities are available, steam cleaned.

53 Fuel tank gauge unit - removal and replacement

1 Working inside the boot detach the spare wheel securing strap and lift away the spare wheel.

2 For safety reasons disconnect the battery positive terminal.

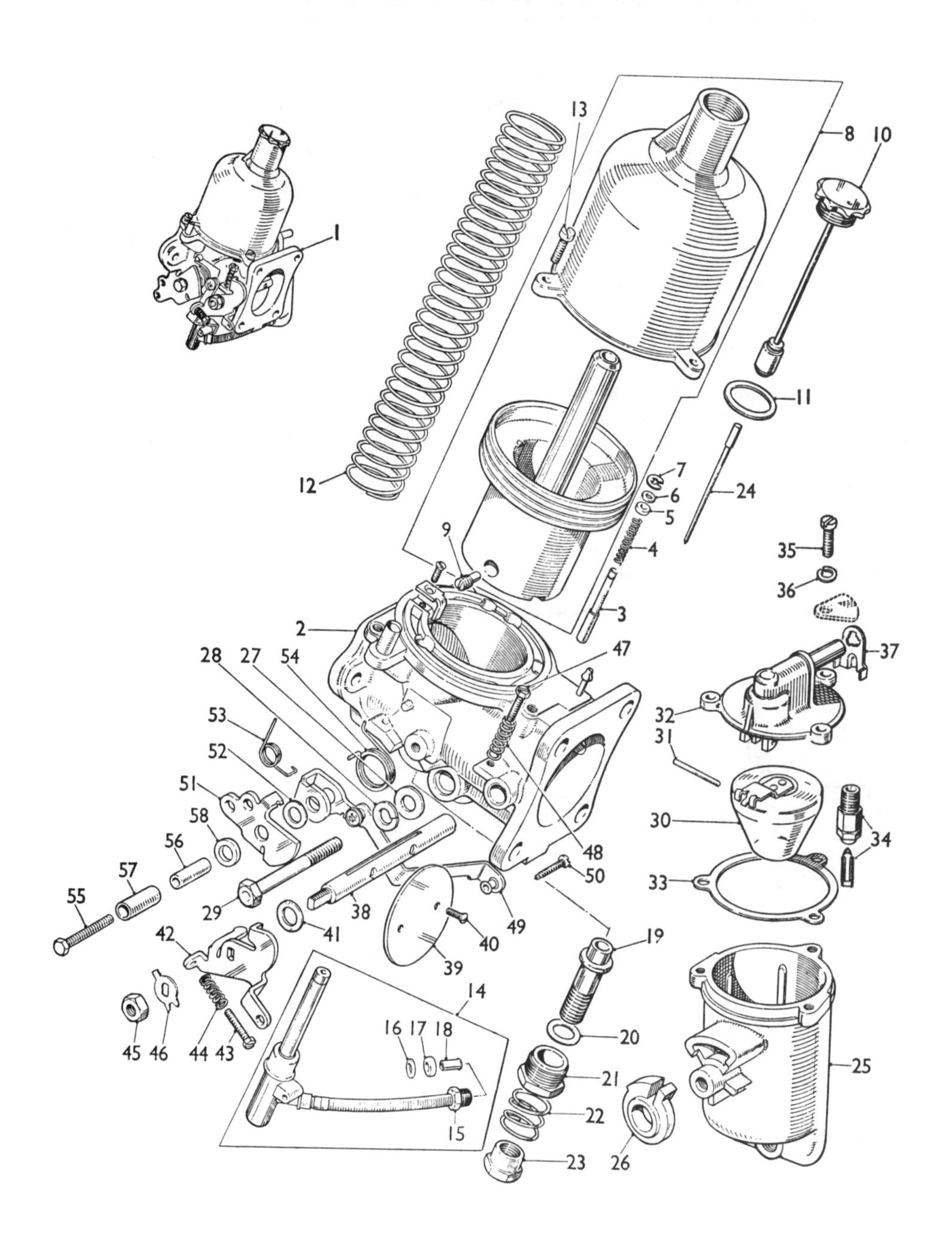

Fig 3.17 COMPONENT PARTS OF SU HS4 AND HS6 CARBURETTOR

1 Carburettor assembly
2 Body
3 Piston lifting pin
4 Spring - pin
5 Neoprene washer
6 Brass washer
7 Circlip - pin
8 Chamber and piston assembly
9 Needle locking screw
10 Piston damper
11 Fibre washer
12 Piston spring
13 Screw - chamber to body
14 Jet assembly
15 Nut
16 Washer
17 Gland
18 Ferrule
19 Jet bearing
20 Brass washer - jet bearing
21 Jet locking nut
22 Jet locking spring
23 Jet adjusting screw
24 Needle
25 Float chamber
26 Adaptor
27 Plain washer
28 Spring washer
29 Bolt - float chamber to body
30 Float
32 Float chamber lid
33 Sealing washer
34 Needle and seat
35 Screw
36 Spring washer
37 Baffle plate lid
38 Throttle spindle
39 Throttle disc
40 Screw - disc to spindle
41 Brass washer - spindle
42 Throttle return lever
43 Cam stop screw
44 Spring - screw
45 Throttle spindle nut
46 Tab washer - nut
47 Throttle adjusting screw
48 Spring - screw
49 Pick-up lever and link
50 Screw - link to jet
51 Cam lever
52 Washer - cam lever
53 Cam lever spring
54 Pick-up lever spring
55 Pivot bolt
56 Pivot bolt tube
57 Outer tube
58 Distance washer

3 Remove the carpeting from the floor of the luggage compartment.
4 Detach the insulated lead from the fuel tank gauge unit terminal.
5 Undo and remove the six screws and spring washers securing the fuel tank gauge unit to the tank and lift away the gauge unit and its joint washer. Take care not to bend the fine wire float lever.
6 Replacement of the unit is a reversal of the above process. To ensure a fuel tight joint, scrape both the tank and sender unit mating faces clean and always use a new joint washer.

54 Throttle control linkage - adjustment

1 It is important that the throttle linkage is not strained causing premature wear. The linkage may be adjusted to allow the toe board to act as a positive stop for the accelerator pedal when the throttles are fully open.
2 To adjust the linkage first slacken the pinch bolt on lever 'A' (Fig 3.18). Note this illustration shows the linkage for a left hand drive model for clarity reasons, on right hand drive cars the lever 'B' is on the right hand side of the accelerator relay shaft.
3 Place a wooden block 'D' 2½ inch thick between the pedal and the toe board and push the pedal down so it rests on and retains the block on the toe board.
4 Adjust the lever 'A' relative to the pedal cross shaft to obtain a clearance of 1/16 inch at point 'X' between lever 'C' and the body flange. Tighten the pinch bolt on lever 'A'.
5 Next slacken the pinch bolt on levers 'A' and 'B' (Fig 3.19) and set the lever 'B' at approximately 45° as shown and retighten the pinch bolt. Make sure that at the same time the throttles are not held open by the idling adjustment screws.
6 Adjust the length of rod 'C' (Fig 3.18) so as to bring lever 'A' parallel with the lever 'B'.
7 Leave the pinch bolt of lever 'A' (Fig 3.18) still slack and press the rod 'D' downwards 1/8 inch to tension the pedal return spring slightly. Retighten the pinch bolt on lever 'A'.
8 Now depress the accelerator pedal fully and check the travel of lever 'E' (Fig 3.18). This must be such that it is at least 20° short of the vertical position when full throttle condition is reached on the carburettors. To achieve this adjust the length of rod 'D'.
9 Check that when the accelerator pedal is fully depressed the carburettor throttles are fully open.
10 If the car is fitted with overdrive it may be necessary to adjust the throttle switch operation. Full information will be found in Chapter 6, Section 19.

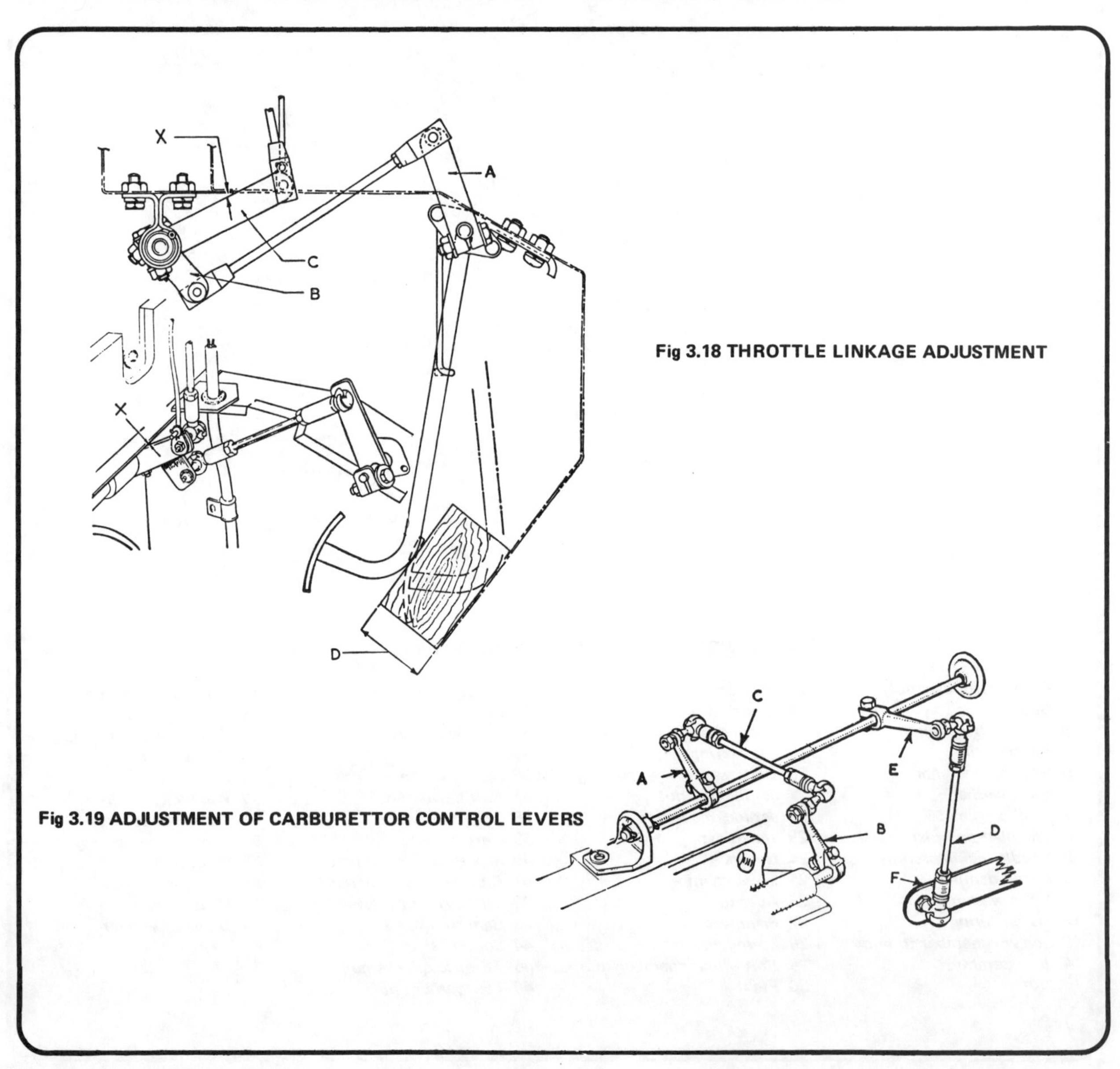

Fig 3.18 THROTTLE LINKAGE ADJUSTMENT

Fig 3.19 ADJUSTMENT OF CARBURETTOR CONTROL LEVERS

Chapter 4 Ignition system

Contents

Specifications

Lucas 12 volt coil system

Austin Healey 100-6 (Series BN4 and BN6)

Distributor	Lucas DM6A
Direction of rotation	Anticlockwise at rotor arm
Contact breaker gap	0.014 to 0.016 inch
Dwell angle	$35^o \pm 3^o$
Condenser capacity	0.2 Mf
Static setting (crankshaft)	6° BTDC
Maximum advance (crankshaft)	35° BTDC
Coil	Lucas HA12
Spark plug	Champion UN12Y
Spark plug gap	0.024 - 0.026 inch

Austin Healey 3000 Mk I and Mk II (Series BN7 and BT7), Mk II (Series BJ7) and Mk III (Series BJ8)

As above but with the following exceptions:

Distributor type	Lucas DM6
Static setting (crankshaft)	6° BTDC
Maximum advance (crankshaft)	36° BTDC

Austin Healey 3000 Mk II and Mk III from engine number 29F/4898

Distributor	
Mk II	Lucas DM6A
Mk II and III	Lucas 25D6 (from engine no 29F3563 serial no 40966B)
Direction of rotation	Anticlockwise at rotor arm
Contact breaker gap	0.014 to 0.016 inch
Dwell angle	$35^o \pm 3^o$
Condenser capacity:	
Mk II	0.2 Mf
Mk II (Series BJ7)	0.1 Mf
Mk III	0.18 - 0.23 Mf
Static setting (crankshaft)	5° BTDC
Later	10° BTDC
Stroboscopic ignition timing (crankshaft)	15° BTDC at 600 rpm
Maximum advance (crankshaft)	35° BTDC
6 port cylinder head	36° BTDC
Coil type	HA12
Resistance	3.1 to 3.5 ohms
Spark plug type	Champion UN12Y
Spark plug gap	0.024 - 0.026 inch

Austin Healey 3000 Mk III

As above but with the following exceptions:

Maximum advance (crankshaft)	34 - 38° at 6400 rpm
Vacuum advance:	
Starts	5 in Hg
Ends	16° at 12 in Hg
Deceleration check (crankshaft)	28° to 32° at 4400 rpm
	22° to 26° at 3400 rpm
	15° to 19° at 2000 rpm
	10° to 16° at 1500 rpm
	2° to 8° at 1100 rpm

6 port cylinder head only

Distributor type	Lucas DM6

1 General description

In order that the engine can run efficiently it is necessary for an electrical spark to ignite the fuel/air mixture in the combustion chamber at exactly the right moment in relation to engine speed and load. The ignition system is based on feeding low tension voltage from the battery to the coil where it is converted to high tension voltage. The high tension voltage is powerful enough to jump the spark plug gap in the cylinders many times a second under high compression pressures, providing that the system is in good condition and that all adjustments are correct.

The ignition system is divided into two circuits. The low tension circuit and the high tension circuit.

The low tension (sometimes known as the primary) circuit consists of the battery, lead to the control box, lead to the ignition switch, lead from the ignition switch to the low tension or primary coil windings (terminal SW), and the lead from the low tension coil windings (coil terminal CB) to the contact breaker points and condenser in the distributor.

The high tension circuit consists of the high tension or secondary coil windings, the heavy ignition lead from the centre of the coil to the centre of the distributor cap, the rotor arm, and the spark plug leads and spark plugs.

The system functions in the following manner. Low tension voltage is changed in the coil into high tension voltage by the opening and closing of the contact breaker points in the low tension circuit. High tension voltage is then fed via the carbon brush in the centre of the distributor cap to the rotor arm of the distributor. The rotor arm revolves inside the distributor cap, and each time it comes in line with one of the four metal segments in the cap, which are connected to the spark plug leads, the opening and closing of the contact breaker points causes the high tension voltage to build up, jump the gap from the rotor arm to the appropriate metal segment and so via the spark plug lead to the spark plug, where it finally jumps the spark plug gap before going to earth.

The ignition is advanced and retarded automatically, to ensure the spark occurs at just the right instant for the particular load at the prevailing engine speed.

The ignition advance is controlled both mechanically and by a vacuum operated system. The mechanical governor mechanism comprises two lead weights, which move out from the distributor shaft as the engine speed rises due to centrifugal force. As they move outwards they rotate the cam relative to the distributor shaft, and so advance the spark. The weights are held in position by two light springs and it is the tension of the springs which is largely responsible for correct spark advancement.

The vacuum control consists of a diaphragm, one side of which is connected via a small bore tube to the side of the carburettor, and the other side to the contact breaker plate. Depression in the inlet manifold and carburettor, which varies with engine speed and throttle opening, causes the diaphragm to move, so moving the contact breaker plate, and advancing or retarding the spark. A fine degree of control is achieved by a spring in the vacuum assembly.

It will be seen from the Specifications at the beginning of this Chapter that depending on the model of car, one of three types of distributor will be fitted. The basic construction of the distributors are similar and the slight differences are described in detail where necessary.

2 Routine maintenance

1 Release the two distributor cap retaining clips and lift off the distributor cap. It is important that the seals of water-repellant silicone grease at the point of entry of the ignition HT leads into the top of the cap are not disturbed. If this seal is disturbed, moisture might find its way into the cap on damp days, possibly causing ignition failure.
2 Smear the cam with a little grease and apply a drop of oil onto the top of the contact breaker lever pivot pin. Do not allow any lubricant to touch the points.
3 Lift off the rotor arm from the top of the cam spindle and allow a few drops of engine grade oil to pass through the passage provided in the cam spindle for the lubrication of the cam bearing and the distributor driveshaft. Do not, however, remove the screw in the centre of the top of the cam spindle. Refit the rotor arm.
4 Lubricate the automatic advance system by squirting a few drops of engine oil into the hole in the contact breaker baseplate.
5 On DM6A distributors only, screw down the grease cap on the distributor driveshaft half of a turn. The grease cap should be refilled when it cannot be screwed any further.
6 The spark plugs must always be correctly set with a gap of 0.025 inch.

3 Contact breaker points - adjustment

1 To adjust the contact breaker points to the correct gap, first pull off the two spring blade clips securing the distributor cap to the distributor body, and lift away the cap. Clean the cap inside and out with a dry cloth. It is unlikely that the six segments will be burned badly or scored, but if they are the cap will have to be renewed.
2 Push the carbon brush located in the top of the cap once or twice to make sure it moves freely.
3 Lift away the rotor arm and gently prise the contact breaker points open to examine the condition of their faces. If they are rough, pitted or dirty, it will be necessary to remove them for resurfacing, or for replacement points to be fitted.
4 Presuming the points are satisfactory, or that they have been cleaned and replaced, measure the gap between the points by turning the engine over until the contact breaker arm is on the peak of one of the six cam lobes.
5 A 0.015 inch feeler gauge should now just fit between the points. If adjustment is necessary, proceed as follows depending

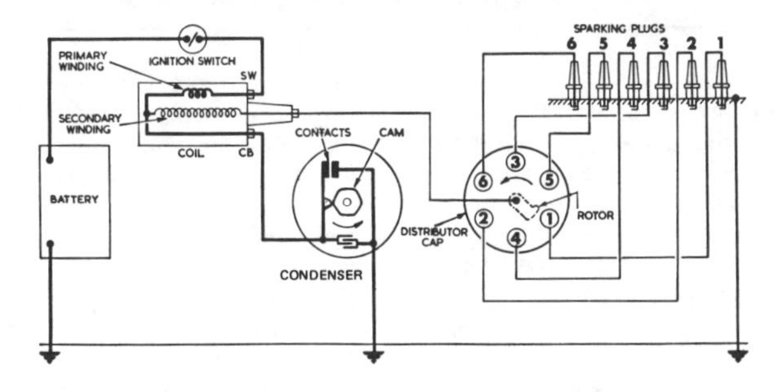

Fig 4.1 DIAGRAMMATIC REPRESENTATION OF THE IGNITION CIRCUIT. LT SYSTEM INDICATED BY THE HEAVIER LINES

Fig 4.2 CONTACT BREAKER POINT ADJUSTMENT

1 Oiling point
2 Feeler gauge
3 Shaft lubricator
4 Contact locking screws
5 Capacitor
6 Contact adjusting screw
7 Micrometer adjuster
8 Contact locking screws

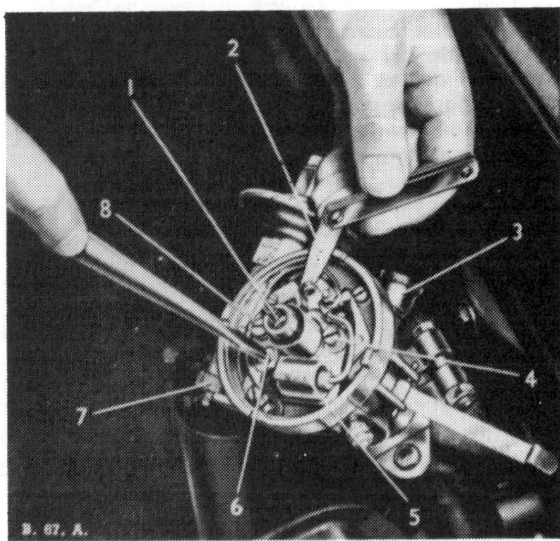

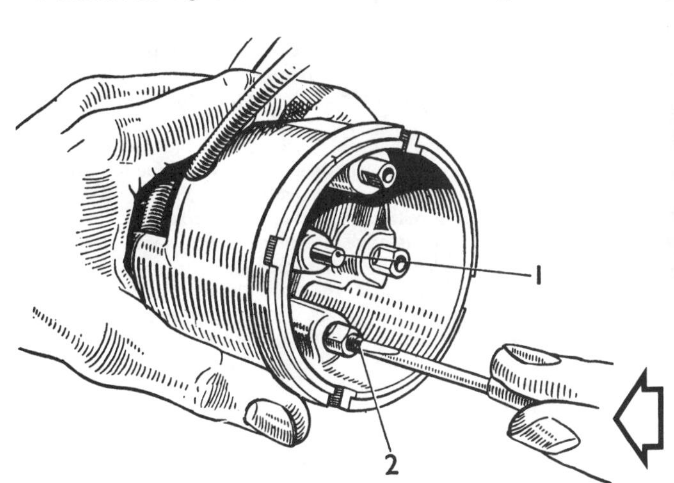

Fig 4.3 DISTRIBUTOR CAP

1 Carbon brush
2 Screw securing cable

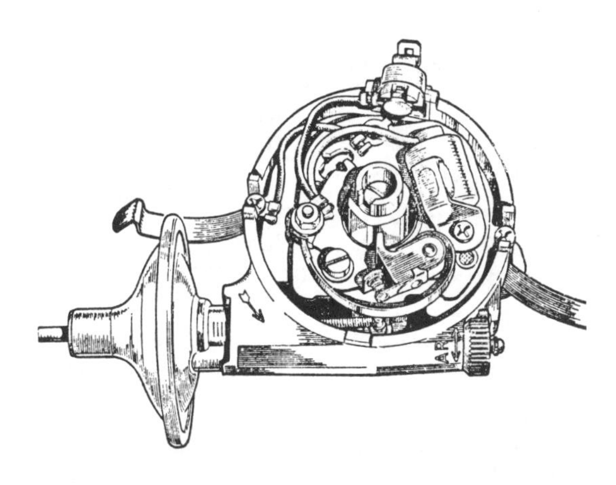

Fig 4.4 VIEW OF DISTRIBUTOR WITH THE CAP REMOVED AND THE CONTACT BREAKER POINTS OPEN

on the type of distributor fitted.
6 DM6A: Slacken the two screws which secure the fixed contact plate. Adjust the setting of the plate by turning the little eccentric screw as necessary until the point gap is correct and then tighten the two securing screws again.
7 25D6: Slacken the one screw which secures the fixed contact plate. Adjust the setting of the plate by inserting a screwdriver in the notched hole at the end of the plate. Turn clockwise to decrease and anticlockwise to increase the gap. Tighten the one securing screw again.
8 Replace the rotor arm and distributor cap and clip the two spring blade retainers into position.

4 Contact breaker points - removal and replacement

1 If the contact breaker points are burnt, pitted or badly worn, they must be removed and either replaced or their faces must be filed smooth.
2 First pull off the two spring blade clips securing the distributor cap to the distributor body, and lift away the cap. Lift away the rotor arm.
3 To remove the points unscrew the terminal nut and remove it together with the steel washer under its head, if fitted. Remove the flanged nylon bush, the condenser head and the low tension lead from the terminal pin.
4 The adjustable contact breaker plate is removed by unscrewing the one holding down screw (25D6) or two holding down screws (DM6A) and removing it complete with spring and flat washer.
5 To reface the points, rub their faces on a fine carborundum stone, or fine emery paper. It is important that the faces are rubbed flat and parallel to each other so that there will be complete face to face contact when the points are closed. One of the points will be pitted and the other will have deposits on it.
6 It is necessary to remove completely the built up deposits, but not necessary to rub the pitted point right down to the stage where all the pitting has disappeared, though obviously, if this is done, it will prolong the time before the operation of refacing the points has to be repeated.
7 To replace the contact breaker points first position the adjustable contact breaker plate over the terminal pin.
8 Secure the contact plate by screwing in the screw (25D6) or screws (DM6A) which should have a spring and flat washer under its head.
9 Then fit the fibre washer over the terminal pin.
10 Next fit the contact breaker arm complete with spring over the terminal pin.
11 Drop the fibre washer over the terminal bolt.
12 Then bend back the spring of the contact breaker arm and fit it over the terminal bolt.
13 Place the terminals of the low tension lead and the condenser over the terminal bolt.
14 Then fit the flanged nylon bush over the terminal bolt with the two leads immediately under its flange.
15 Next fit a steel washer (if originally fitted).
16 Then fit the nut over the terminal bolt and tighten it down.
17 The points are now reassembled and the gap should be set as described in the previous Section.
18 Finally replace the rotor arm and then the distributor cap.

5 Condenser - removal, testing and replacement

1 The purpose of the condenser (capacitor) is to ensure that when the contact breaker points open there is no sparking across them which would waste voltage and cause wear.
2 The condenser is fitted in parallel with the contact breaker points. If it develops a short circuit, it will cause ignition failure as the points will be prevented from interrupting the low tension circuit.
3 If the engine becomes very difficult to start or begins to miss after several miles running and the breaker points show signs of excessive burning, then the condenser must be suspect. A further test can be made by separating the points by hand with the itnition switched on. If this is accompanied by a flash it is an indication that the condenser has failed.
4 Without special equipment the only sure way to diagnose condenser trouble is to replace a suspected unit with a new one and note if there is any improvement.
5 On Lucas distributors to remove the condenser, remove the distributor cap and the rotor arm first.
6 Loosen the outer nut from the contact stud and pull off the condenser lead.
7 Undo the mounting bracket screw and remove the condenser.
8 Replacement is simply a reversal of the removal process. Take particular care that the condenser lead does not short circuit against any portion of the breaker plate.

6 Distributor - removal and refitting

1 To remove the distributor from the engine first turn the engine over by hand until the rotor arm points to the brass segment for No 1 plug lead inside the distributor cap. This will act as a useful datum point for replacement of the distributor later on. Also note the position of the vacuum control unit which will act as a guide for the refitting of the vacuum pipe.
2 Pull the terminals off each of the spark plugs. If the leads are not marked they should be identified so that they are refitted in the correct order. Release the Lucar connector or small nut which holds the low tension lead to the terminal on the side of the distributor and unscrew the high tension lead retaining cap from the coil and remove the lead.
3 Unscrew the union holding the vacuum tube to the distributor housing.
4 Remove the distributor body clamp bolt which holds the distributor clamp plate to the engine and remove the distributor. Note: If it is not wished to disturb the timing then under no circumstances should the clamp pinch bolt which secures the distributor in its relative position on the clamp, be loosened. Providing the distributor is removed without the clamp being loosened from the distributor body, the timing will not be lost.
5 Replacement is a reversal of the above procedure providing that the engine has not been turned in the meantime. If the engine has been turned it will be best to retime the ignition. This will also be necessary if the clamp pinch bolt has been loosened.

7 Distributor - dismantling

1 With the distributor removed from the car and on the bench, first pull off the two spring blade clips securing the distributor cap to the distributor body, and lift away the cap. Lift away the rotor arm. If very tight, lever it off gently with a screwdriver. The two distributors are shown in Figs 4.5 and 4.6.
2 Remove the points first by unscrewing the terminal nut and remove it together with the steel washer under its head, if fitted. Remove the flanged nylon bush, the condenser lead and the low tension lead from the terminal pin.
3 The adjustable contact breaker plate is removed by unscrewing the one holding down screw (25D6) or two holding down screws (DM6A) and removing it complete with spring and flat washer.
4 Remove the condenser from the contact breaker plate by undoing and removing the self-tapping screw and lifting away the condenser.
5 Unhook the vacuum unit spring from its mounting pin on the moving contact breaker plate.
6 Remove the contact breaker plate.
7 Unscrew the two screws and lockwashers which hold the contact breaker baseplate in position and remove the earth lead from the relevant screw. Remember to replace this lead on reassembly.
8 Lift out the contact breaker backplate.
9 Note the position of the slot in the rotor arm drive in relation to the offset drive dog at the opposite end of the

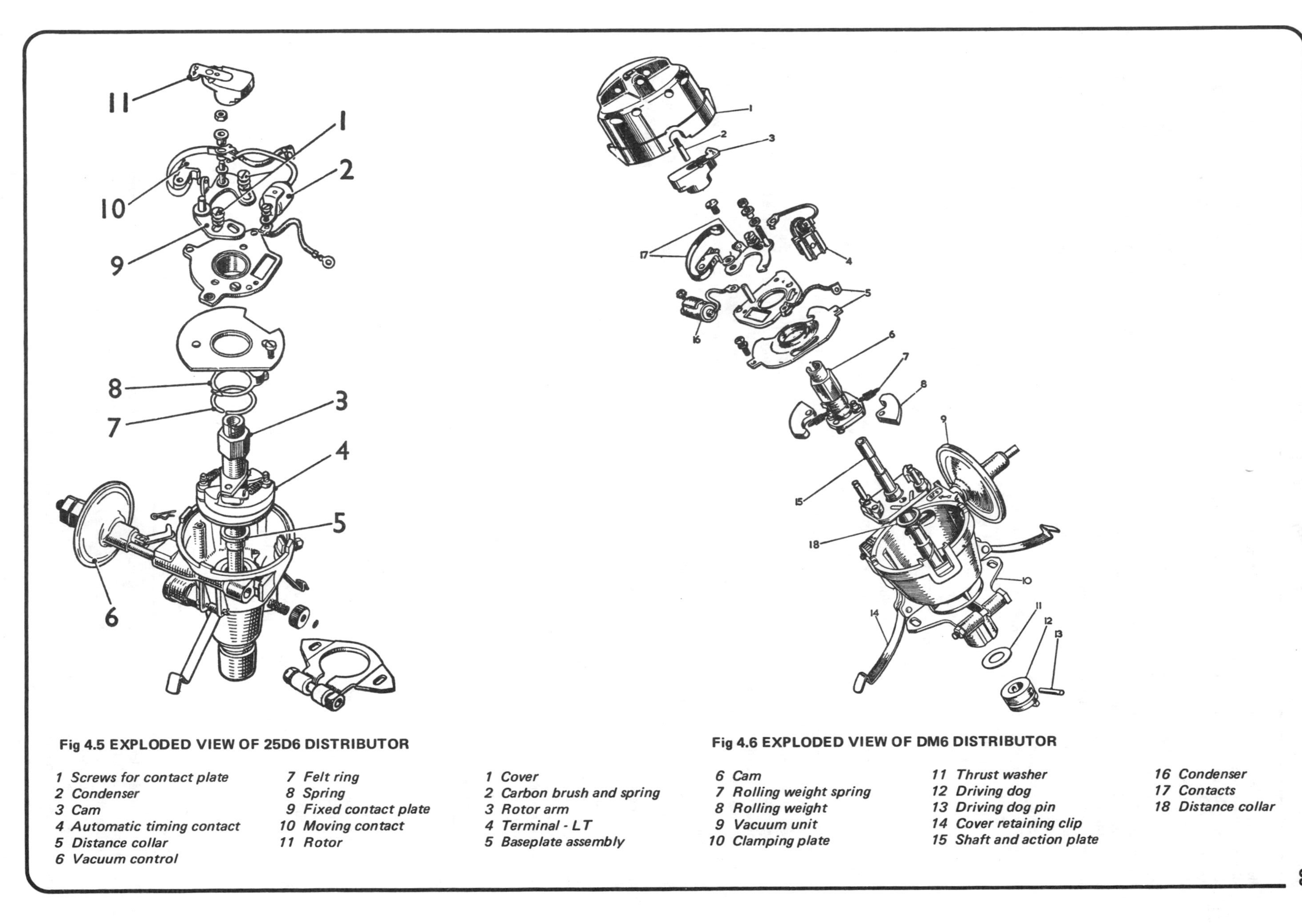

Fig 4.5 EXPLODED VIEW OF 25D6 DISTRIBUTOR

1 Screws for contact plate 2 Condenser 3 Cam 4 Automatic timing contact 5 Distance collar 6 Vacuum control 7 Felt ring 8 Spring 9 Fixed contact plate 10 Moving contact 11 Rotor

Fig 4.6 EXPLODED VIEW OF DM6 DISTRIBUTOR

1 Cover 2 Carbon brush and spring 3 Rotor arm 4 Terminal - LT 5 Baseplate assembly 6 Cam 7 Rolling weight spring 8 Rolling weight 9 Vacuum unit 10 Clamping plate 11 Thrust washer 12 Driving dog 13 Driving dog pin 14 Cover retaining clip 15 Shaft and action plate 16 Condenser 17 Contacts 18 Distance collar

distributor. It is essential that this is reassembled correctly as otherwise the timing will be 180° out.
10 Unscrew the cam spindle retaining screw which is located in the centre of the rotor arm drive, and remove the cam spindle.
11 Lift out the centrifugal weights together with their springs.
12 To remove the vacuum unit spring off the small circlip which secures the advance adjustment nut which should then be unscrewed. With the micrometer adjusting nut removed, release the spring and the micrometer adjusting nut lock spring clip. This is the clip that is responsible for the 'clicks' when the micrometer adjuster is turned, and it is small and easily lost, as is the circlip, so put them in a safe place. Do not forget to replace the lock spring clip on reassembly.
13 It is only necessary to remove the distributor driveshaft or spindle if it is thought to be excessively worn. With a thin punch drive out the retaining pin from the driving tongue collar on the bottom end of the distributor driveshaft. The shaft can then be removed. The distributor is now completely dismantled.

8 Distributor - inspection and repair

1 Check the points as described in Section 4. Check the distributor cap for signs of tracking, indicated by a thin black line between the segments. Replace the cap if any signs of tracking are found.
2 If the metal portion of the rotor arm is badly burned or loose, renew the arm. If slightly burnt, clean the arm with a fine file.
3 Check that the carbon brush moves freely in the centre of the distributor cover.
4 Examine the fit of the breaker plate on the bearing plate and also check the breaker arm pivot for looseness or wear and renew as necessary.
5 Examine the balance weights and pivot pins for wear, and renew the weights or cam assembly if a degree of wear is found.
6 Examine the shaft and the fit of the cam assembly on the shaft. If the clearance is excessive compare the items with new units, and renew either or both, if they show excessive wear.
7 If the shaft is a loose fit in the distributor bush and can be seen to be worn, it will be necessary to fit a new shaft and bush. The single bush is simply pressed out. Note that before inserting a new bush it should be stood in engine oil for at least 24 hours.
8 Examine the length of the balance weight springs and compare them with new springs. If they have stretched they must be renewed.

9 Distributor - reassembly

1 Reassembly is a straight reversal of the dismantling process, but there are several points which should be noted in addition to those already given in the Section on dismantling.
2 Lubricate with Castrol GTX the balance weight and other parts of the mechanical advance mechanism, the distributor shaft, and the portion of the shaft on which the cam bears, during assembly. Do not oil excessively but ensure these parts are adequately lubricated.
3 Check the action of the weights in the fully advanced and fully retarded positions and ensure they are not binding.
4 Tighten the micrometer adjusting nut to the middle position on the timing scale.
5 Finally, set the contact breaker gap to the correct clearance of 0.015 inch.

10 Ignition timing

1 For the correct ignition timing refer to the Specifications at the beginning of this Chapter.
2 Remove the valve rocker cover so that movement of the valves can be observed.
3 Slowly turn the crankshaft in the normal direction of rotation using the starting handle until No 1 piston is at the top of its compression stroke. This is indicated when the exhaust valve of No 6 cylinder is just beginning to close and the inlet valve is just opening.

Fig 4.7 TIMING POINTER SET OPPOSITE TO THE NOTCH IN THE CRANKSHAFT PULLEY (NO 1 PISTON IS AT TDC)

4 Look at the notch on the crankshaft pulley which should now be in line with the pointer on the timing cover as shown in Fig. 4.7. Now refer to Specifications to establish the correct ignition timing for your car. Turn the crankshaft pulley until the notch and pointer are in the correct relative positions for the degree of advancement required.
5 Check the gap of the contact breaker points which should, if necessary, be set to a gap of 0.015 inch which is the position of maximum opening.
6 Set the micrometer adjustment on the distributor to its central position. Place the distributor into its housing and engage the drive dog lugs until the drive gear slots by slowly rotating the rotor arm. It will be observed that the driving dog has an offset so it will only locate correctly one way.
7 Screw in the two bolts that secure the distributor clamp plate to the distributor housing.
8 Rotate the distributor body until the vacuum control unit side of the body is to the rear and the leads are at the cylinder block side of the distributor cap.
9 Again rotate the distributor but in an anticlockwise direction until the contact breaker points are fully closed. Then slowly rotate the distributor in a clockwise direction until the contact breaker points are just commencing to open.
10 Holding the distributor in this position secure the distributor body by tightening up the clamp plate pinch bolt and nut.
11 As a final check make sure that the rotor arm is opposite to the correct segment for No 1 cylinder, which is at the top of the compression stroke.
12 A finer adjustment can be obtained under road test conditions using the micrometer adjustment. This adjustment should not be used for initial setting of the ignition and is only altered if the main setting requires adjustment to meet the characteristics of the grade of petrol being used.
13 The adjuster nut on the side of the distributor alters the ignition timing, it adjusts the timing by 1° for every 11 clicks. There is a considerable amount of latitude for adjustment, but only extremely small movement of the adjustment nut should be made at one time.

Test lamp method

Although the previous method described will give a fairly accurate setting, it is a little difficult to determine the exact spot when the contact breaker points are just opening. To determine this opening point using the test lamp method proceed as follows:
1 Follow the instructions in paragraphs 1 to 8 inclusive, of the

previous part of this Section.
2 Connect one lead of a 12 volt test lamp to the LT terminal on the side of the distributor body and the second lead to earth.
3 Switch on the ignition and slowly rotate the distributor body in an anticlockwise direction until the contact breaker points are fully closed. The lamp will be extinguished.
4 Now rotate the distributor body in a clockwise direction until the light is ignited. This indicates that the contact breaker points have just opened.
5 Holding the distributor in this position, secure the distributor body by tightening up the clamp plate pinch bolt and nut.
6 As a check rotate the rotor arm within its limits in a clockwise and anticlockwise direction whereupon in one position the light should be extinguished and in the other position it should be ignited.
7 Refer to paragraphs 11 and 12 of the earlier part of this Section for the setting of the timing of the later produced engines.

Stroboscopic ignition timing

If a stroboscopic timing light is used to determine the ignition timing proceed as follows:
1 Follow the instructions in paragraphs 1 to 8 inclusive of the first part of this Section.
2 Disconnect the automatic timing control pipe from the side of the distributor.
3 Connect up the stroboscopic light according to the manufacturers' instructions.
4 Start the engine but do not allow the engine speed to exceed 600 rpm, otherwise the centrifugal weights will be in operation giving a false reading.

11 Ignition coil

The ignition coil is a sealed unit and requires no service attention other than making sure that the electrical connections are kept clean and tight. Occasionally wipe the exterior of the coil clean of any oil mist or dirt. This is particularly important between the terminals on the top of the coil.

If the performance of the coil is suspect the easiest method of testing is by substitution for one of known correct functioning or a new one.

12 Spark plugs and leads

1 The correct functioning of the spark plugs is vital for the efficient running of the engine.
2 At intervals of 6000 miles the plugs should be removed, examined, cleaned and if worn excessively, replaced. The condition of the spark plug will also tell much about the overall condition of the engine.
3 If the insulator nose of the spark plug is clean and white, with no deposits, this is indicative of a weak mixture, or too hot a plug (a hot plug transfers heat away from the electrode slowly - a cold plug transfers it away quickly).
4 The plugs fitted as standard are Champion UN12Y. If the top and insulator nose is covered with hard black looking deposits, then this is indicative that the mixture is too rich. Should the plug be black and oily, then it is likely that the engine is fairly worn, as well as the mixture being too rich.
6 If there are any traces of long brown tapering stains on the outside of the white portion of the plug, then the plug will have to be renewed, as this shows that there is a faulty joint between the plug body and the insulator, and compression is being allowed to leak away.
7 Plugs should be cleaned by a sandblasting machine which will free them from carbon more thoroughly than cleaning by hand. The machine will also test the condition of the plugs under compression. Any plug that fails to spark at the recommended pressure should be renewed.
8 The spark plug gap is of considerable importance, as, if it is too large or too small, the size of the spark and its efficiency will be seriously impaired. The spark plug gap should be set to 0.025 inch for the best results.
9 To set it, measure the gap with a feeler gauge, and then bend open, or close, the outer plug electrode until the correct gap is achieved. The centre electrode should never be bent as this may crack the insulation and cause plug failure if nothing worse.
10 When replacing the plugs, remember to use a new plug washer, and replace the leads from the distributor in the correct firing order (see Specifications).
11 The plug leads require no routine attention other than being kept clean and wiped over regularly. At intervals of 12,000 miles, however, pull each lead off the plug in turn and remove them from the distributor by unscrewing the knurled moulded terminal knobs or undoing the securing screws. Water can seep down into these joints giving rise to a white corrosive deposit which must be carefully removed from the brass washer at the end of each cable through which the ignition wires pass.
12 If the HT cable requires renewal the lengths of the lead must be kept the same as those originally fitted as otherwise radio interference will be experienced. Only use 7 mm ignition HT cable.

13 Distributor driving spindle - removal and refitting

1 Refer to Section 6 and remove the distributor.
2 Undo and remove the three setscrews that secure the tachometer housing to the cylinder block and withdraw the housing.
3 Obtain a 5/16 inch UNF bolt approximately 3¼ inches long and screw it into the end of the threaded hole in the end of the drive spindle. Note the position of the drive dog slot.
4 Pull the distributor drive spindle upwards and away from the housing.
5 Whilst the spindle is away from the engine inspect the drive gear teeth for wear and, if evident, a new spindle should be obtained.
6 To refit the drive spindle, if the engine crankshaft position has not been disturbed, proceed direct to paragraph 9, otherwise proceed as follows.
7 Remove the valve rocker cover and with the starting handle turn the engine until No 1 piston is at the top of its compression stroke. This is indicated by the No 6 cylinder exhaust valve just closing and the inlet valve just opening.
8 Turn the crankshaft very carefully until the notch in the crankshaft pulley flange is in line with the pointer on the timing chain cover as shown in Fig 4.7.
9 Screw the 5/16 inch UNF bolt into the threaded hole in the distributor drive and replace the drive in the cylinder block so that the centrally cut slot takes up the 'twenty-to-two' position as shown in Fig 4.8.

Fig 4.8 THE DISTRIBUTOR DRIVE SHOWING SLOT IN THE 'TWENTY TO TWO' POSITION

10 Replace the tachometer housing whilst at the same time rotating the external drive dog until it mates up with the slot in the distributor drive spindle.
11 The smaller segment of the offset dog located within the tachometer housing must be in the downward position. Tighten the three tachometer housing securing set bolts and spring washers.
12 Refer to Section 6 and refit the distributor.

14 Ignition system - fault finding

By far the majority of breakdown and running troubles are caused by faults in the ignition system either in the low tension or high tension circuits.

15 Ignition system - fault symptoms

There are two main symptoms indicating faults: either the engine will not start or fire, or the engine is difficult to start and misfires. If it is a regular misfire ie: the engine is running on only two or three cylinders the fault is almost sure to be in the secondary, or high tension, circuit. If the misfiring is intermittent, the fault could be in either the high or low tension circuits. If the car stops suddenly, or will not start at all it is likely that the fault is in the low tension circuit. Loss of power and overheating, apart from faulty carburation settings, are normally due to faults in the distributor or incorrect ignition timing.

16 Fault diagnosis - engine fails to start

1 If the engine fails to start and the car was running normally when it was last used, first check there is fuel in the petrol tank. If the engine turns over normally on the starter motor and the battery is evidently well charged, then the fault may be in either the high or low tension circuit. First check the HT circuit. Note: If the battery is known to be fully charged, the ignition light comes on, and the starter motor fails to turn the engine CHECK THE TIGHTNESS OF THE LEADS OF THE BATTERY TERMINALS and also the secureness of the earth lead to its CONNECTION TO THE BODY. It is quite common for the leads to have worked loose, even if they look and feel secure. If one of the battery terminal posts gets very hot when trying to work the starter motor this is a sure indication of a faulty connection to that terminal.
2 One of the commonest reasons for bad starting is wet or damp spark plugs, leads and distributor. Remove the distributor cap. If condensation is visible internally, dry the cap with a rag and also wipe over the leads. Replace the cap.
3 If the engine still fails to start, check that current is reaching the plugs, by disconnecting each plug lead in turn at the spark plug end, and holding the end of the cable about 3/16 inch away from the cylinder block. Spin the engine on the starter motor by pressing the rubber button on the starter motor solenoid switch (under the bonnet).
4 Sparking between the end of the cable and the block should be fairly strong with a regular blue spark. (Hold the lead with rubber to avoid electric shocks.) If current is reaching the plugs, then remove them and clean and regap to 0.025 inch. The engine should now start.
5 Spin the engine as before, when a rapid succession of blue sparks between the end of the lead and the block indicates that the coil is in order, and that either the distributor cap is cracked; the carbon brush is stuck or worn; the rotor arm is faulty; or the contact points are burnt, pitted or dirty. If the points are in bad shape, clean and reset them as described in Section 3.
6 If there are no sparks from the end of the lead from the coil, then check the connections of the lead to the coil and distributor lead, and if they are in order, check out the low tension circuit starting with the battery.
7 Switch on the ignition and turn the crankshaft so the contact breaker points have fully opened. Then with either a 20 volt voltmeter or bulb and length of wire, check that current from the battery is reaching the starter solenoid switch. No reading indicates that there is a fault in the cable to the switch, or in the connections at the switch or at the battery terminals. Alternatively the battery earth lead may not be properly earthed to the body.
8 If in order, check that the current is reaching the two-way fuse unit A1 terminal. Connect the voltmeter between the fuse unit A1 terminal and earth. If there is no reading this indicates a faulty cable or loose connection between the solenoid switch and the fuse unit. Remedy and the engine will start.
9 Check with the voltmeter between the control box terminal A1 and earth. No reading indicates a fault in the control box. Fix a new control box and start the car.
10 If in order, then check the current is reaching the lighting switch by connecting the voltmeter between the switch input terminal A (brown and blue cable) and earth. No reading indicates a faulty cable or loose connection.
11 Check with the voltmeter between the fuse unit A3 terminal (white cable) and earth. No reading indicates a broken cable or loose connection.
12 If in order connect the voltmeter between the ignition coil terminal (SW) and earth. This cable is coloured white. No reading indicates a faulty cable or loose connection.
13 Connect the voltmeter between the ignition coil terminal (CB) and earth. No reading indicates a faulty ignition coil.
14 If in order connect the voltmeter between the distributor low tension terminal in the side of the distributor and earth. If no reading then check the wire for loose connections etc. If a reading is obtained then the final check is the low tension circuit across the breaker points. No reading means a broken condenser which, when replaced, will enable the car to start.

17 Fault diagnosis - engine misfires

1 If the engine misfires regularly, run it at a fast idling speed, and short out each of the plugs in turn by placing a short screwdriver across from the plug terminal to the cylinder. Ensure that the screwdriver has a WOODEN or PLASTIC INSULATED HANDLE.
2 No difference in engine running will be noted when the plug in the defective cylinder is short circuited. Short circuiting the working plugs will accentuate the misfire.
3 Remove the plug lead from the end of the defective plug and hold it about 3/16 inch from the block. Restart the engine. If the sparking is fairly strong and regular the fault must lie in the spark plug.
4 The plug may be loose, the insulation may be cracked, or the points may have burnt away giving too wide a gap for the spark to jump. Worse still, one of the points may have broken off. Either renew the plug, or clean it, reset the gap and then test it.
5 If there is no spark at the end of the plug or if it is weak and intermittent, check the ignition lead from the distributor to the plug. If the insulation is cracked or perished, renew the lead. Check connections at the distributor cap.
6 If there is still no spark, examine the distributor cap carefully for tracking. This can be recognised by a very thin black line running between two or more electrodes, or between an electrode and some other part of the distributor. These lines are paths which now conduct electricity across the cap thus letting it run to earth. The only answer is a new distributor cap.
7 Apart from the ignition timing being incorrect, other causes of misfiring have already been dealt with under the section dealing with the failure of the engine to start (Section 16).
8 If the ignition timing is too far retarded, it should be noted that the engine will tend to overheat and there will be quite a noticeable drop in power. If the engine is overheating and the power is down, and the ignition timing is correct, then the carburettors should be checked, as it is likely that this is where the fault lies. See Chapter 3 for details of this.

White deposits and damaged porcelain insulation indicating overheating

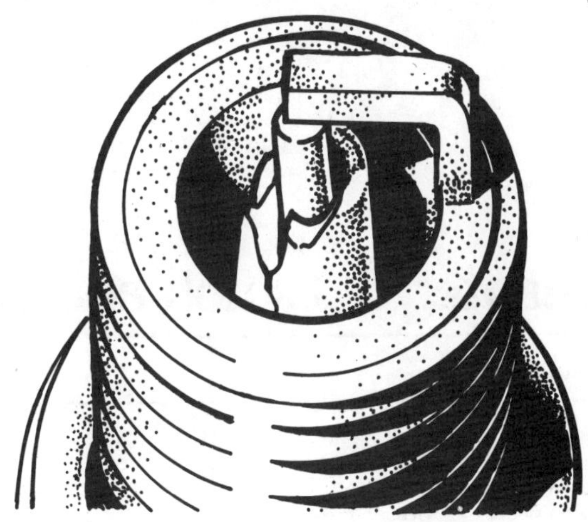

Broken porcelain insulation due to bent central electrode

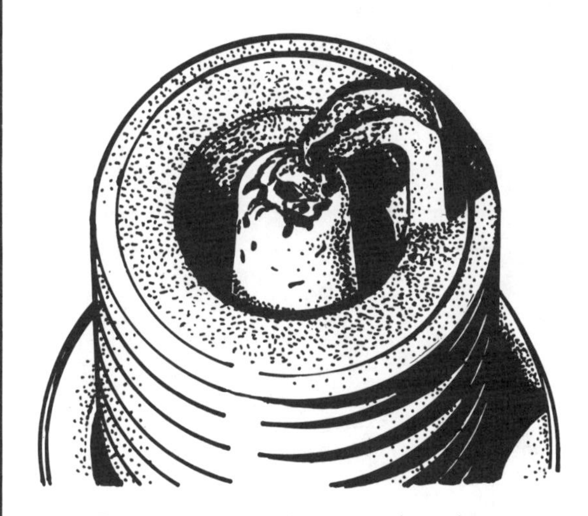

Electrodes burnt away due to wrong heat value or chronic pre-ignition (pinking)

Excessive black deposits caused by over-rich mixture or wrong heat value

Mild white deposits and electrode burnt indicating too weak a fuel mixture

Plug in sound condition with light greyish brown deposits

Chapter 5 Clutch and actuating mechanism

Contents

Specifications

Austin Healey 100-6 (Series BN4 and BN6)

Borg and Beck. Single dry plate

Diameter	9 inch
Total friction area	36.5 sq in
Friction lining thickness	0.150 inch
Release bearing type	Carbon graphite or copper carbon graphite
Number of springs	9
Total axial spring pressure	1215 to 1305 lb
Distance thrust race to thrust plate	0.10 inch
Thrust plate travel to fully released position	0.42 to 0.47 inch

Austin Healey 3000 Mk I and Mk II (Series BN7 and BT7) Mk II (Series BJ7) and Mk III (Series BJ8)

Borg and Beck. Single dry plate

Diameter	10 inch
Total friction area	39 sq in
Friction bearing thickness	0.150 inch
Release bearing type	Carbon graphite or copper carbon graphite
Number of springs	12
Colour of springs	Yellow/light green
Total axial spring pressure	1620 to 1740 lb
Distance thrust race to thrust plate	0.10 inch
Thrust plate travel to fully released position	0.42 to 0.47 inch

Austin Healey 3000 Mk II and Mk III from engine no 29F/4818

Borg and Beck DS.G diaphragm spring

Clutch plate diameter	9.63 inch
Facing material	Wound yarn
Number of damper springs	6
Damper spring load	110 to 120 lb
Damper spring colour	Dark grey/light green
Clutch release bearing	Graphite (MY3D)
Clutch fluid	Castrol Girling Brake Fluid

1 General description

The clutch unit fitted to earlier produced cars was of the Borg and Beck single plate dry disc type which is hydraulically actuated and automatically adjusts for wear.

The clutch assembly comprises a steel cover which is bolted and dowelled to the rear face of the flywheel and contains the pressure plate, pressure plate springs, release levers and clutch disc or driven plate.

The pressure plate, pressure springs and release levers are all attached to the clutch assembly cover. The clutch disc is free to slide along the splined first motion shaft and is held in position between the flywheel and the pressure plate by the pressure of the pressure plate springs.

Friction lining material is rivetted to the clutch disc and it has a spring cushioned hub to absorb transmission shocks.

The clutch is actuated hydraulically. The pendant clutch pedal is connected to the clutch master cylinder and hydraulic fluid reservoir by a short pushrod. The master cylinder and hydraulic reservoir are mounted on the engine side of the bulkhead in front of the driver.

Depressing the clutch pedal moves the piston in the master cylinder forwards so forcing hydraulic fluid through the clutch hydraulic pipe to the slave cylinder.

The piston in the slave cylinder moves forward on the entry of the fluid and actuates the clutch release arm by means of a short pushrod. The opposite end of the release arm is forked and is located behind the release bearing.

As this pivoted clutch release arm moves backwards, it bears against the release bearing pushing it forwards to bear against the release bearing thrust plate and three clutch release levers. These levers are also pivoted so as to move the pressure plate backwards against the pressure of the pressure plate springs, in this way disengaging the pressure plate from the clutch disc.

When the clutch pedal is released, the pressure plate springs force the pressure plate into contact with the high friction linings on the clutch disc, at the same time forcing the clutch disc against the flywheel and so taking the drive up.

As the friction linings on the clutch disc wear, the pressure plate automatically moves closer to the disc to compensate. This makes the inner ends of the release levers travel further towards the gearbox which decreases the release bearing clearance but not the clutch free pedal travel, as unless the master cylinder has been disturbed this is automatically compensated for.

On later cars a Borg and Beck DS.G diaphragm spring clutch was fitted. The clutch comprises a steel cover which is bolted and dowelled to the rear face of the flywheel and contains the pressure plate and clutch disc or driven plate. The pressure plate, diaphragm spring and release plate are all attached to the clutch assembly cover.

The clutch disc as with the earlier type is free to slide along the splined first motion shaft and is held in position between the flywheel and the pressure plate by the pressure of the diaphragm spring.

The friction lining material is rivetted to the clutch disc and has a spring cushioned hub to absorb transmission shocks.

The clutch is actuated hydraulically again in the same manner as the earlier type.

The piston in the slave cylinder moves forward on the entry of the fluid from the hydraulic master cylinder and actuates the clutch release arm by means of a short pushrod. The opposite end of the release arm is located behind the release bearing.

As the pivoted clutch release arm moves it bears against the release bearing pushing it forwards to bear against the release plate, so moving the centre of the diaphragm spring inwards. The spring is sandwiched between two annular rings which act as fulcrum points. As the centre of the spring is pushed in, the outside of the spring is pushed out, so moving the pressure plate backwards and disengaging the pressure plate from the clutch disc.

When the clutch pedal is released, the diaphragm spring forces the pressure plate into contact with the high friction linings on the clutch disc and at the same time pushes the clutch disc a fraction of an inch forward on its splines so engaging the clutch disc with the flywheel. The clutch disc is now firmly sandwiched between the pressure plate and the flywheel so the drive is taken up.

As the friction linings on the clutch disc wear, the pressure plate automatically moves closer to the disc to compensate. There is therefore no need to periodically adjust the clutch.

2 Routine maintenance

1 Routine maintenance consists of checking the level of the hydraulic fluid in the master cylinder reservoir and topping up with a recommended grade of hydraulic brake fluid if the level falls. This level should be to within ¾ inch of the filler neck. Do not overfill.

2 If it is noted that the level of the liquid has fallen then an immediate check should be made to determine the source of the leak.

3 Before checking the level of the fluid in the master cylinder reservoir, carefully clean the cap and the body of the reservoir unit with a clean rag so to ensure that no dirt enters the system when the cap is removed. On no account should paraffin or any other cleaning solvent be used in case the hydraulic fluid becomes contaminated.

4 Check that the vent hole in the top is clear.

3 Clutch system - bleeding

1 Gather together a clean jam jar, a nine inch length of rubber tubing which fits tightly over the bleed nipple in the slave cylinder, a tin of hydraulic brake fluid and someone to help.

2 Check that the master cylinder is full. If it is not fill it and cover the bottom two inches of the jar with hydraulic fluid.

3 Remove the rubber dust cap if fitted from the bleed nipple on the slave cylinder and open the bleed nipple one turn.

4 Place one end of the tube securely over the nipple and insert the other end in the jam jar so that the tube orifice is below the level of the fluid.

5 The assistant should now pump the clutch pedal up and down slowly until air bubbles cease to emerge from the end of the tubing. He should also check the reservoir frequently to ensure that the level of the hydraulic fluid does not drop too low so letting air into the system.

6 When no more air bubbles appear, tighten the bleed nipple on the down stroke.

7 Replace the rubber dust cap over the bleed nipple.

4 Clutch pedal - removal and replacement

1 Upon inspection it will be seen that the clutch and brake pedal linkages are mounted in a common bracket and therefore has to be removed as one unit before separating the clutch pedal from the bracket.

2 Working inside the car, withdraw the split pin and remove the clevis pin to release the clutch pedal from the clutch master cylinder pushrod.

3 Withdraw the split pin and remove the clevis pin to release the brake pedal from the brake master cylinder pushrod.

4 Open the bonnet and locate the six bolts securing the pedal bracket to the engine bulkhead. Partially unscrew the six bolts until there is sufficient room to allow the clutch and brake pedal linkage bracket to be removed from the inside of the car. It is not necessary to remove the bolts completely.

5 Disconnect the pedal return springs from the clutch and brake pedals.

6 Undo and remove the nut that secures the clutch and brake pedal pivot shaft and carefully withdraw the pivot shaft.

7 Lift away the clutch and brake pedal levers and recover the distance piece.

8 If excessive pedal movement on the shaft other than in the normal direction is evident inspect the lever bush for wear and if evident it is possible to drift out the old bush and fit a new one.
9 Refitting is the reverse sequence to removal. Lubricate the pedal pivot bushes with Castrol GTX.

5 Clutch - removal

1 Remove the gearbox as described in Chapter 6, Section 3.
2 Remove the clutch assembly by unscrewing the six bolts that hold the clutch cover assembly to the rear face of the flywheel. Unscrew the bolts diagonally, half a turn at a time so as to prevent distortion to the cover flange.
3 Make sure that as the six bolts are gradually unscrewed, the cover assembly rides up the dowels otherwise if all the bolts are removed the cover assembly could suddenly release from the dowels and fly off.
4 With the bolts and spring washers removed, lift the clutch assembly from the ends of the dowels. The driven plate or clutch disc will fall out at this stage as it is not attached to either the clutch cover assembly or the flywheel.
5 Note which way round the driven plate or clutch disc is fitted as it is important that upon refitting it is assembled the right way round. The smaller end of the boss should face the flywheel.

6 Clutch - replacement

1 It is important that no oil or grease gets onto the clutch disc friction linings, or the pressure plate and flywheel faces. It is advisable to handle the clutch with clean hands and wipe down the pressure plate and flywheel faces with a dry rag before reassembly begins.
2 Place the clutch disc against the flywheel with the shorter end of the hub facing the flywheel.
3 Replace the clutch cover assembly loosely on the dowels. Replace the six bolts and spring washers and tighten them finger tight so that the clutch disc is gripped but can still be moved.
4 The clutch disc must now be centralised to allow the gearbox input shaft to pass through the splines in the centre of the driven plate hub.
5 Centralisation can be carried out quite easily by inserting a round bar or long screwdriver through the hole in the centre of the clutch, so that the end of the bar rests in the small hole in the end of the crankshaft containing the input shaft bearing bush.
6 Centralisation is easily judged by removing the bar and viewing the driven hub in relation to the hole in the release bearing contact plate. When the hub appears exactly in the centre of the release bearing plate hole all is right. Alternatively, if an old Healey input shaft can be used, this will eliminate all the guesswork.
7 Tighten the clutch bolts firmly in a diagonal sequence to ensure that the cover plate is pulled down evenly, and without distortion of the flange.
8 Mate the engine and gearbox, bleed the slave cylinder if the pipe was disconnected and check the clutch for correct operation.

7 Clutch - dismantling, reassembly and inspection

1 In the normal course of events clutch dismantling and reassembly is the term used for simply fitting a new clutch pressure plate and friction disc. Under no circumstances should the clutch unit be dismantled. If a fault develops in the pressure plate assembly an exchange replacement unit must be fitted.
2 If a new clutch disc is being fitted it is false economy not to renew the release bearing at the same time. This will preclude having to replace it at a later date when wear on the clutch linings is still very small.
3 Examine the clutch disc friction linings for wear or loose rivets and the disc for rim distortion, cracks and worn splines.
4 It is always best to renew the clutch driven plate as an assembly to preclude further trouble.
5 Check the machine faces of the flywheel and the pressure plate. If either is badly grooved or shows signs of cracking or overheating it should be either machined until smooth as applicable or replaced with a new item. If the pressure plate is cracked or split it must be renewed.

8 Clutch slave cylinder - removal, dismantling, examination and reassembly

1 Before removing the clutch slave cylinder take off the clutch master cylinder reservoir cap and place a piece of polythene over the top of the reservoir. Screw the cap down tightly over the polythene.
2 Wipe the area around the flexible hose where it enters the slave cylinder body to remove any dust or dirt. Slacken the flexible hose at this point but do not attempt to remove it as it will only damage the flexible hose.
3 Extract the split pin and remove the clevis pin to release the pushrod from the clutch fork and lever.
4 Undo and remove the two set bolts and spring washers that secure the slave cylinder to the clutch housing.
5 The slave cylinder can now be removed from the flexible hydraulic hose by unscrewing it from the hose. Note that the thicker washer on the hose connection is nearest to the cylinder.
6 Wrap a piece of clean non-fluffy rag around the exposed end to stop dirt ingress into the system.
7 Clean the outside of the cylinder before dismantling. Remove the rubber dust cap and shake the piston, seal, seal filler and spring out of the cylinder. These component parts are shown in Fig 5.3. Clean all the components thoroughly with hydraulic fluid and then dry them off.
8 Carefully examine the rubber components for signs of wear, swelling, distortion and splitting and check the piston and the cylinder wall for wear and score marks. Replace any parts that are found faulty.
9 To reassemble the clutch slave cylinder thoroughly wet all internal parts with clean hydraulic fluid and insert the spring (3) (Fig 5.3) cup filler (4), piston cup (5) and piston (6) into the slave cylinder bore.
10 Secure the rubber boot (9) to the pushrod (10) with the small boot clip (7).
11 Ease the rubber boot over the lip of the slave cylinder and seat it into the groove securing with the large boot clip.
12 To refit the slave cylinder is the reverse sequence to removal but it will be necessary to bleed the hydraulic system and full details of this will be found in Section 3 of this Chapter. Do not forget to remove the polythene from under the cap of the clutch master cylinder.

9 Clutch master cylinder - removal and refitting

1 Drain the hydraulic fluid from the clutch hydraulic system by attaching a length of suitable size plastic tubing to the bleed screw on the slave cylinder. Place the other end in a clean jam jar. Open the bleed screw one turn and depress the clutch pedal. Tighten the bleed screw and allow the pedal to return. Repeat this procedure until the system has been drained.
2 Wipe the master cylinder hydraulic pipe connection with a clean non-fluffy rag and disconnect the union. Wrap the end in a piece of clean rag to stop dirt ingress or fluid dripping onto the paintwork. Plug the master cylinder union connection to stop accidental dirt entry into the master cylinder.
3 Extract the split pin from the master cylinder to piston pushrod clevis pin, remove the plain washer and withdraw the clevis pin. Separate the pushrod from the clutch pedal.
4 Remove the fixing bolt and spring washer from the underside

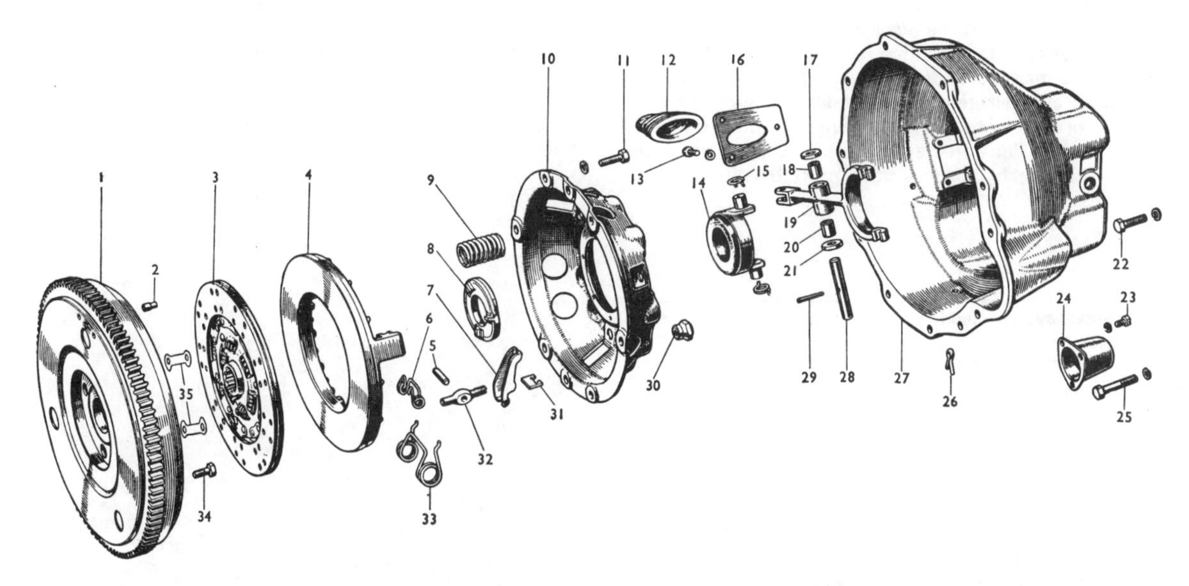

Fig 5.1 EXPLODED VIEW OF EARLY TYPE CLUTCH

1 Flywheel
2 Locating peg
3 Clutch plate with lining
4 Pressure plate
5 Release lever pin
6 Release lever retainer
7 Release lever
8 Release lever plate
9 Pressure plate spring
10 Clutch cover
11 Cover setpin
12 Fork and lever seal
13 Retaining plate screw
14 Release bearing
15 Release bearing retainer spring
16 Seal retaining plate
17 Fork and lever thrust washer
18 Fork and lever shaft bush
19 Clutch fork and lever
20 Fork and lever shaft bush
21 Fork and lever thrust washer
22 Clutch to gearbox setpin
23 Starter cover screw
24 Cover
25 Clutch to gearbox setpin
26 Split pin for drain hole
27 Clutch housing
28 Fork and lever shaft
29 Taper pin
30 Eye bolt nut
31 Release lever strut
32 Eye bolt
33 Anti-rattle spring
34 Flywheel to crankshaft bolt
35 Lockwashers

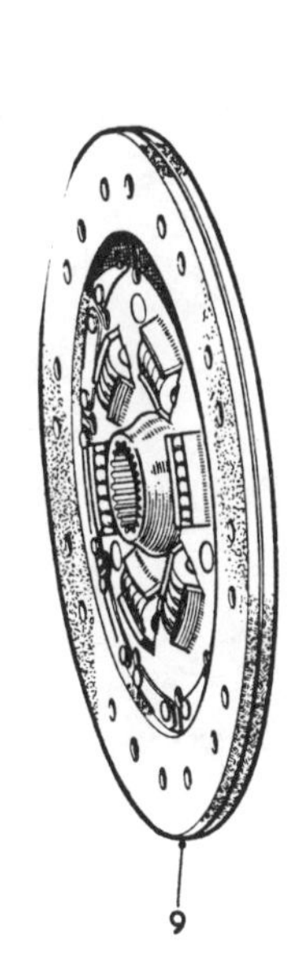

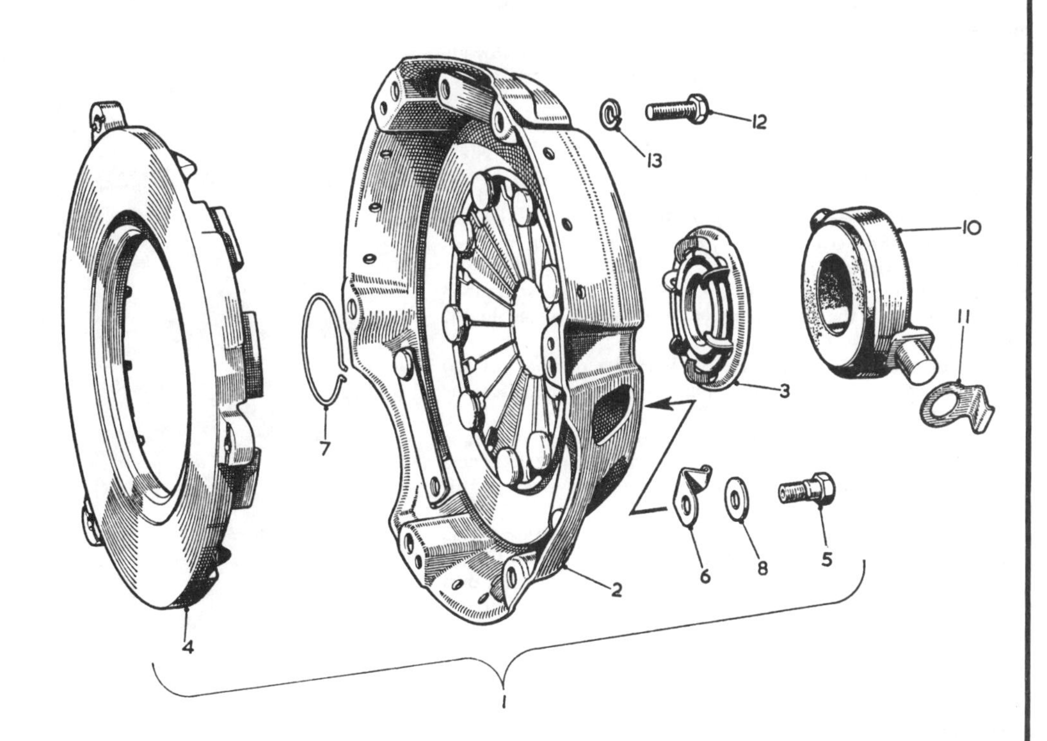

Fig 5.2 EXPLODED VIEW OF LATER TYPE DIAPHRAGM CLUTCH

1 Cover assembly
2 Cover with strap and diaphragm spring
3 Plate - release
4 Plate - pressure
5 Bolt - strap
6 Clip - pressure plate
7 Circlip - release plate
8 Washer - strap
9 Plate assembly - driven
10 Bearing assembly - release
11 Retainer - bearing
12 Screw - clutch to flywheel
13 Washer for screw - spring

of the master cylinder and the nut and spring washer from the top of the master cylinder mounting flange to the bulkhead, and carefully ease the master cylinder away from the bulkhead.
5 The master cylinder refitting procedure is the reverse to removal but care must be taken when offering up to the bulkhead that the pushrod is in line with the clutch pedal. Once connections have been made the hydraulic system must be bled and the clutch road tested.

10 Clutch master cylinder (early type) - dismantling, examination and reassembly

1 The numbers in the text refer to Fig 5.4. Pull off the rubber dust cover (13) which exposes the circlip (11) which must be removed so the pushrod complete with metal retaining washer can be pulled out of the master cylinder.
2 Pull the piston (9) and valve assembly as one unit from the master cylinder.
3 The next step is to separate the piston and valve assemblies. With the aid of a small screwdriver prise up the inner leg of the piston return spring retainer (8) which engages under a shoulder in the front of the piston (9) and holds the retainer (8) in place.
4 The retainer (8), spring (7) and valve assembly (4, 5, 6) can then be separated from the piston.
5 To dismantle the valve assembly compress the spring (7) and move the retainer (8) which has an offset hole to one side in order to release the valve stem (4) from the retainer (8).
6 With the seat spacer (6) and curved valve seal washer (5) removed, the rubber seals can be removed.
7 Clean and carefully examine all parts, especially the piston cup and rubber washers for signs of distortion, swelling, splitting or other wear and check the piston and cylinder for wear and scoring. Renew any parts that are suspect. It is recommended that new rubber seals are always fitted.
8 Rebuild the piston and valve assembly in the following sequence, ensuring that all parts are thoroughly wetted with clean brake fluid.

a) Fit the piston seal to the piston (9) so that the larger circumference of the lip will enter the cylinder bore first.
b) Fit the valve seal to the valve (4) in the same way as in (a).
c) Place the valve spring seal washer (5) so that its convex face abuts against the valve stem flange (4) and then fit the seat spacer (6) and spring (7).
d) Fit the spring retainer (8) to the spring (7) which must then be compressed so the valve stem (4) can be re-inserted in the retainer (8).
e) Replace the front of the piston (9) in the retainer (8) and then press down the retaining leg so it locates under the shoulder at the front of the piston (9).
f) With the valve assembly well lubricated with clean hydraulic fluid carefully insert it in the master cylinder bore taking care that the rubber seal is not damaged or the lip reversed as it is pushed into the bore.
g) Fit the pushrod (12) and washer in place and secure with the circlip (11). Smear the sealing areas of the dust cover with Girling Grease and pack the cover with rubber grease to act as a dust trap, and fit to the master cylinder body. The master cylinder is now ready for refitting to the car.

11 Clutch master cylinder (later type) - dismantling, examination and reassembly

The component parts of the later type master cylinder are shown in Fig 5.5, where it will be seen that the internal components are the same as for the earlier type. For full service information refer to Section 10 of this Chapter.

12 Clutch fork and release bearing - removal and replacement

1 With the gearbox and engine separated to provide access to the clutch, attention can be given to the release bearing and fork and lever located in the gearbox bellhousing as shown in Fig 5.1.
2 To remove the clutch release bearing, ease back the two spring clips located at the ends of the release bearing carrier and lift away the release bearing (photos).

12.2A

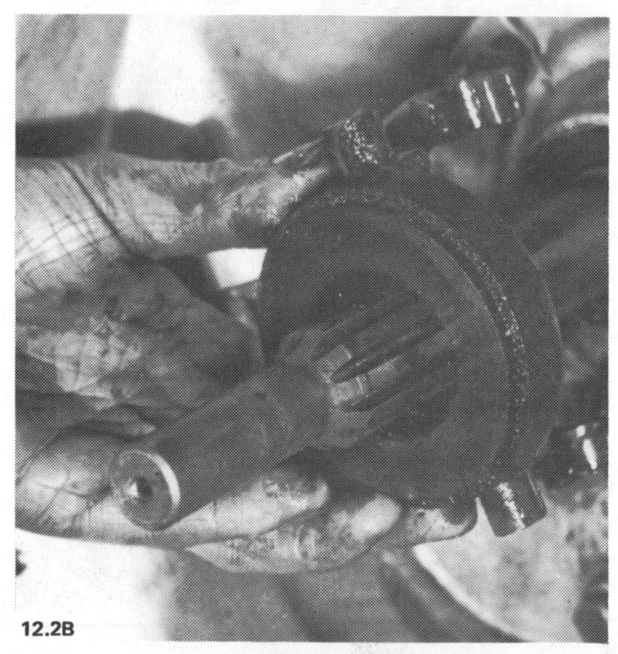

12.2B

3 Undo and remove the three bolts and spring washers that secure the fork and lever rubber seal to the side of the bellhousing. Lift away the seal and retaining plate.
4 Undo and remove the shaft blanking screw from the end of the fork and lever shaft.
5 Using a small parallel pin punch, drive out the taper pin located in the hub of the fork and lever securing the fork and lever to the shaft (photo).

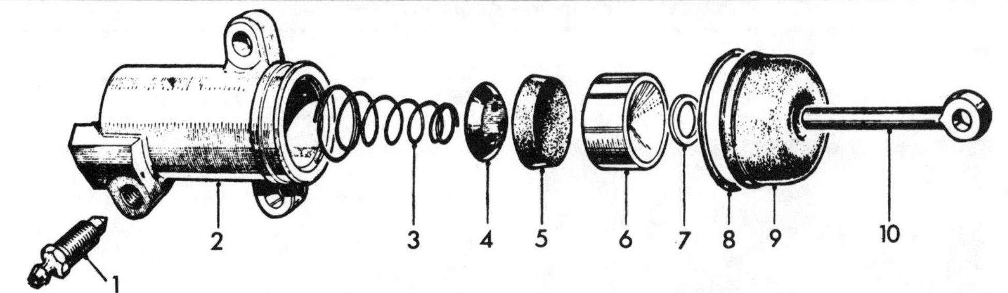

Fig 5.3 CLUTCH SLAVE CYLINDER COMPONENTS

1 Bleeder screw
2 Body
3 Return spring
4 Cup filler
5 Piston cup
6 Piston
7 Boot clip (small)
8 Boot clip (large)
9 Dust excluder boot
10 Pushrod

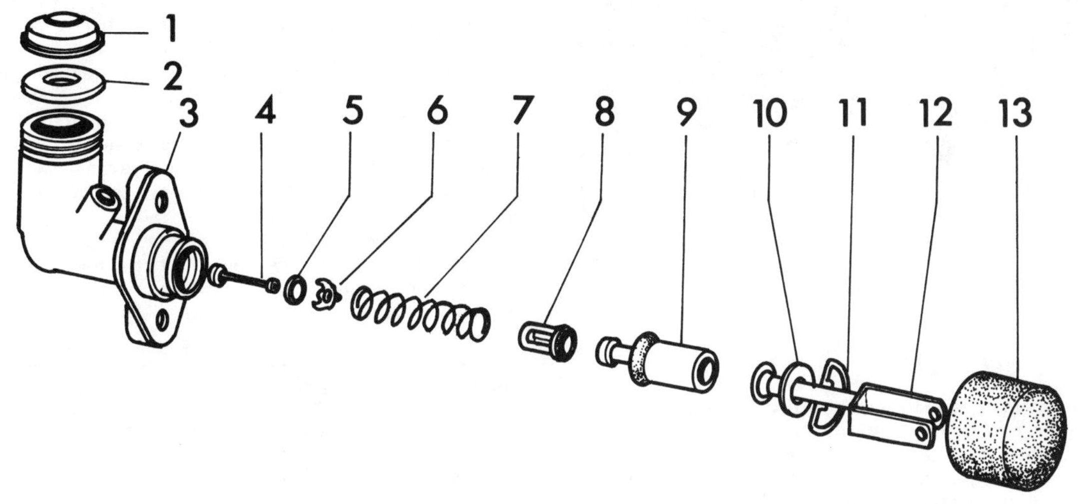

Fig 5.4 CLUTCH MASTER CYLINDER COMPONENT PARTS – EARLY TYPE

1 Filler cap
2 Washer
3 Master cylinder
4 Valve stem
5 Spring washer
6 Valve spacer
7 Return spring
8 Return spring retainer
9 Plunger
10 Dished washer
11 Circlip
12 Fork
13 Dust cover

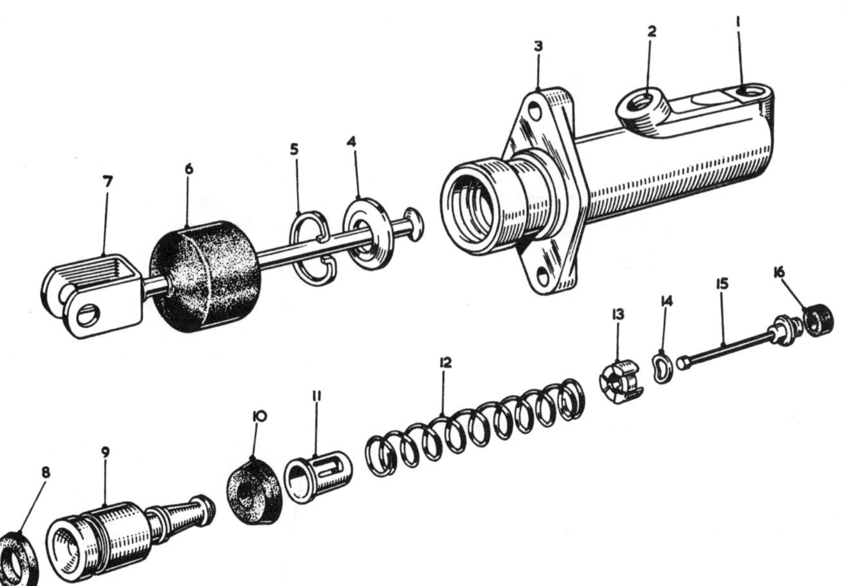

Fig 5.5 CLUTCH MASTER CYLINDER COMPONENT PARTS – LATER TYPE

1 Fluid inlet
2 Fluid outlet
3 Master cylinder
4 Dished washer
5 Circlip
6 Dust cover
7 Pushrod
8 End seal
9 Plunger
10 Plunger seal
11 Return spring retainer
12 Return spring
13 Valve spacer
14 Spring washer
15 Valve stem
16 Valve seal

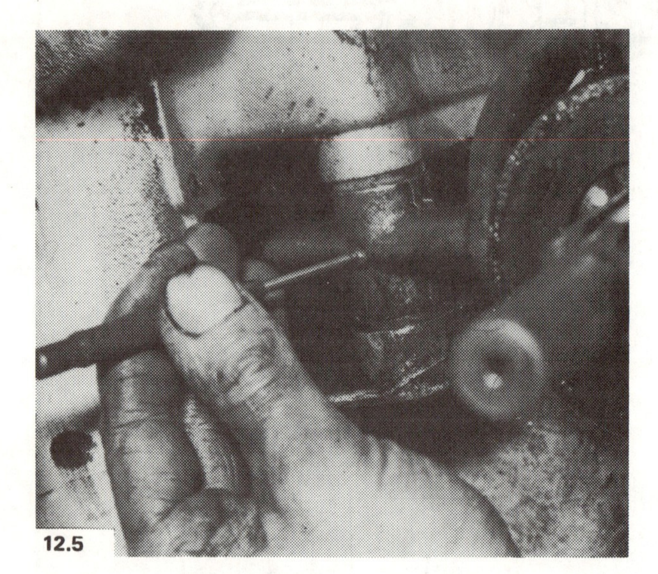
12.5

6 With a suitable sized soft metal drift carefully drive out the fork and lever shaft from its bushes and lift away the clutch fork and lever together with the two circular thrust washers.
7 Inspect the bushes and shaft for signs of wear, which if evident, new parts should be fitted.
8 To remove the old bushes, carefully drift out using a small diameter soft metal drift from their location in the bellhousing.
9 To refit the new bushes they should be carefully drifted in using a suitable sized drift and then if available use a parallel reamer of a suitable matching diameter to the fork and lever shaft. In most cases it will not be found necessary to ream these two bushes to final size unless they have been inadvertantly damaged during refitting.
10 Reassembling the clutch fork and lever assembly is the reverse sequence to removal.

13 Clutch faults

There are four main faults to which the clutch and release mechanism are prone. They may occur by themselves or in conjunction with any of the other faults. They are clutch squeal, slip, spin and judder.

14 Clutch squeal - diagnosis and cure

1 If, on taking up the drive or when changing gear, the clutch squeals, this is a sure indication of a badly worn clutch release bearing.
2 As well as regular wear due to normal use, wear of the clutch release bearing is much accentuated if the clutch is ridden or held down for long periods in gear, with the engine running. To minimise wear of this component the car should always be taken out of gear at traffic lights and for similar hold-ups.
3 The clutch release bearing is not an expensive item, but difficult to get at.

15 Clutch slip - diagnosis and cure

1 Clutch slip is a self-evident condition which occurs when the clutch friction plate is badly worn, oil or grease have got onto the flywheel or pressure plate faces, or the pressure plate itself is faulty.
2 The reason for clutch slip is that due to one of the faults above, there is either insufficient pressure from the pressure plate, or insufficient friction from the friction plate to ensure solid drive.
3 If small amounts of oil get onto the clutch, they will be burnt off under the head of the clutch engagement, and in the process, gradually darken the linings. Excessive oil on the clutch will burn off leaving a carbon deposit which can cause quite bad slip, or fierceness, spin and judder.
4 If clutch slip is suspected, and confirmation of this condition is required, there are several tests which can be made.
5 With the engine in second or third gear and pulling lightly up a moderate incline, sudden depression of the accelerator pedal may cause the engine to increase its speed without any increase in the road speed. Easing off on the accelerator will then give a definite drop in engine speed without the car slowing.
6 In extreme cases of clutch slip the engine will race under normal acceleration conditions.
7 If slip is due to oil or grease on the linings a temporary cure can sometimes be effected by squirting tetrachloride into the clutch. The permanent cure is, of course, to renew the clutch driven plate and trace and rectify the oil leak.

16 Clutch spin - diagnosis and cure

1 Clutch spin is a condition which occurs when there is a leak in the clutch hydraulic actuating mechanism; there is an obstruction in the clutch either in the primary gear splines or in the operating lever itself; or the coil may have partially burnt off the clutch linings and have left a resinous deposit which is causing the clutch disc to stick to the pressure plate or flywheel.
2 The reason for clutch spin is that due to any, or a combination of, the faults just listed, the clutch pressure plate is not completely freeing from the centre plate even with the clutch pedal fully depressed.
3 If clutch spin is suspected, the condition can be confirmed by extreme difficulty in engaging first gear from rest, difficulty in changing gear, and very sudden take up of the clutch drive at the fully depressed end of the clutch pedal travel as the clutch is released.
4 Check the clutch master and slave cylinders and the connecting hydraulic pipe for leaks. Fluid in one of the rubber boots fitted over the end of either the master or slave cylinder is a sure sign of a leaking piston seal.
5 If these points are checked and found to be in order then the fault lies internally in the clutch, and it will be necessary to remove the clutch for examination.

17 Clutch judder - diagnosis and cure

1 Clutch judder is a self-evident condition which occurs when the gearbox or engine mountings are loose or too flexible; when there is oil on the face of the clutch friction plate; or when the clutch pressure plate has been incorrectly adjusted.
2 The reason for clutch judder is that due to one of the faults just listed, the clutch pressure plate is not freeing smoothly from the friction disc, and is snatching.
3 Clutch judder normally occurs when the clutch pedal is released in first or reverse gears, and the whole car shudders as it moves backwards or forwards.

Chapter 6 Gearbox and overdrive

Contents

Specifications

Type	4 speed, synchromesh on 2nd, 3rd and top
Type of gear	Helical constant mesh
Type of overdrive	Laycock de Normanville electrially operated

All models except Austin Healey 3000
Mk I and II (Series BN7, BT7)
Mk II (Series BJ7)
Mk III (Series BJ8)

Gear ratios:

First	3.076 : 1
Second	1.913 : 1
Third	1.333 : 1
Overdrive, third	1.037 : 1
Fourth	Direct
Overdrive, fourth	.778 : 1
Reverse	4.16 : 1

Overall gear ratios:

Standard box:

First	12.027 : 1
Second	7.48 : 1
Third	5.212 : 1
Fourth	3.91 : 1
Reverse	16.4 : 1

Including overdrive:

First	12.6 : 1
Second	7.84 : 1
Third	5.47 : 1
Third and overdrive	4.24 : 1
Fourth	4.1 : 1
Fourth and overdrive	3.19 : 1
Reverse	17.1 : 1

Layshaft bearing

Type	Needle roller
Number of rollers	46
Length of roller	1.551 in (39.6 mm)
Diameter of roller	3.118 in (3 mm)

Mainshaft bearing

Make	R and M
Type	MJ35
Size	1.38 x 3.15 x 0.827 in (35 x 80 x 21 mm)

First motion shaft bearing

Make	R and M
Type	IMJ40G
Size	1.58 x 3.55 x 0.905 in (40 x 90 x 23 mm)

Austin Healey 3000
Mk I and II (Series BN7, BT7)
Mk II (Series BJ7)
Mk III (Series BJ8)

Overall gear ratios:

		From engine no 10897 with overdrive and engine no 11342 without overdrive	Mk III
First	2.93 : 1	2.83 : 1	2.637 : 1
Second	2.053 : 1	2.06 : 1	2.071 : 1
Third	1.309 : 1	1.31 : 1	1.306 : 1
Fourth	Direct		Direct
Overdrive	.822 : 1		8.20 : 1
Reverse	3.78 : 1	3.72 : 1	3.391 : 1

Standard box (3.545 : 1 axle):

		From engine no 11342	Mk III
First	10.386 : 1	10.209 : 1	9.348 : 1
Second	7.877 : 1	7.302 : 1	7.341 : 1
Third	4.640 : 1	4.743 : 1	4.629 : 1
Fourth	3.545 : 1		3.545 : 1
Reverse	13.400 : 1	13.127 : 1	12.021 : 1

Including overdrive (3.909 : 1 axle):

		From engine no 10897	Mk III
First	11.453 : 1	11.257 : 1	10.308 : 1
Second	8.025 : 1	8.052 : 1	8.095 : 1
Third	5.116 : 1	5.120 : 1	5.105 : 1
Third and overdrive	4.195 : 1	4.198 : 1	4.188 : 1
Fourth	3.909 : 1	3.909 : 1	3.909 : 1
Fourth and overdrive	3.205 : 1	3.205 : 1	3.207 : 1
Reverse	14.776 : 1	14.541 : 1	13.255 : 1

Layshaft bearing

Type	Needle roller
Number of rollers	46
Length of roller	1.551 in (39.6 mm)
Diameter of roller	3.118 in (3 mm)
Endfloat	0.021 in (0.30 mm)

Mainshaft bearing

Make	R and M
Type	MJ35
Size	1.39 x 3.15 x 0.827 in (35 x 80 x 21 mm)
End float	Nil

First motion shaft bearing

Make	R and M
Type	IMJ40G
Size	1.58 x 3.55 x 0.905 in (40 x 90 x 23 mm)

Synchromesh hubs

End float

2nd speed	0.007 in (0.18 mm)
3rd and 4th speeds	0.031 in (0.8 mm)

Overdrive

Overdrive ratio	0.802 : 1
Output shaft selective washers	0.146 in ± 0.0005 (3.708 mm ± 0.01)
	0.151 in ± 0.0005 (3.835 mm ± 0.01)
	0.156 in ± 0.0005 (3.962 mm ± 0.01)
	0.160 in ± 0.0005 (4.064 mm ± 0.01)

	0.161 in ± 0.0005 (4.089 mm ± 0.01)
Sunwheel end float	0.008 to 0.014 in (0.20 to 0.35 mm)
Sunwheel and float selective washers	0.113 to 0.114 in (2.8 to 2.9 mm)
	0.107 to 0.108 in (2.72 to 2.74 mm)
	0.101 to 0.102 in (2.56 to 2.59 mm)
	0.095 to 0.096 in (2.4 to 2.44 mm)
	0.089 to 0.090 in (2.26 to 2.28 mm)
	0.083 to 0.084 in (2.1 to 2.13 mm)
	0.077 to 0.078 in (1.9 to 1.98 mm)
Pump plunger spring free length	2 inch (5.08 mm)
Pump spring rate	11 lb in (12.7 kg cm)
Pump operating pressure	470 to 490 lb/sq in (33.04 to 34.45 kg/cm^2)
Solenoid cross shaft end float	0.008 to 0.010 inch (0.20 to 0.25 mm)
Clutch spring length:	
Inner	4½ inch (115 mm)
Outer	4¼ inch (108 mm)
Relay type	SB 40
Throttle switch type	RTS1
Gear switch	SS10
Solenoid unit	TGS1

Oil capacities

Oil capacity (standard box)	5 pints (2.8 litres)
Oil capacity (overdrive fitted)	6¼ Imp pints (3.6 litres)

Torque wrench settings

Mainshaft nut	183 lb ft (25.3 kg m)
Bellhousing bolts	35 lb ft (4.83 kg m)
Coupling flange nut (overdrive)	100 to 130 lb ft (13.83 to 17.97 kg m)

1 General description

The gearbox fitted to models covered by this manual contains four forward and one reverse gear. Overdrive is fitted as an extra.

The gearbox comprises helical cut constant mesh gears on shafts which are free to rotate in ball or roller bearings. Gear selection is obtained by moving the gear change lever forwards or rearwards which causes selectors to engage the gears. Top gear is a direct drive, third and second gears are constant mesh whilst first and reverse are obtained by sliding spur pinions.

On later produced Mk II BN7 and BT7 cars the gearbox had a centrally mounted remote control change speed lever in place of the side mounted box which was previously fitted. This modified gearbox together with the accompanying body and gearbox cover modifications were introduced from car numbers BN7 16039 and BT7 15881.

The Laycock de Normanville electrically operated overdrive is fitted to the rear of the gearbox and further details of this unit will be found in Section 13 of this Chapter.

2 Routine maintenance

1 Every 3000 miles lift up the floor covering and take out the inspection panel in the top right hand side of the gearbox cover giving access to the combined filler plug and dipstick. Wipe the area around the combined dipstick and filler and remove the dipstick. Wipe it clean of oil and re-insert it fully. Withdraw it again and note the level. Top up if necessary using Castrol GTX.

2 Every 6000 miles drain the oil from the gearbox. This is best done when the car has just completed a run so that the oil is still warm. Wipe the area around the drain plug/s and remove the drain plug/s. Allow the oil to drain into a container having a 6¼ pint capacity for a full five minutes. Replace the drain plug/s and refill with 5 pints (standard gearbox) or 6¼ pints (with overdrive).

3 Gearbox - removal and replacement

1 The gearbox complete with overdrive, if fitted, can be removed in unit with the engine through the engine compartment or from underneath the body complete with engine and front suspension as described in Chapter 1. Alternatively, the gearbox complete with overdrive, if fitted, can be separated from the engine end plate and lowered from the car. The latter is the best method to adopt.

2 Drain the gearbox oil as detailed in Section 2, paragraph 2 and place the rear wheels on ramps or high axle stands. Even better when available, raise the car on a lift or place over a pit.

3 For safety reasons switch off the battery master switch which is to be found inside the rear luggage compartment.

4 Working inside the car, remove the seat cushions and release the clips that secure the padded arm rest to the central tunnel.

5 Unclip and remove the carpeting and felt from the short gearbox tunnel giving access to the tunnel.

6 Unscrew and remove the twelve setscrews that secure the tunnel to the body of the car and lift away the tunnel and its carpeting.

7 Unscrew and remove the six setscrews (three on each side) which secure the carpet covered bulkhead and remove the bulkhead.

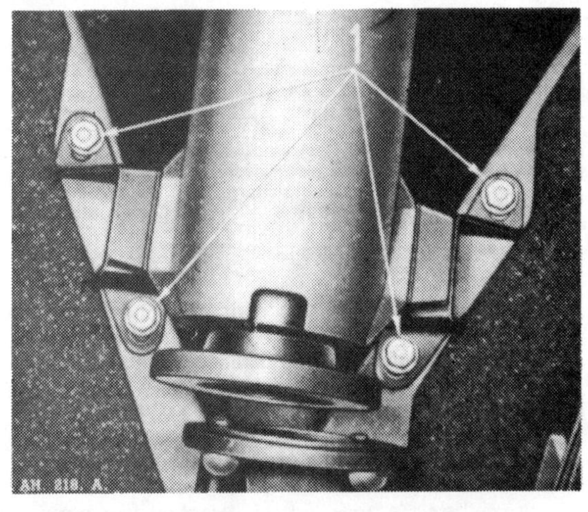

Fig 6.1 GEARBOX REAR UPPER BRACKET

1 Setpins

8 With a scriber or file mark the propeller shaft and gearbox/overdrive flanges so that they may be correctly refitted in their original position.
9 Using a flat chisel tap back the propeller shaft flange securing nuts locking washers (photo).
10 Undo and remove the bolts and carefully lower the propeller shaft to the ground (photo).
11 Refer to Fig 6.1 and unscrew and remove the four bolts with plain and spring washers from the gearbox rear mounting brackets.
12 Unscrew the speedometer cable knurled nut at its connection to the gearbox/overdrive unit.
13 If overdrive is fitted, undo the bolt securing the cable clip to the gearbox switch and release it at its terminal on the switch (photo).
14 Refer to Fig 6.2 and working under the car, remove the bolts (1) and the nuts (2, 3) so to release the stabiliser bar.

Fig 6.2 GEARBOX LOWER RETAINING BOLTS

1 Setpins
2 Stabiliser adjustment nut
3 Securing pin

15 Detach the clutch slave cylinder from the gearbox bellhousing by removing the two securing bolts (photo). Extract the split pin and remove the clevis pin connecting the slave cylinder pushrod to the clutch operating lever. Tie the slave cylinder out of the way using string or wire.
16 Release the heavy duty cable from the terminal on the rear of the starter motor by undoing and removing the nut and spring washer.
17 Undo and remove the two starter motor securing bolts and manoeuvre the starter forwards below the oil filter and lift away from the engine compartment.
18 Place a jack or support stand under the gearbox bellhousing and a further one under the engine sump.
19 Unscrew the nuts, bolts and set bolts that secure the bellhousing to the engine backplate.
20 Help from an assistant is now beneficial to prevent any accidents. Very carefully withdraw the gearbox first motion shaft from the rear of the engine. The gearbox must not be allowed to hang on the first motion shaft otherwise the clutch or shaft could be damaged. It should be remembered that the gearbox and overdrive unit are heavy (up to 135 lbs) so care must be taken to prevent it rolling off the jack whilst it is being removed.
21 Refitting the gearbox/overdrive unit is the reverse sequence to removal but the following additional points should be noted.

a) Make sure that the clutch disc is centralised correctly before attempting to refit the gearbox/overdrive unit if the clutch has been overhauled.
b) Do not tighten the bellhousing bolts until the set pins are in position.
c) Refill the gearbox/overdrive unit as described in Section 2, paragraph 2, with Castrol GTX.

4 Gearbox - dismantling (early type)

All numbers in brackets refer to Fig 6.3 unless otherwise stated.
1 Remove the oil level dipstick from the top of the gearbox main casing (42).
2 Undo the drain plug (41) located on the underside of the gearbox and allow the oil to drain into a container having a capacity of at least 5 pints. If overdrive is fitted, undo the overdrive unit drain plug (Fig 6.7) and drain the unit. The total capacity of the gearbox plus overdrive is 6¼ pints.
3 Unscrew the speedometer drive from the right hand side of the rear extension housing.
4 With the gearbox on the bench in its normal mounted position with the side cover (53) accessible, unscrew and remove the thirteen bolts and spring washers securing the side cover to the gearbox housing. Lift away the side cover and paper gasket (photo).
5 Note that there are two dowels accurately locating the side cover to the main casing. These are to be found at the top right and bottom left hand side of the main casing.
6 Unscrew and remove the seven short and one long bolt with spring washers securing the clutch bellhousing to the main casing (42) (photo).
7 Lift away the clutch bellhousing and paper gasket from the front of the main casing (photo).
8 Recover the bearing spring plate (33) and bearing plate (34) to be found on the front face of the first motion shaft bearing outer track (photo).
9 Recover the three selector ball bearings and springs (30) which will be released once the side cover has been removed (photo). On a gearbox without overdrive, undo and remove the eight bolts and spring washers that secure the rear extension housing to the gearbox main casing and slide off the rear extension and its paper gasket. It may be necessary to use a soft faced hammer to start the movement of the extension housing.
10 If overdrive is fitted refer to Section 14 and separate the overdrive unit from the adaptor plate on the rear of the main casing (photo).
11 Draw the overdrive unit rearwards away from the adaptor plate. Note that in this photograph the two long studs in the adaptor plate may be seen and these are used for drawing the overdrive unit into place on refitting.
12 Lift away the overdrive unit pump operating cam from the mainshaft (photo).
13 Separate the overdrive adaptor plate from the rear main casing by undoing the securing nuts and spring washers. This operation is fully described in Section 14 (photo).
14 Lift away the overdrive unit adaptor plate (photo).
15 With a pair of side cutters cut the wires locking the selector fork retaining screws to the selector forks. Unscrew the selector fork retaining screws (photo).
16 Using a suitable diameter soft metal drift, carefully drive out the reverse fork selector rod (62) towards the rear of the main casing. Note that this rod has two notches at the forward end (photo).
17 Lift out the reverse selector fork (72) and recover the selector plunger (68), selector plunger spring (69), detent plunger (70) and detent plunger spring (71) (photo).
18 Recover the interlock ball bearing (29) which is located between the centre selector rod and the reverse selector rod location (photo).
19 Again with the soft metal drift carefully drive out the first and second selector rod (60) towards the rear of the main casing. Note that this rod has three notches at the forward end and the selector fork retaining screw dowel hole about two thirds of the way down from the front.
20 Using the soft metal drift carefully drive out the third and fourth gear selector rod (59) towards the rear of the main casing. Note that this rod has three notches at the forward end and the selector fork retaining screw dowel hole about one third of the way down from the front.

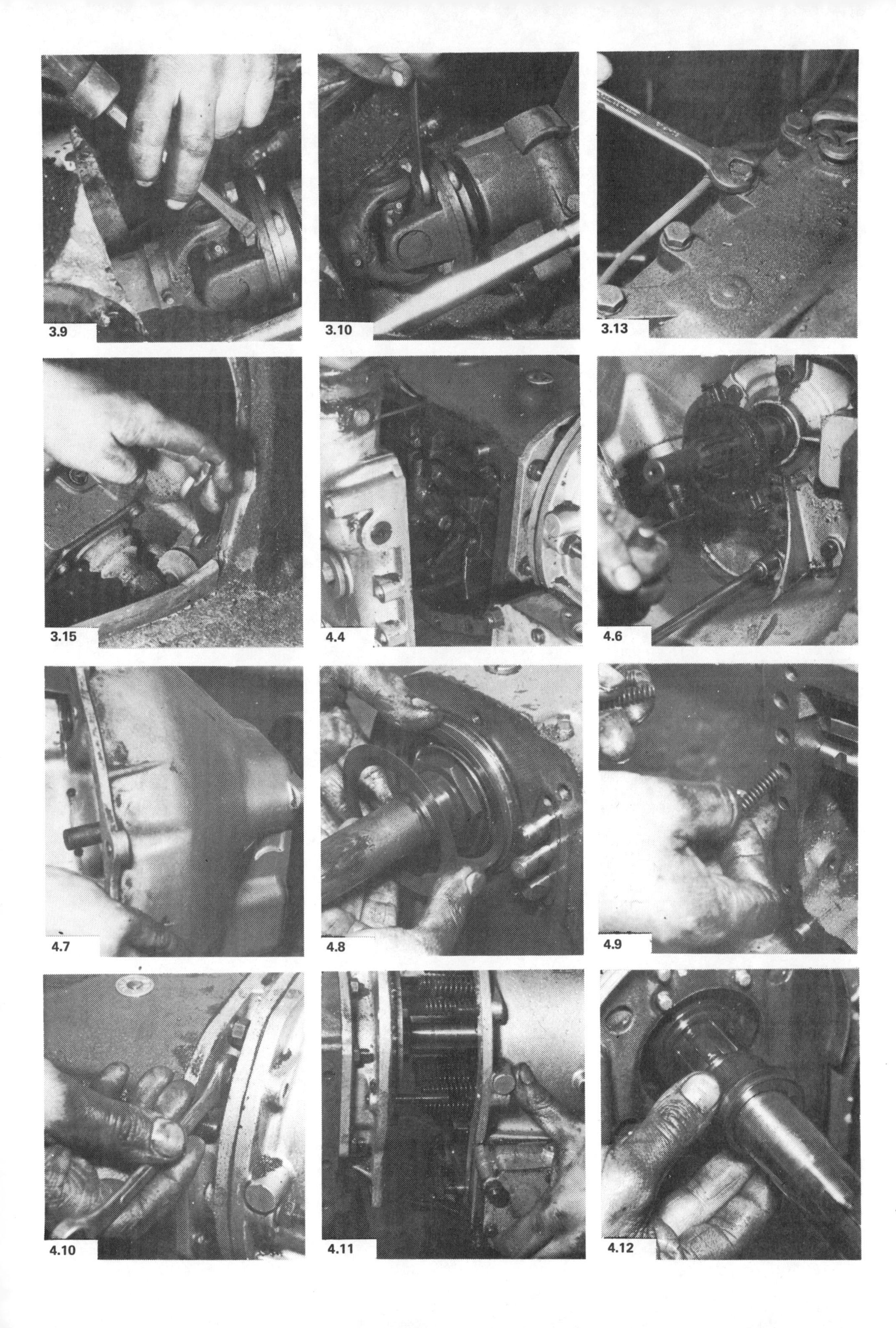

3.9

3.10

3.13

3.15

4.4

4.6

4.7

4.8

4.9

4.10

4.11

4.12

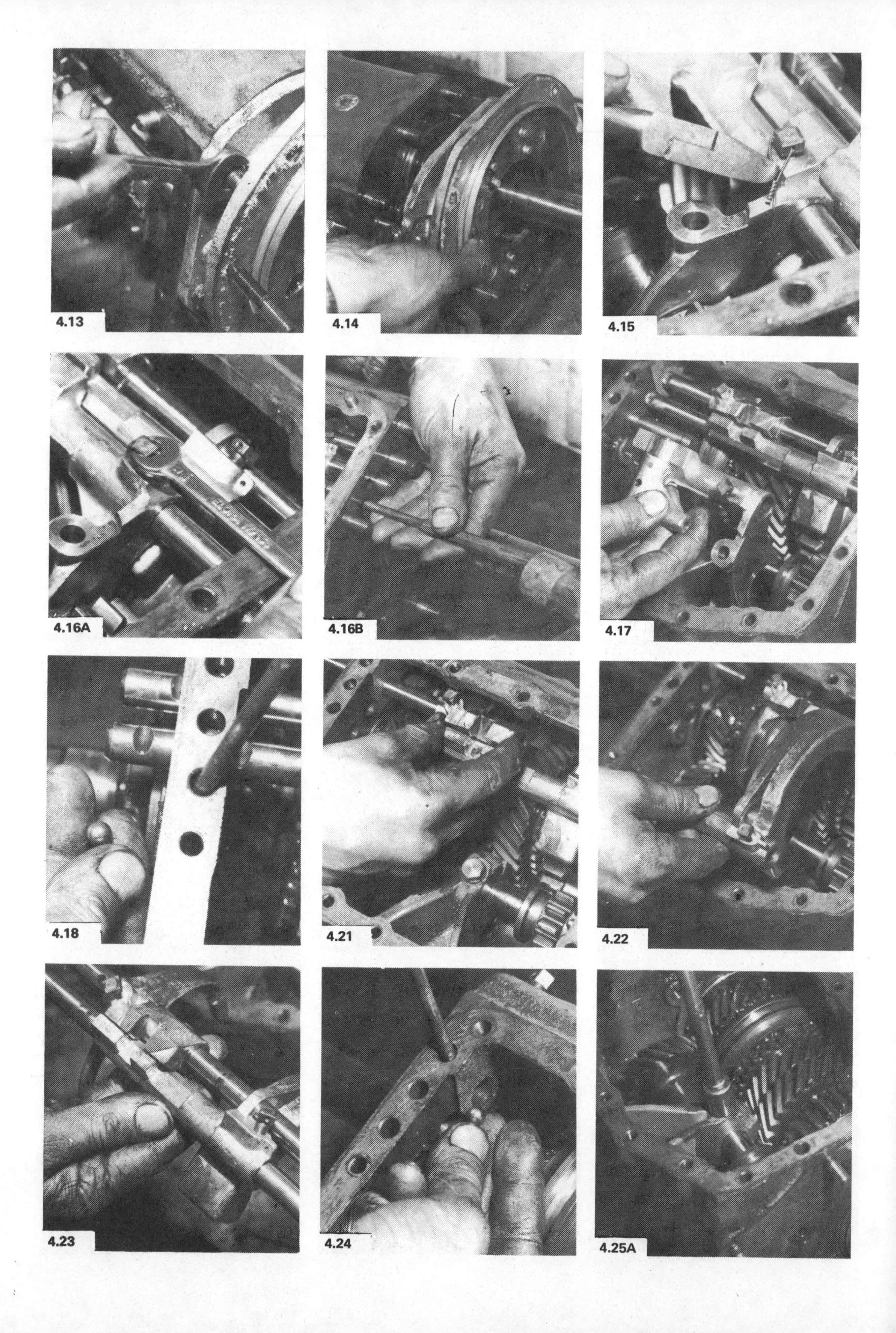
4.13 4.14 4.15
4.16A 4.16B 4.17
4.18 4.21 4.22
4.23 4.24 4.25A

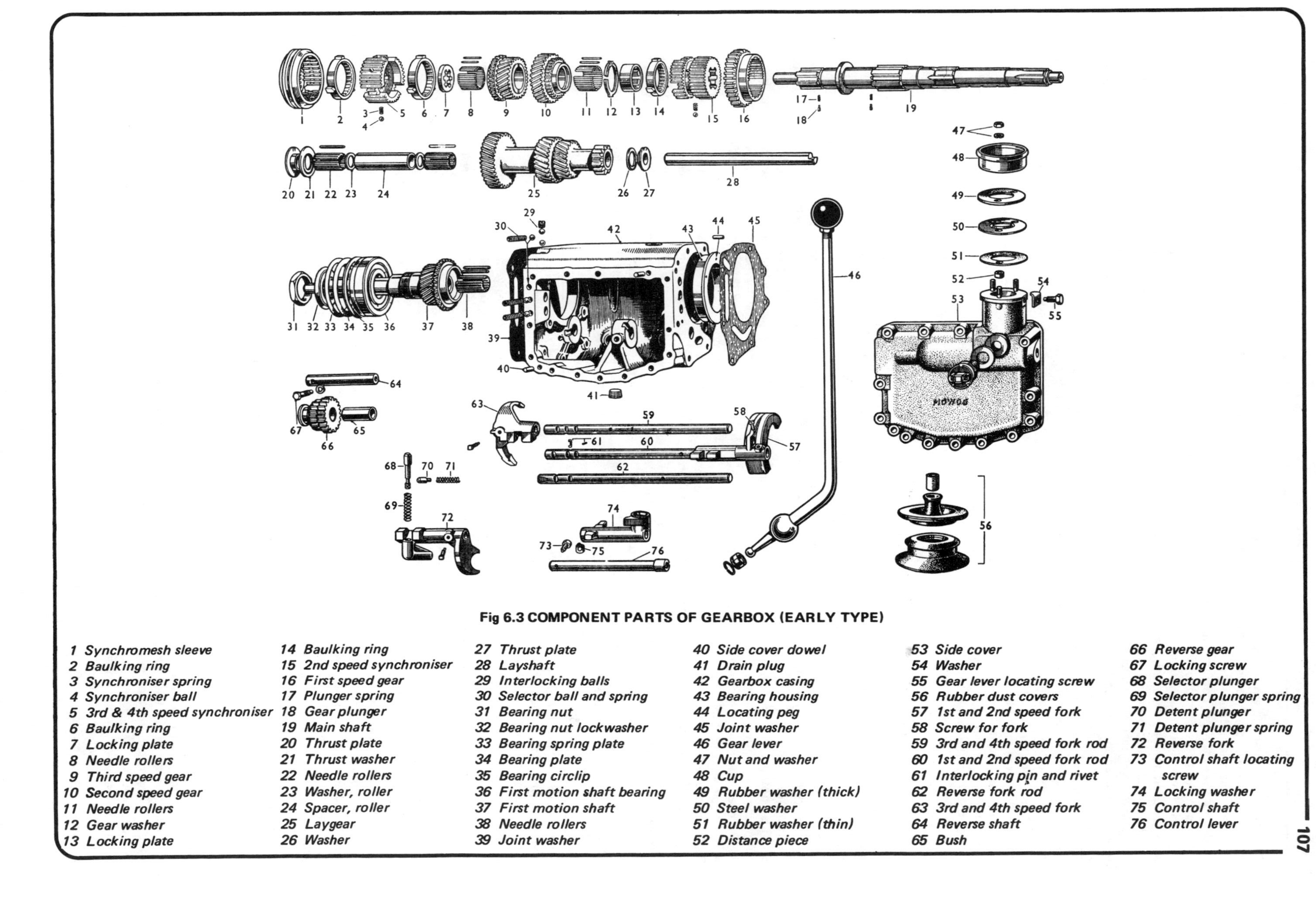

Fig 6.3 COMPONENT PARTS OF GEARBOX (EARLY TYPE)

1 Synchromesh sleeve
2 Baulking ring
3 Synchroniser spring
4 Synchroniser ball
5 3rd & 4th speed synchroniser
6 Baulking ring
7 Locking plate
8 Needle rollers
9 Third speed gear
10 Second speed gear
11 Needle rollers
12 Gear washer
13 Locking plate
14 Baulking ring
15 2nd speed synchroniser
16 First speed gear
17 Plunger spring
18 Gear plunger
19 Main shaft
20 Thrust plate
21 Thrust washer
22 Needle rollers
23 Washer, roller
24 Spacer, roller
25 Laygear
26 Washer
27 Thrust plate
28 Layshaft
29 Interlocking balls
30 Selector ball and spring
31 Bearing nut
32 Bearing nut lockwasher
33 Bearing spring plate
34 Bearing plate
35 Bearing circlip
36 First motion shaft bearing
37 First motion shaft
38 Needle rollers
39 Joint washer
40 Side cover dowel
41 Drain plug
42 Gearbox casing
43 Bearing housing
44 Locating peg
45 Joint washer
46 Gear lever
47 Nut and washer
48 Cup
49 Rubber washer (thick)
50 Steel washer
51 Rubber washer (thin)
52 Distance piece
53 Side cover
54 Washer
55 Gear lever locating screw
56 Rubber dust covers
57 1st and 2nd speed fork
58 Screw for fork
59 3rd and 4th speed fork rod
60 1st and 2nd speed fork rod
61 Interlocking pin and rivet
62 Reverse fork rod
63 3rd and 4th speed fork
64 Reverse shaft
65 Bush
66 Reverse gear
67 Locking screw
68 Selector plunger
69 Selector plunger spring
70 Detent plunger
71 Detent plunger spring
72 Reverse fork
73 Control shaft locating screw
74 Locking washer
75 Control shaft
76 Control lever

21 Lift out the first and second speed selector fork (57) together with the third and fourth speed selector fork (63) (photo).
22 To remove the first and second speed fork it will be necessary to draw the selector fork boss towards the lower end of the main casing so to clear the gear (photo).
23 This photograph shows the relative positions of the first and second speed fork also the third and fourth speed fork when installed on the selector rods in the main casing (photo).
24 Recover the second interlock ball bearing (29) which is located between the centre selector rod and the third and fourth speed fork rod (photo).
25 Unscrew and remove the reverse shaft locating screw (67) and spring washer (photo). Using a soft metal drift push the reverse idler shaft into the main casing (photo).
26 Lift out the reverse idler shaft and reverse idler gear (66) noting which way round the reverse idler gear is fitted (photo).
27 Using the soft metal drift tap out the layshaft (28) towards the front of the main casing. Note that the milled slot is at the rear of the main casing (photo).
28 Lift away the layshaft from the front of the main casing (photo).
29 The laygear (25) should be allowed to drop into the bottom of the main casing (photo).
30 Using a soft metal drift with one end tapered, locate the taper on the front face of the mainshaft bearing housing and tap out the complete mainshaft through the rear of the main casing (photo).
31 Should the mainshaft prove difficult to move use a soft faced hammer and tap on the front face of the spigot on the front end of the mainshaft.
32 This photograph shows the mainshaft being drawn rearwards out of the main casing.
33 Recover the sixteen spigot rollers (38) from the rear of the first motion shaft (37) (photo).
34 Using the soft metal drift with one end tapered, locate the taper on the rear face of the outer track of the first motion shaft bearing (36) and tap out the first motion shaft assembly complete with bearing from the front of the main casing (photo).
35 Lift away the first motion shaft assembly (photo).
36 The layshaft gear (25) may now be lifted out from the bottom of the main casing (photo).
37 Recover the layshaft gear thrust washers (21, 26) and thrust plates (20, 27) from inside the main casing (photo).
38 The four gear assemblies have now been separated from the gearbox casing. Inspect the condition of the gears, splines, bushes and thrust washers to indicate whether or not further dismantling is necessary.

5 Gearbox - examination and renovation

1 Carefully clean and then examine all the components for general wear, distortion, slackness of fit and damage to machined faces and threads.
2 Examine the gearwheels for excessive wear and chipping of the teeth. Obtain new parts as necessary. Refit the layshaft unit thrust washers and plates and with feeler gauges determine the end float which should be not greater than 0.012 inch. If this limit is exceeded new thrust washers and plates should be obtained. It is most probable that after 50,000 miles new thrust washers and plates will be necessary.
3 Examine the layshaft for signs of wear where the layshaft gear unit needle roller bearings bear, and check the layshaft gear unit on a new shaft, if available, for worn bearings. These rollers may be removed and new sets fitted to the inside of the layshaft gear unit.
4 The three synchroniser rings (2, 6, 14) (Fig 6.3) are bound to be badly worn and it is false economy not to renew them. New rings will improve the smoothness and speed of the gearchange considerably.
5 The roller bearings (38) located in the rear of the first motion shaft (37) may be worn so check by refitting into position and placing the first motion shaft on the mainshaft. If sideways movement is evident, a new set of rollers should be obtained.
6 Examine the condition of the first motion shaft bearing (36) and mainshaft bearing. Check them for roughness or looseness between the inner and outer races and for general wear. Normally they should be renewed on a gearbox that is being rebuilt.
7 Inspect the faces of the selector forks (57, 63, 72) for signs of wear, and if evident, new forks should be obtained.
8 Check the reverse idler gear (66) on the shaft (64) for wear of the bush (65). If this bush is worn, drift it out and fit a new one using a drift of suitable diameter so as not to damage the bush face.
9 If the gearbox is to be left for any length of time it is a good idea to wire the respective components together wherever possible so that confusion is avoided at a later date.

6 First motion shaft - dismantling and reassembly

1 The first motion shaft assembly may be dismantled by first holding it vertically between soft faces in a vice. Bend back the lockwasher (32) and undo the large nut (31). Note that it has a left hand thread.
2 Lift away the nut and lockwasher.
3 Place the first motion shaft on the top of the vice with the outer track of the race resting on soft faces.
4 Using a soft faced hammer, drift the first motion shaft through the race inner track. The strain placed on the bearing does not matter as the bearing would not be removed unless it was being renewed. Alternatively use a three legged universal puller.
5 Lift away the race from the first motion shaft noting that the circlip groove on the outer track is offset towards the front. Remove the circlip (35) from the outer track of the bearing.
6 To assemble the first motion shaft place the race against soft metal (old shell bearings suitably straightened) on the top of the jaws of the vice and using a drift located in the mainshaft spigot bearing hole in the rear of the first motion shaft, drift the shaft into the bearing. Make sure the bearing is the correct way round.
7 Refit the lockwasher and secure the retaining nut, not forgetting it has a left hand thread. Bend over the locking washer to secure the nut.

7 Mainshaft - dismantling and reassembly

1 Slide the top and third gear hub and interceptors (1-5) (Fig 6.3) from the forward end of the mainshaft (photo).
2 With a small parallel pin punch or screwdriver carefully depress the plunger locating the third gear locking plate (7) and rotate the plate so as to line up the splines.
3 Slide the locking plate up off the front end of the mainshaft (photo).
4 Remove the plunger (18) and spring (17) from the mainshaft (photo).
5 Slide the third speed gear (9) and its 32 rollers (8) from the mainshaft (photo).
6 Place two pieces of soft metal in the jaws of a firm bench vice ready for holding the mainshaft.
7 Place the front end of the mainshaft between two soft faces of the vice so that it is parallel with the top of the bench.
8 Bend back the lockwasher securing the nut and with a large ring spanner undo the nut. Note: This nut has a left hand thread. Remove the nut and lockwasher from the mainshaft. On gearboxes having overdrive as well there will be no speedometer drive gear, locking washer and nut on the end of the mainshaft.
9 Lift the speedometer drive gear bearing with housing (43) and distance collar. Remove the speedometer drive gear key from the mainshaft.
10 Slide the first and second speed hub (15) second speed interceptor and the first speed gear (16) rearwards from the mainshaft. Note which way round the first speed gear is fitted.
11 Should it be necessary the first speed gear may be withdrawn from the hub. Wrap the assembly in a rag so as to catch the ball

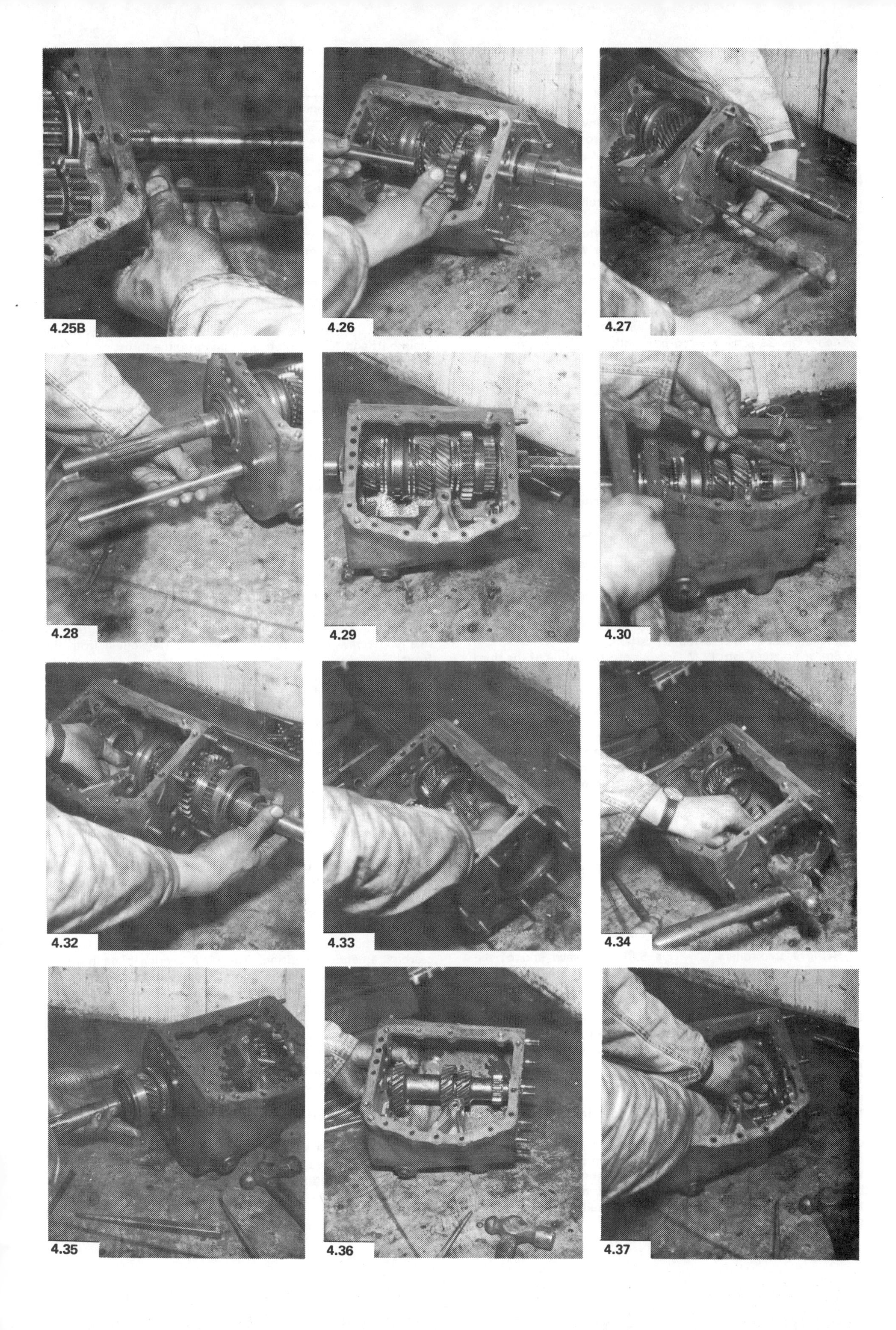
4.25B
4.26
4.27
4.28
4.29
4.30
4.32
4.33
4.34
4.35
4.36
4.37

bearings and springs and slide the first speed gear (16) from the hub (15). Recover the ball bearings and springs.
12 With a small parallel pin punch or screwdriver depress the second speed gear locking collar plunger and rotate the collar (13) to line up the splines. Slide the collar from the shaft and extract the two halves of the second gear washer (12).
13 Recover the spring and plunger.
14 The second speed gear (10) and its 33 rollers (11) may now be withdrawn from the mainshaft.
15 Dismantling the mainshaft is now complete.
16 Inspect all gears for signs of damage or wear. If a constant mesh gear tooth is damaged it is likely that the corresponding gear on the laygear will also be damaged. If this is the case a new laygear will also be necessary.
17 To assemble the mainshaft once all new parts have been obtained first smear the mainshaft with Castrol LM Grease and assemble the 33 second speed gear rollers (11) onto the mainshaft.
18 Very carefully slide the second gear (15) over the rollers.
19 Refit the plunger and spring followed by the two halves of the second speed gear washer (12) and slide the collar (13) onto the splines.
20 Depress the plunger and push the collar (13) into position whilst at the same time locating the lugs of the washer in the cut-outs of the collar.
21 Rotate the collar so as to bring the splines out of line.
22 Replace the balls and springs in the second and first speed hub (15). Depress the balls and slide the first speed gear on the hub the correct way round as shown in Fig 6.4 and then fit the assembly to the shaft.
23 Refit the bearing distance collar followed by the key and housing (43).
24 Insert the speedometer drive gear key and refit the gear, locking washer and nut.
25 Tighten the nut (left hand thread) and lock by bending over the locking washer in at least one position.
26 Smear a little Castrol LM Grease onto the inside of the third gear (9) and position the 32 rollers inside the gear. Carefully slide the gear with rollers onto the mainshaft.
27 Refit the plunger and spring followed by the third speed locking collar (7) (photo).
28 Rotate the collar so as to bring the splines out of line.
29 Refit the balls and springs to the top and third speed hub and slide the striking dog into position on the hub.
30 Slide the hub, striking dog and interceptors onto the mainshaft.
31 The mainshaft is now completely assembled and is ready for refitting to the gearbox main casing.

8 Laygear - dismantling and reassembly

1 Remove the rollers (22) (Fig 6.3) from each end of the laygear (25) and keep them in their sets.
2 Lift out the distance tube (24) and two washers (23). The components are shown in this photo.
3 To reassemble the laygear first insert the distance tube (24) and place one washer (23) at each end of the tube.
4 Smear the rollers with Castrol LM Grease and position them in each end of the gear.
5 So as to retain the rollers in position obtain a length of round bar the same outside diameter as the layshaft and the length of the laygear plus thrust washers and plates. Insert this rod into the laygear.
6 It should be noted that on gearboxes fitted to engines numbered 11342 and onwards for a limited period the laygear was fitted with plain bushes but this was subsequently replaced by a laygear assembly with needle roller bearings. This modification commenced at the same time as slightly redesigned mainshaft gears and new gears may only be interchanged in complete sets.

9 Gearbox - reassembly

1 Fit the thrust washers (21, 26) and plates (20, 27) to the laygear (25) and place the gear in the bottom of the main casing. Make sure that it is in the correct way round.
2 Carefully insert the mainshaft assembly into the main casing, feeding it through the rear.
3 Smear some Castrol LM Grease into the spigot bore at the rear of the first motion shaft (37) and insert the 16 rollers (38).
4 Insert the first motion shaft bearing outer track into its bore in the front of the main casing and with a soft metal drift tap the bearing into position. Take care that the rollers do not become dislodged as the mainshaft front end spigot engages with the rollers.
5 Invert the gearbox so that the laygear meshes with the mainshaft and first motion shaft and ease the thrust washer tags into their grooves.
6 Carefully push the layshaft through the housing and gear and withdraw the retaining bar as the shaft pushes it out of the gear.
7 Note that the cut-away portion of the shaft must be aligned to fit the groove in the bellhousing provided to prevent the layshaft from turning.
8 Refit the reverse idler gear (66), the correct way round into the gearbox and insert the shaft (64) in such a manner that the dowel locking setscrew hole lines up with the setscrew hole in the casting web.
9 Replace the dowel setscrew (67) and washer making sure that the dowel lines up with the hole in the reverse idle shaft (64).
10 Position the third and fourth gear selector fork (63) onto the synchromesh sleeve (1).
11 Replace the first and second gear selector fork (57) in its location on the first speed gear (16).
12 Insert the first and second gear selector rod (60). Note this rod has three notches at the forward end and the selector fork retaining screw about two thirds of the way down from the front.
13 Fit one interlock ball bearing (29) above the first and second shifter shaft (60) and insert the third and fourth gear selector rod. Note this rod is the one with three notches at the forward end and the selector fork retaining screw dowel hole about one third of the way down from the front.
14 Place the second interlock ball (29) and hold it in position with some grease.
15 Fit the reverse gear selector fork (72) and refit the selector rod (62) together with the detent plungers and springs (68, 69, 70, 71).
16 Position the selector forks and rods in such a manner that the dowel holes are lined up with the forks and then refit the selector setscrews. Ensure that the dowel ends engage in the drilling in the selector rod.
17 Lock the selector setscrews by using soft iron wire threaded through the setscrew head and hole in the selector fork.
18 Wipe the mating faces of the gearbox main casing and the extension housing. Fit a new paper gasket (45) and replace the gearbox extension housing. The plain bearing plate must be fitted against the bearing. Secure the extension housing with the eight bolts and spring washers.
19 If an overdrive unit is fitted slide the adaptor plate together with its bearing and paper joint washer along the mainshaft.
20 Fit and tighten down the eight set pins securing the adaptor plate to the gearbox.
21 Refit the distance piece which covers the space between the rear main bearing and the groove for the circlip. Refit the circlip.
22 Refit the overdrive unit as described in Section 14 of this Chapter.
23 Insert the three selector balls into the holes in the gearbox main casing and the springs in the holes in the side cover.
24 Wipe the mating faces of the side cover (53) and main casing (42) and fit a new paper gasket.
25 Secure the side cover (53) with the 13 bolts and spring washers. Note that the longer bolt must be fitted in the top right

7.1

7.3

7.5

7.27

Fig 6.4 FIRST GEAR SYNCHRONISER

1 Gear
2 Baulking ring
3 Synchroniser

8.2

hand position.

26 Wipe the mating faces of the clutch bellhousing and the main case and fit a new gasket.

27 Refit the bearing spring plate (33) and bearing plate (34) to the outer track of the first motion shaft ball race.

28 Place the gearbox bellhousing onto the main casing and secure with the seven short and one long bolt with spring washers and tighten in a diagonal manner to a torque wrench setting of 35 lb ft.

29 Refit the speedometer drive and the dipstick.

10 Gearbox side cover - removal, overhaul and refitting (early type)

1 Undo the drain plug (41) (Fig 6.3) located on the underside of the gearbox and allow the oil to drain into a container having a capacity of at least five pints. If overdrive is fitted, undo the overdrive unit drain plug (Fig 6.7) and drain the unit. The total capacity of the gearbox plus overdrive is 6¼ pints.

2 Detach the switch electric cable. Unscrew and remove the thirteen bolts and spring washers securing the side cover to the gearbox housing. Lift away the side cover and paper gasket.

3 Take care that the three selector balls and springs do not fall out of the front of the main casing.

4 Undo and remove the three nuts and spring washers (47) (Fig 6.3) from the studs on the gear change lever cup on the side cover (53).

5 Lift away the cup (48) together with the three washers (49, 50, 51) and the three distance pieces (52) located on the three studs.

6 Lift away the gear change lever with the bush on the end of the lever ball.

7 If it is necessary to remove the control lever first remove the switch on the side cover.

8 Remove the plug on the rear end of the side cover.

9 Bend back the tab washer (75) and undo the control shaft locating screw (73).

10 Draw the control shaft (76) rearwards through the hole in the rear of the side cover whilst sliding the control lever in the centre of the side cover.

11 Check the control shaft movement in the side cover and if excessive side movement is evident a new side cover must be obtained.

12 Check that the control shaft locating screw dowel (73) is a good fit in the drilling in the control shaft. If wear is evident obtain a new control shaft and locating screw.

13 Reassembly and refitting is the reverse sequence to removal.

11 Gearbox - dismantling, overhaul and reassembly (later type)

The procedure for the dismantling, overhaul and reassembly of the later type of gearbox is basically identical to that for the earlier type. Upon inspection of Fig 6.5 and Fig 6.6 it will be seen that the gear change system is centrally mounted on the top of the gearbox instead of on the side of the gearbox.

This type of gearbox was introduced from car numbers BN7 16039 and BT7 15881.

The main differences in working on the gearbox are listed below for reference:

1 The dipstick is located in the gearbox top cover.

2 If it is necessary to remove the reverse selector plunger from the reverse striking fork, extract the split pin so as to release the plunger and spring which in turn will release the detent plunger and spring.

3 Information on dismantling, overhaul and reassembly of the top cover will be found in Section 12 of this Chapter.

Fig 6.5 GEARBOX COMPONENT PARTS (WITHOUT OVERDRIVE) – LATER TYPE

1 Clutch housing
2 Fork and lever shaft bush
3 Buffer pad
4 Oil seal
5 Bolt - long
6 Bolt - short
7 Spring washer
8 Clutch fork and lever
9 Fork and lever shaft
10 Clutch withdrawal fork screw
11 Taper pin
12 Thrust washer for fork and lever
13 Fork and lever seal
14 Seal retaining plate
15 Retaining plate screw
16 Spring washer
17 Starter end cover
18 End cover screw
19 Spring washer
20 Gearbox case
21 Oil drain plug
22 Interlock ball hole plug
23 Case to clutch housing joint
24 Oil level indicator
25 Rubber grommet
26 Gearbox top cover
27 Cover oil seal
28 Cover plug
29 Cover to gearbox joint
30 Bolt - long
31 Bolt - short
32 Spring washer
33 Gearbox breather
34 Gearbox extension casing
35 Casing taper plug
36 Speedometer pinion thrust button
37 Oil seal
38 Bearing
39 Bearing washer
40 Coupling flange
41 Spring washer
42 Flange nut
43 Casing to gearbox joint
44 Bolt - casing to gearbox
45 Spring washer
46 Drive gear
47 Bearing for drive gear
48 Bearing circlip
49 Bearing plate
50 Bearing plate (spring)
51 Bearing nut
52 Lockwasher
53 Roller for drive gear
54 Mainshaft
55 Mainshaft bearing
56 Bearing housing
57 Locating peg
58 Bearing circlip
59 Bearing plate
60 Bearing plate (spring)
61 Top and third sliding hub with striking dog
62 Sliding hub interceptor
63 Sliding hub ball
64 Ball spring
65 Third speed gear
66 Gear roller
67 Locking plate
68 Gear plunger
69 Plunger spring
70 Second speed gear
71 Gear roller
72 Gear washer
73 Locking plate
74 Gear plunger
75 Plunger spring
76 First speed gear with first and second sliding hub
77 Sliding hub interceptor
78 Sliding hub ball
79 Ball spring
80 Mainshaft distance collar
81 Reverse gear
82 Gear bush
83 Gear shaft
84 Shaft retaining screw
85 Spring washer
86 Layshaft
87 Layshaft gear unit
88 Gear unit roller
89 Roller washer
90 Roller spacer
91 Gear unit thrust plate - front
92 Gear unit thrust plate - rear
93 Gear unit thrust washer - front
94 Gear unit thrust washer - rear
95 Top and third shifter shaft
96 Shaft interlocking ball
97 Top and third striking fork
98 Screw for striking fork
99 Shifter shaft ball
100 Ball spring
101 First and second shifter shaft
102 Shaft interlock pin
103 Interlocking pin rivet
104 First and second striking fork
105 Screw for striking fork
106 Shifter shaft ball
107 Ball spring
108 Reverse shifter shaft
109 Reverse striking fork
110 Screw for striking fork
111 Shifter shaft ball
112 Ball spring
113 Reverse selector plunger
114 Plunger spring
115 Detent plunger
116 Detent plunger spring
117 Remote control shaft
118 Change speed lever shaft
119 Selector lever
120 Selector lever and change speed lever socket screw
121 Spring washer
122 Selector lever and change speed lever socket key
123 Change speed lever
124 Lever bush
125 Circlip for bush
126 Rollpin
127 Ball and retaining spring
128 Spring washer
129 Circlip
130 Change speed lever knob
131 Locknut for knob
132 Plunger retaining plug
133 Plug washer
134 Plunger
135 Plunger spring
136 Speedometer gear
137 Key for gear
138 Locknut for gear
139 Lockwasher for gear
140 Speedometer pinion
141 Pinion bearing
142 Washer for bearing
143 Pinion distance collar
144 Pinion oil seal
145 Reverse switch hole plug
146 Clutch housing bolt - lon
147 Spring washer
148 Clutch housing bolt - sh
149 Spring washer
150 Nut
151 Clutch housing dowel bolt
152 Spring washer for dowel bolt
153 Nut for dowel bolt
154 Speedometer drive adaptor box

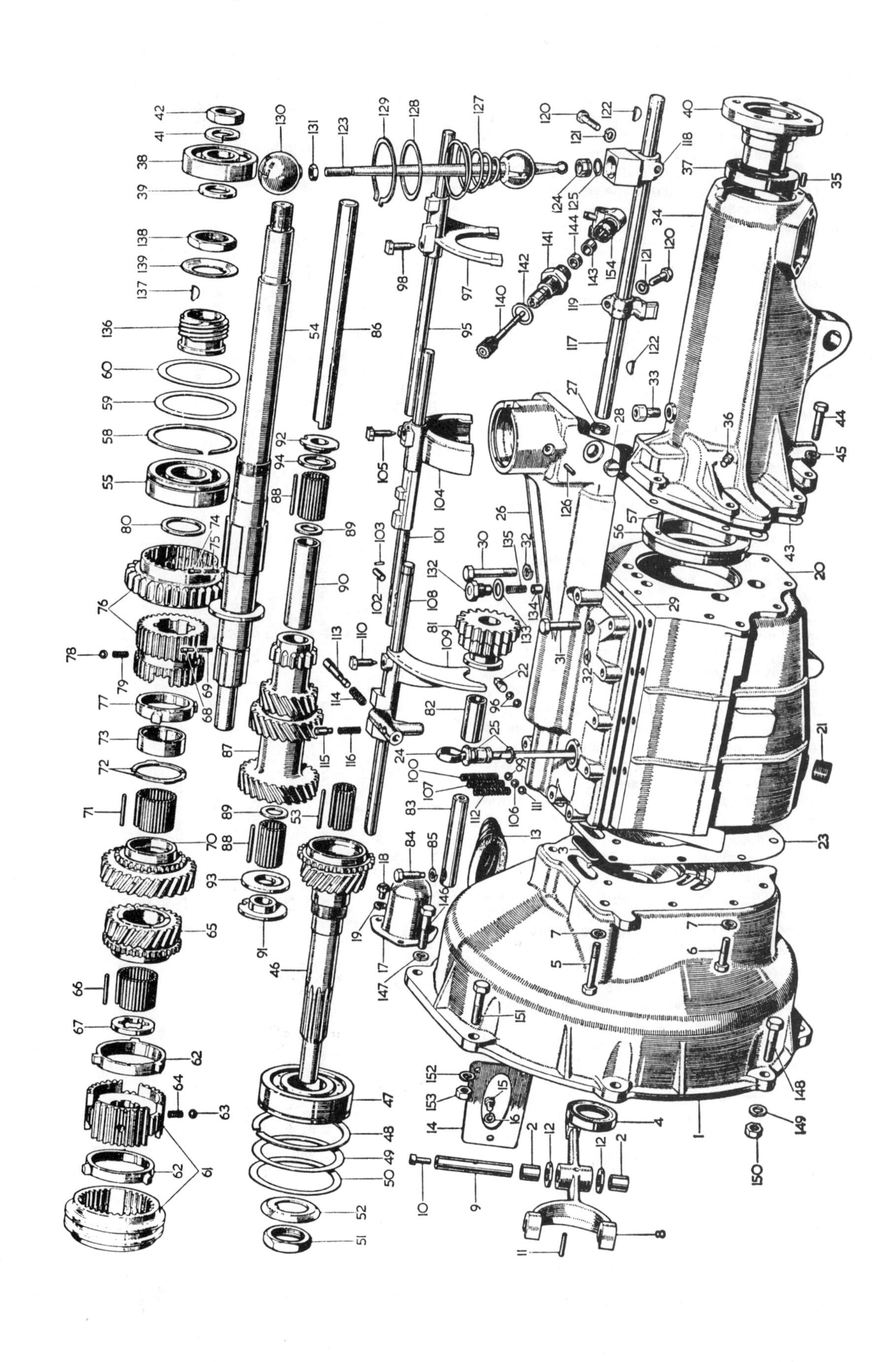

12 Gearbox top cover - dismantling, overhaul and reassembly (later type)

1 Undo and remove the 12 bolts securing the top cover to the gearbox main casing. Note that the two cover securing bolts nearest the change speed lever turret are longer.
2 Carefully lift away the top cover and recover the three detent springs positioned in the gearbox casing and not lifted up from their drillings and accidentally dropped into the gearbox.
3 To remove the gear change lever from the top cover, release the circlip, lift away the washer and conical spring from the change lever turret.
4 Using a parallel pin punch drift out the two roll pins in turn into the 3/16 inch diameter holes on each side of the change speed turret. This will cause them to move into the bore of the lever ball.
5 Lift out the lever and retrieve the roll pins from the ball end.
6 Remove the remote control shaft and plug under the turret.
7 Undo and remove the two locking bolts and spring washers securing the selector lever and change speed shaft lever to the change speed shaft.
8 Move the selector lever and change speed shaft lever up the change speed shaft and withdraw the two Woodruff keys.
9 Draw the change speed shaft through the rear of the top cover.
10 If oil has been leaking from either the front or rear of the top cover the old oil seal should be removed and a new one fitted. The lip of the seal faces inwards.
11 Reassembly and refitting is the reverse sequence to removal. Take care that the fine lips of the oil seals are not damaged by the keyways which can have sharp edges.

13 Overdrive - general description

The overdrive unit is attached to the extension on the rear of the gearbox by eight studs and nuts, and takes the form of an hydraulically operated epicyclic gear. Overdrive operates on third and fourth speeds to provide fast cruising at lower engine revolutions. The overdrive 'in-out' switch on the right hand side of the steering wheel actuates a solenoid attached to the side of the overdrive unit. In turn the solenoid operates a valve which opens the hydraulic circuit which pushes the cone clutch into contact with the annulus when overdrive is engaged.

During high speed motoring, the engine speed is decreased with the engagement of overdrive so that with continual use of the unit there will be an increase in engine life.

A special switch called an inhibitor switch is incorporated in the electrical circuit and prevents the engagement of overdrive in reverse, first or second gears. The switch is located on the top of the gearbox top cover. The overdrive unit will lock solid if engaged in reverse gear.

The normal minimum engagement speeds are top gear 40 mph and third gear 30 mph whilst the minimum disengagement speed in top is under the control of the driver who must take care not to over rev the engine at high speeds. For third speed the disengagement should occur at a maximum of 70 mph.

The overdrive unit operates in the following manner. The operating gears in the overdrive are of epicyclic design and comprise a sunwheel which meshes with three planet gears carried in a circular metal carrier. These planet gears mesh with an annulus which has internal teeth. The planet carrier is attached to the input shaft which is the output shaft of the gearbox. The annulus is an integral part of the output shaft.

When the driver selects overdrive, hydraulic pressure is built up by a plunger type pump which operates from a cam splined to the input shaft, ie: the output shaft from the gearbox. The hydraulic pressure built up by the pump forces the two pistons against bridge pieces which are themselves attached to the thrust ring. The thrust ring is pushed forwards by the pistons so engaging the clutch with the brake ring, with sufficient force to hold the sunwheel firmly at rest.

Fig 6.6 GEARBOX COMPONENT PARTS (WITH OVERDRIVE) — LATER TYPE

1 Clutch housing
2 Fork and lever shaft bush
3 Buffer pad
4 Oil seal
5 Bolt - long
6 Bolt - short
7 Spring washer
8 Clutch fork and lever
9 Fork and lever shaft
10 Clutch withdrawal fork, screw
11 Taper pin
12 Thrust washer for fork and lever
13 Fork and lever seal
14 Seal retaining plate
15 Retaining plate screw
16 Spring washer
17 Starter end cover
18 End cover screw
19 Spring washer
20 Gearbox case
21 Oil drain plug
22 Interlock ball hole plug
23 Case to clutch housing joint
24 Oil level indicator
25 Rubber grommet
26 Gearbox top cover
27 Cover oil seal
28 Cover to gearbox joint
29 Bolt - long
30 Bolt - short
31 Spring washer
32 Overdrive switch
33 Joint for switch
34 Drive gear
35 Bearing for drive gear
36 Circlip for bearing
37 Plate for bearing
38 Plate for bearing (spring)
39 Nut for bearing
40 Lockwasher
41 Drive gear roller
41 Mainshaft
43 Mainshaft bearing
44 Bearing housing
45 Locating peg
46 Bearing circlip
47 Plate for bearing
48 Plate for bearing (spring)
49 Mainshaft circlip
50 Bearing abutment collar
51 Abutment collar retaining ring
52 Shim
53 Top and third sliding hub with striking dog
54 Sliding hub interceptor
55 Sliding hub ball
56 Ball spring
57 Third speed gear
58 Roller for gear
59 Locking plate
60 Gear plunger
61 Plunger spring
62 Second speed gear
63 Roller for gear
64 Gear washer
65 Locking plate
66 Gear plunger
67 Plunger spring
68 First speed gear with first and second sliding hub
69 Sliding hub interceptor
70 Sliding hub ball
71 Ball spring
72 Mainshaft distance collar
73 Reverse gear
74 Gear bush
75 Gear shaft
76 Shaft retaining screw
77 Spring washer
78 Layshaft
79 Layshaft gear unit
80 Gear unit roller
81 Roller washer
82 Roller spacer
83 Gear unit thrust plate - front
84 Gear unit thrust plate - rear
85 Gear unit thrust washer - front
86 Gear unit thrust washer - rear
87 Top and third shifter shaft
88 Shaft interlocking ball
89 Top and third striking fork
90 Screw for striking fork
91 Shifter shaft ball
92 Ball spring
93 First and second shifter shaft
94 Shaft interlocking pin
95 Interlocking pin rivet
96 First and second striking fork
97 Screw for striking fork
98 Shifter shaft ball
99 Ball spring
100 Reverse shifter shaft
101 Reverse striking fork
102 Screw for striking fork
103 Shifter shaft ball
104 Ball spring
105 Reverse selector plunger
106 Plunger spring
107 Detent plunger
108 Detent plunger spring
109 Remote control shaft
110 Change speed lever rocket
111 Selector lever
112 Selector lever and change speed lever socket screw
113 Spring washer
114 Selector lever and change speed lever socket key
115 Change speed lever
116 Lever bush
117 Circlip for bush
118 Roll pin
119 Ball end and retaining spring
120 Washer for spring
121 Circlip
122 Change speed lever knob
123 Locknut for knob
124 Plunger retaining plug
125 Plug washer
126 Plunger
127 Plunger spring

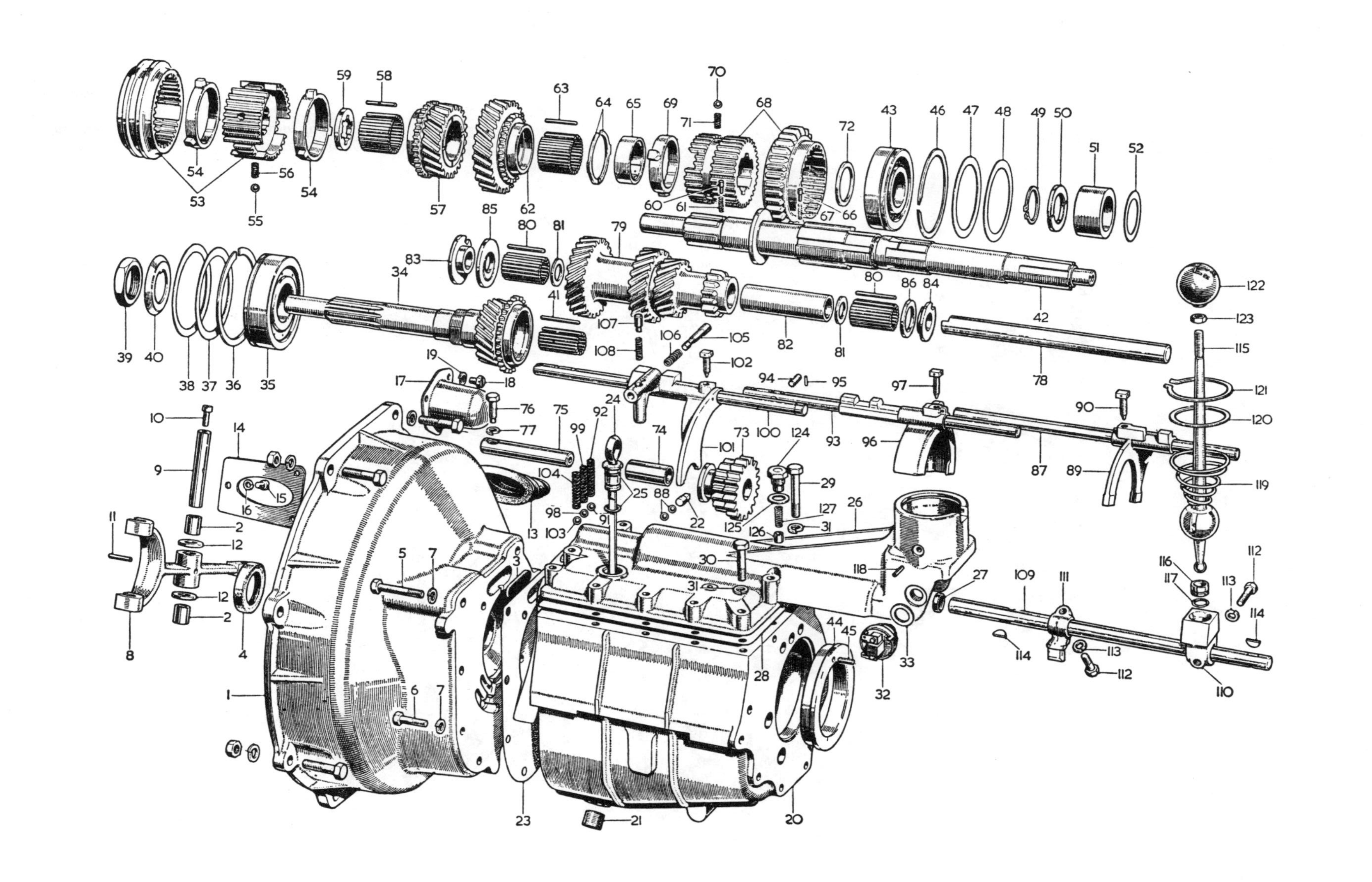
70
59
58
63
64
65
69
68
72
43
46
47
48
49
50
51
52
71
56
54
54
53
55
57
85
62
60
61
67
66
80
81
79
83
34
80
86
84
122
123
42
41
107
106
105
82
81
78
115
39
40
38
37
36
35
108
102
19
17
18
94
95
97
121
24
10
76
75
99
92
90
120
14
77
100
93
73
124
96
101
74
87
89
9
104
25
29
119
16
15
88
98
127
26
11
22
2
103
13
125
31
12
126
116
112
5
7
3
30
118
27
109
111
117
113
12
31
114
2
8
4
44
45
114
33
113
28
112
1
6
7
32
110
23
21
20

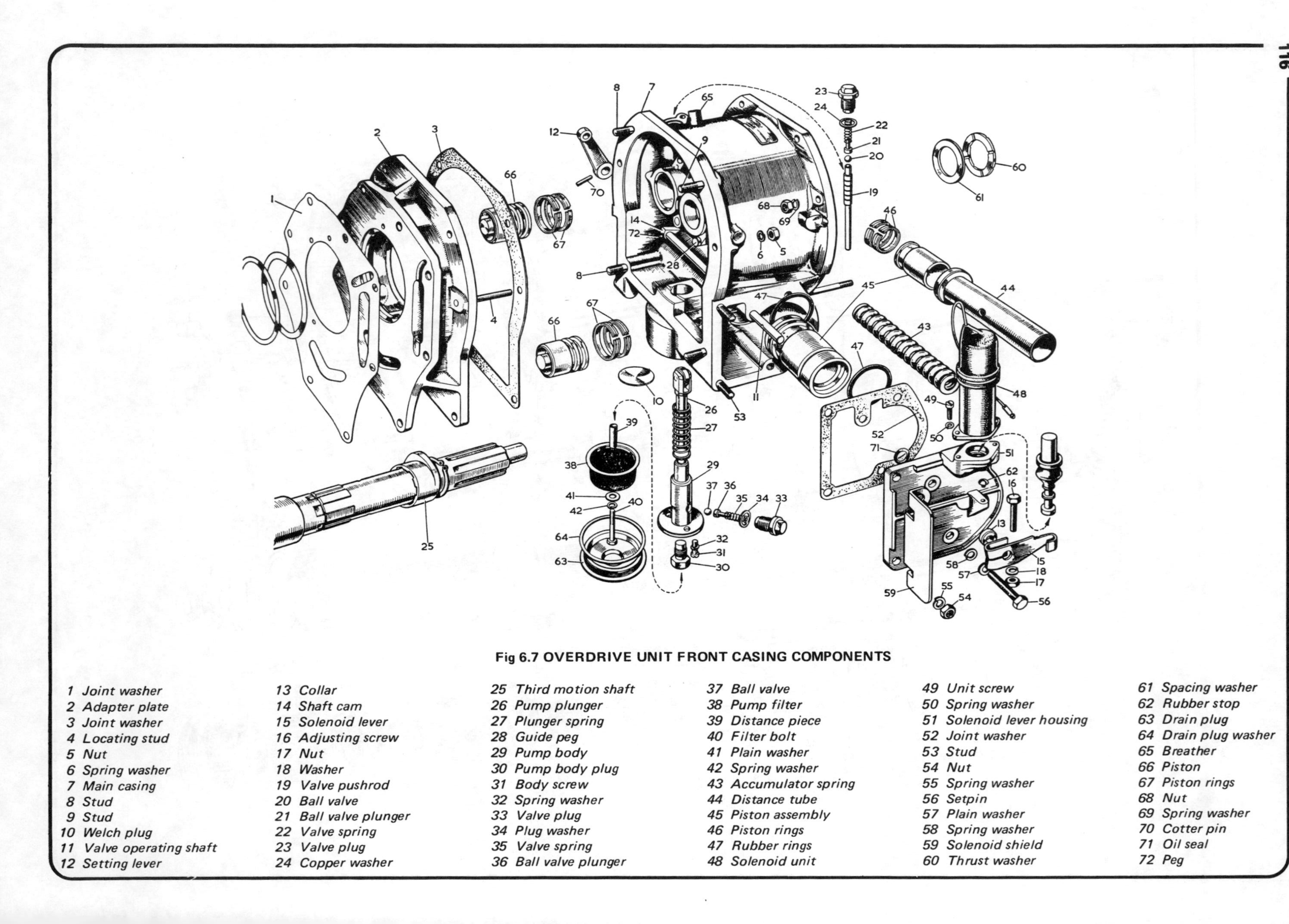

Fig 6.7 OVERDRIVE UNIT FRONT CASING COMPONENTS

1 Joint washer
2 Adapter plate
3 Joint washer
4 Locating stud
5 Nut
6 Spring washer
7 Main casing
8 Stud
9 Stud
10 Welch plug
11 Valve operating shaft
12 Setting lever
13 Collar
14 Shaft cam
15 Solenoid lever
16 Adjusting screw
17 Nut
18 Washer
19 Valve pushrod
20 Ball valve
21 Ball valve plunger
22 Valve spring
23 Valve plug
24 Copper washer
25 Third motion shaft
26 Pump plunger
27 Plunger spring
28 Guide peg
29 Pump body
30 Pump body plug
31 Body screw
32 Spring washer
33 Valve plug
34 Plug washer
35 Valve spring
36 Ball valve plunger
37 Ball valve
38 Pump filter
39 Distance piece
40 Filter bolt
41 Plain washer
42 Spring washer
43 Accumulator spring
44 Distance tube
45 Piston assembly
46 Piston rings
47 Rubber rings
48 Solenoid unit
49 Unit screw
50 Spring washer
51 Solenoid lever housing
52 Joint washer
53 Stud
54 Nut
55 Spring washer
56 Setpin
57 Plain washer
58 Spring washer
59 Solenoid shield
60 Thrust washer
61 Spacing washer
62 Rubber stop
63 Drain plug
64 Drain plug washer
65 Breather
66 Piston
67 Piston rings
68 Nut
69 Spring washer
70 Cotter pin
71 Oil seal
72 Peg

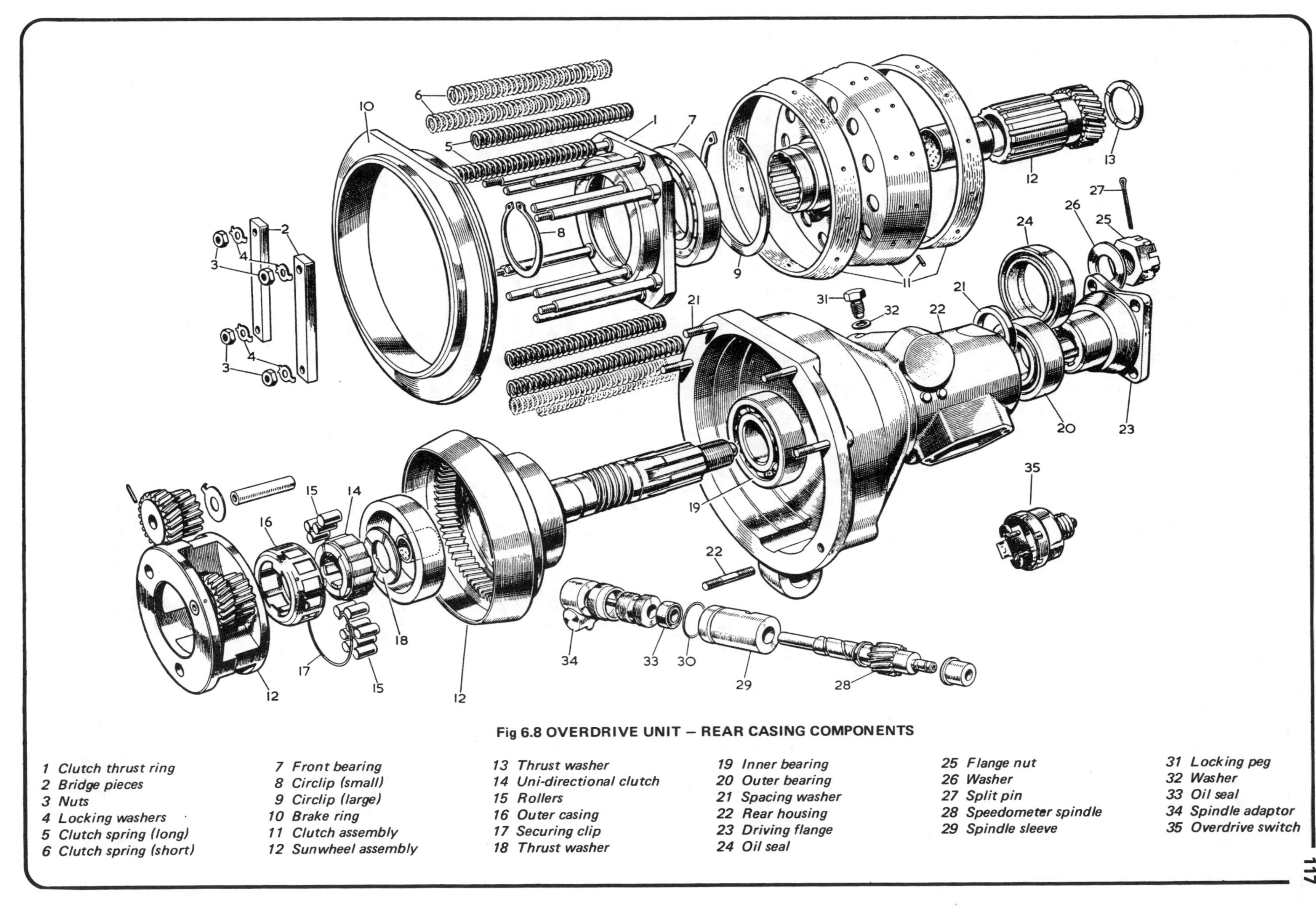

Fig 6.8 OVERDRIVE UNIT – REAR CASING COMPONENTS

1 Clutch thrust ring
2 Bridge pieces
3 Nuts
4 Locking washers
5 Clutch spring (long)
6 Clutch spring (short)
7 Front bearing
8 Circlip (small)
9 Circlip (large)
10 Brake ring
11 Clutch assembly
12 Sunwheel assembly
13 Thrust washer
14 Uni-directional clutch
15 Rollers
16 Outer casing
17 Securing clip
18 Thrust washer
19 Inner bearing
20 Outer bearing
21 Spacing washer
22 Rear housing
23 Driving flange
24 Oil seal
25 Flange nut
26 Washer
27 Split pin
28 Speedometer spindle
29 Spindle sleeve
31 Locking peg
32 Washer
33 Oil seal
34 Spindle adaptor
35 Overdrive switch

The planet carrier is now able to rotate with the input shaft allowing the planet wheels in the planet carrier assembly to rotate about their own axes so driving the annulus at a faster speed than the input shaft is rotating.

Oil is drawn by the pump through a wire mesh filter and delivers it to the operating valve through an hydraulic accumulator, the amount of pressure being controlled by a pressure relief valve built into the accumulator. The oil is free to pass between the gearbox and overdrive unit and there is one common level for both units. The oil level is indicated by a level plug on the side of the gearbox.

Whenever the oil is drained the two drain plugs, one on the underside of the gearbox and the other on the underside of the overdrive unit must be removed but it is usual practice not to change the oil during normal servicing but to top up. Only recommended grades of oil must be used and it is important that under no circumstances must oil anti-friction additives be used otherwise the overdrive unit will not operate correctly.

Cleanliness is very important so do not remove the drain plug without first wiping the surrounding area. Whenever the oil is drained for service/repair work on either the gearbox or the overdrive always clean the mesh filter.

The overdrive is normally a very reliable unit and trouble is usually due to either the solenoid sticking; a fault in the hydraulic system due to dirt ingress; insufficient oil; or incorrect solenoid operating lever adjustment.

14 Overdrive - removal and replacement

1 It is necessary to remove the overdrive from the car in order to attend to the following: the hydraulic lever setting; the relief valve; the non-return valve; the solenoid; and the operating valve.

2 If the unit as a whole requires overhaul it must be removed from the car together with the gearbox as described in Chapter 6, Section 3.

3 To separate the overdrive from the gearbox, undo the eight nuts from the ¼ inch diameter studs (noting the extra length of two of the studs) to separate the main overdrive casing adaptor plate from the gearbox rear extension. Carefully pull the overdrive off the end of the mainshaft.

4 To mate the overdrive and gearbox, start by placing the overdrive in an upright position and then line up the splines on the clutch and planet carrier by eye, turning them anticlockwise only, with the aid of a long thick screwdriver. Make certain that the spring clip is correctly positioned in the groove in the mainshaft and that it does not protrude over the mainshaft splines.

5 Under normal circumstances if everything is in line the gearbox mainshaft should enter the overdrive easily. If trouble is experienced do not try to force the components together but separate them and re-align the components. Place the gearbox in top gear while refitting.

6 As the mainshaft is fed into the overdrive, gently rotate the input shaft to and fro to help the mating of the mainshaft into the splines. At the same time make certain that the lowest portion of the cam on the mainshaft will rest against the pump and that as the gearbox extension and overdrive come together, the end of the mainshaft enters into the needle roller bearing in the tail shaft.

7 The remainder of the replacement procedure is a straightforward reversal of the removal sequence.

15 Overdrive - dismantling, overhaul and reassembly

1 To enable a satisfactory overhaul to be completed there are several special tools that will be required. Full details of these are given as and where they are needed and they should be obtained before work commences.

2 It is recommended that before the unit is dismantled the exterior is thoroughly cleaned and dried as it is important that no dirt gets into any of the internal parts of the unit.

3 Place the unit on a clean bench and lift away the long and short clutch springs (5, 6) (Fig 6.8) noting their locations so that they may be refitted correctly. Place these in a clean jam jar for safe keeping.

4 Bend back the tab washers (4) (Fig 6.8) and unscrew the four nuts (3). Lift away the nuts, tab washers and two bridge pieces (2).

5 Undo the six nuts securing the rear housing to the main casing. Lift away the nuts and spring washers from the studs.

6 The two casings may now be separated. Lift away the brake ring (10) (Fig 6.8).

7 Working from the forward end of the sunwheel assembly first remove the steel thrust washer followed by the phosphor bronze thrust washer (60, 61) (Fig 6.7). Lift out the clutch sliding member (11) (Fig 6.8) complete with the thrust ring (1) and bearing (7).

8 Withdraw the sunwheel and the planet carrier assembly (12).

9 Unscrew and remove the operating valve plug (23) (Fig 6.7) and copper washer (24). Lift out the spring (22), plunger (21) and ball bearing (20) followed by the valve (19).

10 Using a pair of pliers carefully grip the operating pistons (66) on the centre bosses and rotate whilst pulling the pistons out of the front casing (7).

11 Undo and remove the two nuts and spring washers (54, 55) (Fig 6.7) securing the solenoid shield (59) to the solenoid lever housing (51) and lift away the solenoid shield.

12 Undo and remove the two bolts and spring washers (49, 50) (Fig 6.7) securing the solenoid (48) to the solenoid lever housing (51). Lift away the solenoid, carefully easing the plunger from the yoke of the valve operating lever (15).

13 Slacken the clamp bolt (16) and retaining nut (17) on the valve solenoid operating lever (15) and withdraw the lever (15) and distance collar (13) from the shaft (11).

14 Undo and remove the two bolts with plain and spring washers (56, 57, 58) (Fig 6.7) that secure the solenoid lever housing (51) to the main casing (7). Lift away the solenoid lever housing (51) followed by the spring (43) and spacer tube assembly (44).

15 If it is necessary to remove the accumulator sleeve (45) and piston assembly, a special tool is required. Refer to Fig 6.9 and insert tool number L182 into the accumulator sleeve and tighten the lower wing nut. Withdraw the accumulator sleeve and piston assembly by applying a rotary pull to the upper bolt of the tool. There is no other way of removing these parts without causing damage to their very fine surface finish.

16 The next part to be removed is the pump return valve which is positioned in the cavity in the main body casing once the solenoid bracket assembly (51) is removed. Undo and lift out the hexagonal plug (33) (Fig 6.7) and washer (34) followed by the spring (35), plunger (36) and ball bearing (37). It is important that these parts are removed before the pump is removed from the main casing.

17 Undo and remove the drain plug (63) (Fig 6.7). Lift away the filter (38) having first released the securing bolt (40) with spring and plain washers (41, 42). Recover the distance piece (39).

18 To remove the pump another special tool is required having a part number L183A/1, undo and remove the two retaining screws with spring washers (31, 32) and the base plug (30). Screw the short threaded portion of the spindle of the special tool into the pump body. Locate the adaptor in position against the casing and tighten the wing nut which will cause the body (29) to be withdrawn from the main casing. This operation may be seen in Fig 6.10. Lift out the plunger (26) and spring (27).

19 Using a pair of circlip pliers remove the circlip (8) from its groove in the forward end of the clutch hub and taking care not to damage the clutch member or friction lining, driving the clutch member from the thrust ring (1) (Fig 6.8) and bearing (7) using a soft metal drift and hammer.

20 With a pair of circlip pliers remove the large diameter circlip (9) and using a vice and piece of suitable diameter tube press the bearing (7) from the thrust ring (1).

21 Should it be necessary to remove the uni-directional clutch special assembly ring tool number L178 is necessary. Position

Fig 6.9 USE OF TOOL TO WITHDRAW ACCUMULATOR SLEEVE AND PISTON

Fig 6.10 SPECIAL PUMP REMOVAL TOOL

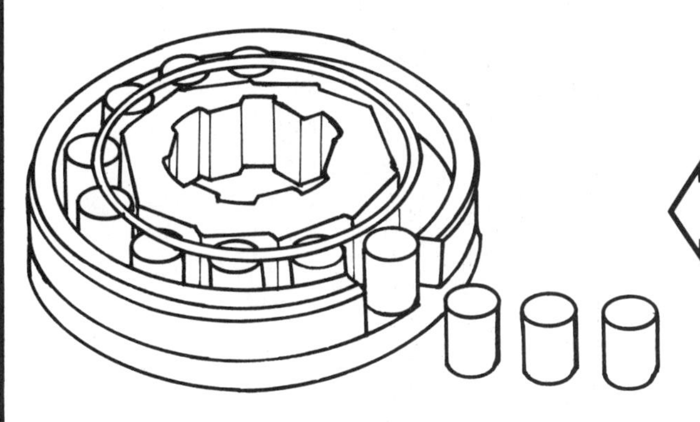

Fig 6.11 ASSEMBLY OF UNI-DIRECTIONAL CLUTCH

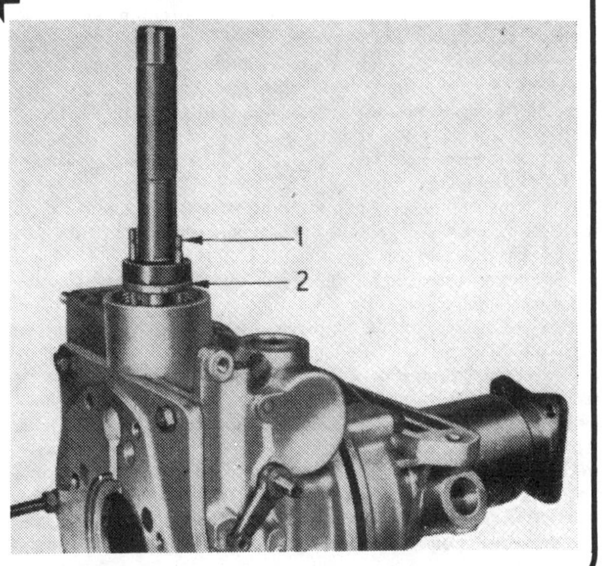

Fig 6.12 REFITTING PUMP BODY

1 Guide pins *2 Pump body*

the assembly ring over the front face of the annulus and lift the inner member of the uni-directional clutch (14) up to it. This operation is shown in Fig 6.11.

22 Then remove the assembly ring and allow the rollers to come out followed by the hub so exposing the spring (17).

23 Lift away the phosphor bronze thrust washer (18) that is fitted between the uni-directional clutch and the annulus.

24 Undo and remove the speedometer dowel screw and spring washer. Withdraw the speedometer drive bearing and pinion (28). Note the O ring (30) on the bearing centre outer circumference.

25 Extract the split pin (27) (Fig 6.8) securing the castellated nut (25) to the annulus. Undo the castellated nut and remove followed by the thick plain washer (26). Slide the coupling flange (23) from the splines on the annulus.

26 Using a press or a hammer and block of wood on the end of the annulus with the castellated nut (25) replaced to protect the threads, remove the annulus from the rear casing (22).

27 The front bearings (19) should remain in position on the annulus, but if it must be removed a suitable two legged puller or a press should be used. Note the position of the spring washer which should be located on a shoulder in front of the annulus splines.

28 If the oil seal (24) (Fig 6.8) has shown signs of leaking or the bearing (20) is to be renewed, prise the oil seal from the rear casing (22) making a note of which way round it fits. The bearing (20) may be drifted from the rear casing using a long soft metal drift.

29 Thoroughly clean all the component parts and then examine them carefully. Check that the oil pump plunger (26) (Fig 6.7) and body (29) are not worn and that the spring has not contracted (free length should not be less than 2 inches). Examine the piston rings (67) (Fig 6.7) on the operating pistons (66) and renew them if worn. Check that the cylinder bores are free from score marks and wear. Check all the ball bearings for roughness when turned or for looseness between inner and outer tracks. Examine the splines for burrs and wear, and the rollers of the uni-directional clutch for chips and flat spots.

30 Renew the clutch linings if they are burnt or worn and carefully examine the main (7) (Fig 6.7) and rear (22) (Fig 6.8) casings for cracks, split threads or other damage. Renew the steady bush if it is worn and examine the gear teeth for cracks, chips and general wear. Examine the sealing balls for ridges which will prevent them seating properly and check the free lengths of the springs, the measurements being given in the Specifications.

31 Assembly of the unit can commence after any damaged or worn parts have been exchanged and new gaskets and seals obtained.

32 The first part to be refitted is the pump assembly for which tool L184 is required to ensure accurate refitting. This is shown in Fig 6.12. Screw the two guide pegs of the tool into the holes in the pump bottom face.

33 Refit the spring (27) to the pump plunger (26) (Fig 6.7) and insert this into the pump body (29). Insert the pump assembly into the casing (7) positioning the flange of the body (29) over both guide pegs of the tool L184 and locating the flat of the pump plunger against the guide peg in the front casing adjacent to the central guide bushes.

34 Drift the pump body home using the drift, this being part of tool L184. Undo and remove the two guide pegs and fit the two retaining screws and spring washers (31, 32).

35 Refit the base plug (30). Reassemble the filter assembly (38-42) to the base plug (30). Replace the drain plug and washer (63, 64) to the underside of the main casing and tighten fully to prevent subsequent oil leaks.

36 Insert the non-return valve ball bearing (37) (Fig 6.7) into its drilling. Note that this ball bearing has a diameter of ¼ inch. Using a soft metal drift of diameter slightly less than ¼ inch, tap the ball bearing lightly so as to seat it in its drilling. Insert the plunger (36), spring (35) and plug (33) with a new copper washer (34) fitted under its head. Check that the copper washer seats on its location correctly to ensure no oil leaks.

37 Refit the piston (45) into the sleeve taking care that the piston rings (46) are not damaged. With the sleeve upright push the piston down until the wings are resting on the top of the bore. Using two thumbs compress each ring whilst a second person pushes the piston down.

38 Insert the accumulator spring (43) into the tube (44) and fit the accumulator tube (44) into the recess in the accumulator sleeve and then carefully push into the casing, easing the sealing rings (47) into the bore.

39 Fit a new O ring over the operating shaft (11) if one was originally fitted and using a new joint washer (52) refit the solenoid lever housing (51), ensuring that the accumulator spring (43) locates over a dowel in the solenoid lever housing.

40 Tighten the two screws with spring and plain washers (56, 57 58) evenly and then the two nuts with spring washers (54, 55) once the solenoid shield has been refitted onto the two short studs (53).

41 Fit the new rings (67) (Fig 6.7) to the piston (66) and lubricate with oil. Insert the pistons into their respective bores carefully easing the piston rings into the bores in the casing. Note that the centre bosses of the pistons face outwards to the front of the casing.

42 Insert the operating valve (19) (Fig 6.7) into its bore in the casing ensuring that it is the correct way up with the hemi-spherical end engaging on the flat of the small cam (14) on the operating shaft (11).

43 Drop in the ball bearing (20) (5/16 inch diameter) followed by the plunger (21) with the larger diameter innermost and the spring (22). Fit a new copper washer (24) onto the operating valve plug (23) and refit the plug. Ensure the copper washer is seating correctly on its location on the plug and tighten securely.

44 If any new parts are to be fitted to the rear casing or annulus a special tail shaft end float setting gauge will be necessary. This is number L190A and will enable the thickness of the spacing washer (21) to be determined so that the bearings (19, 20) will not have excessive end float or pre-load. This gauge is shown in

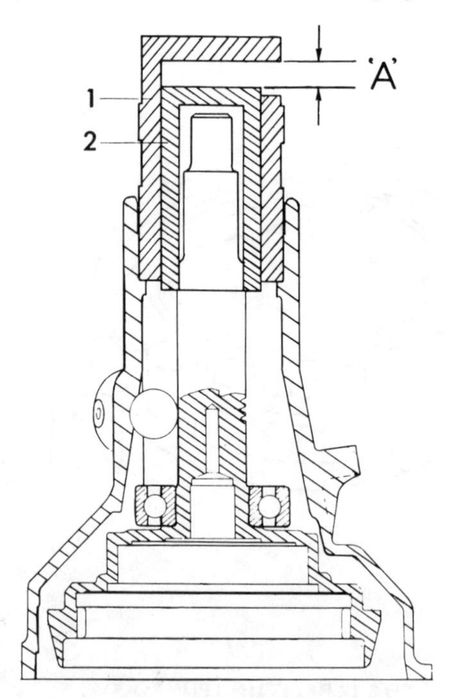

Fig 6.13 DETERMINATION OF TAILSHAFT END FLOAT USING TOOL NUMBER L190A

1 Outer member - spacing washer gauge
2 Inner member - spacing washer gauge
A Thickness of required spacer

Fig 6.13 and comprises an inner member which rests against the end of the annulus output shaft. The outer part rests on the rear bearing abutment in the rear casing.

45 Using a drift of suitable diameter to locate the outer track of the front bearing (19) insert the bearing into the rear casing until the outer track buts against a shoulder in the casing. Next, using a press or a drift of suitable diameter insert the annulus into the front bearing which has been inserted into the casing.

46 Referring to Fig 6.13 fit the gauge over the output shaft of the annulus until the outer member (1) contacts the rear bearing shoulder in the rear casing. Gently press down the inner member (2) and using feeler gauges determine dimension 'A'. Select a spacing washer (21) of the same thickness as the measurement just made using the feeler gauges. A range of washers is available in the following sizes: 0.146, 0.151, 0.156, 0.161, 0.166 inch. Remove the setting gauge.

47 Fit the previously selected washer (21) onto the annulus output shaft and using a drift of suitable diameter drive the rear bearing (20) into position in the rear casing. Also use a drift to refit the oil seal (24) ensuring that the lip is facing inwards. Lubricate the oil seal inner face.

48 Refit the rear coupling flange (23) into the splines of the annulus output shaft followed by the plain washer (26) and castellated nut (25). Tighten the nut and secure with a new split pin (27).

49 Fit a new O ring (30) to the bearing (29) and insert the speedometer pinion gear (28) into the bearing. Insert the bearing and gear assembly into the casing and to ensure correct meshing rotate the annulus. Align the holes in the casing and the bush and fit the dowel screw with a new copper washer under its head. Tighten the dowel screw securely.

50 Refit the spring (17) (Fig 6.8) into the roller cage (16) of the uni-directional clutch. Insert the inner member (14) into the cage (16) and engage it into the other end of the spring (17). Also engage the slots of the inner member with the tongues on the roller cage ensuring that the spring is able to rotate the cage, so moving the rollers (15) when they are refitted up the inclined faces of the inner member (14). The cage should be spring loaded in an anti-clockwise direction when looking at it from the front.

51 Insert this assembly into the special assembly ring, tool number L178, with its front end facing downwards and insert the rollers through the milled slot in the tool shown in Fig 6.11. It will be necessary to turn the uni-directional clutch in a clockwise direction until all rollers are in place.

52 Refit the uni-directional clutch assembly to the annulus, having first inserted the thrust washer (18) (Fig 6.8). The assembly tool will allow the rollers to enter into the annulus without falling out or jamming. If the tool is not available a strong elastic band should be wrapped around the cage and then lifted to allow each roller to be inserted.

53 Before the planet carriet is refitted, the gears must be specially set. Turn each planet gear in turn until a dot mark on one of the teeth of the large gear is positioned radially outwards as shown in Fig 6.14. Locate the phospor bronze washer in its recess in the planet carrier and insert the sunwheel.

54 Ensure that the sunwheel meshes correctly with the planet gears at the same time keeping the dot marks in their originally set position.

55 Insert the planet carrier and sunwheel assembly into the annulus.

56 Obtain a piece of metal bar the same diameter as the output shaft of the gearbox and insert it into the sunwheel until the rod engages the planet carrier and uni-directional clutch splines.

57 The end float of the sunwheel must be checked to ensure that it is within the limits of between 0.008 and 0.014 inch. To do this slide an additional thrust washer of known thickness over the previously inserted metal rod until it rests on the top of the sunwheel followed by the original phosphor bronze thrust washer and the steel thrust washer.

58 Fit the brake ring (10) (Fig 6.8) into the front casing and using a soft faced hammer tap it firmly into position. Carefully slide the front casing over the metal rod and position it up to the rear casing assembly.

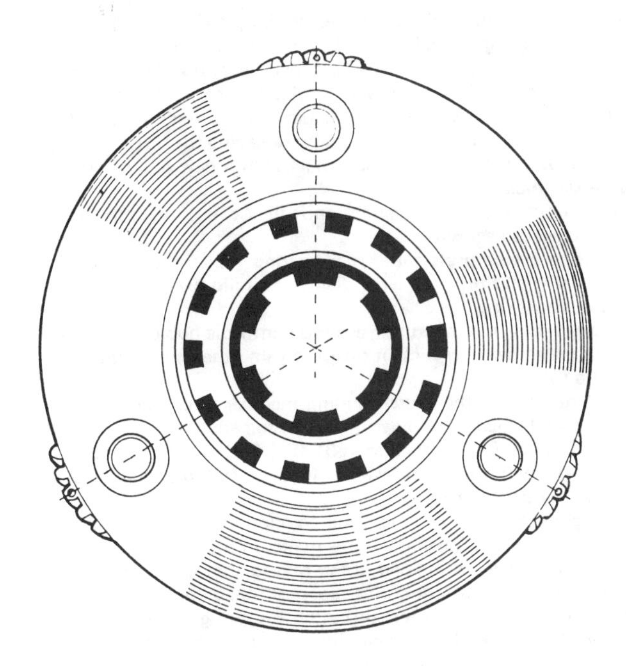

Fig 6.14 BEFORE ASSEMBLING THE PLANET CARRIER TO THE SUNWHEEL, ROTATE THE PLANET WHEELS UNTIL THE PUNCH MARKS ARE IN THE POSITION SHOWN

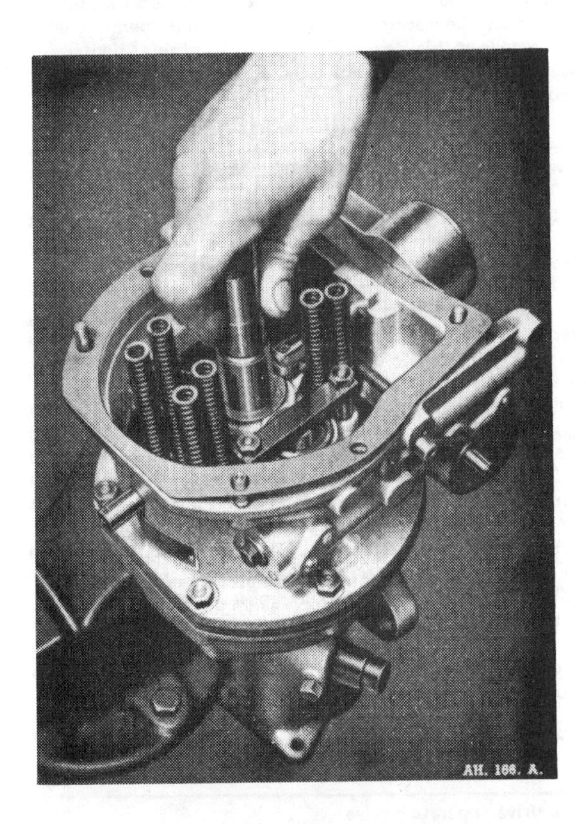

Fig 6.15 IF A SPARE MAINSHAFT IS AVAILABLE, OR USING A PIECE OF SUITABLE ROD, CENTRALISE THE GEARS

59 As an additional thrust washer has been fitted a gap between the two casings should now be evident. Using feeler gauges measure this gap. The thickness of the extra thrust washer MINUS the end float of the sunwheel is the required dimension.
60 If this indicated end float is outside the limit specified in paragraph 57, it must be adjusted by replacing the STEEL thrust washer at the front of the sunwheel with a new one of greater or lesser thickness.
61 Remove the front casing and thrust washers previously positioned as necessary.
62 Using a vice press the thrust bearing (7) (Fig 6.8) into the thrust ring and refit the large diameter circlip (9) into its groove in the thrust ring.
63 Next press the thrust ring assembly into the hub of the clutch sliding member and lock in position using the smaller diameter circlip (8).
64 Carefully fit the clutch sliding member over the sunwheel splines and engage the inner linings of the annulus assembly. Fit the phosphor bronze washer (60) (Fig 6.7) on the top of the sunwheel and the steel washer (61) of suitable thickness as previously determined.
65 Lightly smear a little jointing compound onto both sides of the brake ring flange and tap this home on the main casing.
66 Fit the main casing and brake ring to the rear casing taking care that the thrust ring pins are positioned through the four corresponding holes in the main casing. Fit spring washers onto the six studs followed by the retaining nuts and tighten securely in a diagonal manner.
67 Replace the two operating piston bridge pieces (2) (Fig 6.8) and secure on the studs using new tab washers (4) and nuts (3). Bend over the locking tabs. Slide on the distance collar (13) (Fig 6.7) onto the lever shaft (11).
68 Refit the operating lever (15) to the lever shaft (11) and fully tighten the locknut (17).
69 Fit the solenoid to its mounting on the solenoid lever housing and secure with the two setscrews and spring washers (49, 50).
70 It will now be necessary to set the solenoid operating lever and full details of this are given in Section 16 of this Chapter.
71 Reassembly is now complete. Do not forget to refill the gearbox and overdrive unit with oil once it has been refitted.

16 Overdrive unit - operating lever adjustment

1 If the overdrive does not engage, or will not release when it is switched out, providing the solenoid is not at fault, the trouble is likely to be that the operating lever is out of adjustment. Adjustment can be made without removing the overdrive.
2 To one end of a shaft passing through the overdrive casing is secured a setting lever having a 3/16 inch hole in its outer end as shown in Fig 6.16. The other end of the shaft is attached to a solenoid lever as shown in Fig 6.17.
3 Switch on the ignition and set the overdrive switch to energise the solenoid. The hole in the setting lever should align with a similar hole in the casing which will indicate that the operating valve is fully open. To check this try and insert a 3/16 inch diameter rod through both holes (Fig 6.16). If it is not possible adjustment is necessary. Switch off the ignition.
4 Undo the six solenoid bracket cover securing bolts and spring washers (later type). These parts are shown in Fig 6.22 (items 35, 36). Lift away the cover (37) and joint washer (80).
5 Slacken the clamp nut and bolt and adjust the lever until the plunger bolt is resting on the rubber stop whilst the rod is locking the setting lever. There must be between 0.008 and 0.010 inch end float on the cross shaft.
6 Later models. Refit the solenoid bracket cover and joint washer and secure with the six bolts and spring washers.

17 Overdrive - operating valve

Should the overdrive unit not function correctly and the fault be diagnosed from Section 21 that the operating valve be at fault it may be removed and checked as follows.
1 Refer to Section 3, paragraphs 4 to 6 inclusive and remove the propeller shaft tunnel.
2 Switch on the ignition but do not start the engine. Activate the overdrive control switch several times so as to operate the solenoid thus releasing any residual oil pressure. Wipe the area around the valve plug free of dust.
3 Unscrew and remove the operating valve plug (Fig 6.18) and copper washer.
4 Using a paper clip which has been straightened and the end bent to a small hook, withdraw the spring. The plunger may be removed using a small magnet or magnetised screwdriver. Also remove the ball bearing.
5 Using the other end of the paper clip with a slight kink in it, carefully insert it into the centre of the valve and withdraw the valve.
6 Clean the removed parts in petrol and allow to dry. Locate the small drilling near to the base of the valve and check that it is free of dirt.
7 Inspect the ball bearing for signs of pitting which, if evident, indicates that a new ball bearing should be obtained. It has a diameter of 5/16 inch.
8 If the ball bearing is satisfactory reseat it by placing the ball bearing in a block of soft wood. Invert the valve, place on top of the ball bearing and lightly tap the end. If it is tapped too hard the drilling in the side of the valve or in the end may be closed.
9 Reassembling the valve is the reverse sequence to removal.

18 Overdrive - pump non-return valve

1 Access to the non-return valve located in the bottom of the overdrive is simply gained. First drain the oil from the gearbox and overdrive.
2 Remove the solenoid from the side of the overdrive unit.
3 Slacken off the clamping bolt in the operating lever and remove the lever complete with the solenoid plunger.
4 Lift away the distance collar from the valve operating shaft.
5 Unscrew the two nuts from the studs securing the solenoid bracket to the overdrive unit body. Lift away the two nuts and spring washers.
6 Unscrew the two bolts identified with red paint marks. Do not remove the bolts before the nuts (paragraph 5) are removed.
7 As the two bolts are slackened the tension on the accumulator spring will be released.
8 Lift away the solenoid bracket.
9 Unscrew the valve cap (Fig 6.19) and lift out the spring, plunger and ball.
10 Clean the valve ball and seat with a non-fluffy rag. Reseat the ball by tapping it on its seat with a light hammer and drift.
11 Reassembly is the reverse sequence to removal. Make sure that the soft copper washer located between the valve cap and pump housing is nipped up tightly to prevent subsequent oil leaks.
12 Refer to Section 16 and reset the valve operating lever.

19 Throttle switch - adjustment

It is important that the throttle switch is correctly adjusted as otherwise the overdrive will disengage when the car slows down with the throttle closed. This will be accompanied by a noticeable braking effect.

Normally the switch will require adjustment only when the carburettor or accelerator controls have been reset. To check and adjust the switch setting proceed as follows.
1 Connect a 12 volt 2.2 watt test lamp across terminal 'A' (Fig 6.20) and a convenient earth point on the car.
2 It should be observed that the bulb lights when the overdrive and the ignition are both switched on and the gear lever is set in the third or top gear position.
3 When the overdrive is switched off the bulb should remain alight with the throttle still closed.

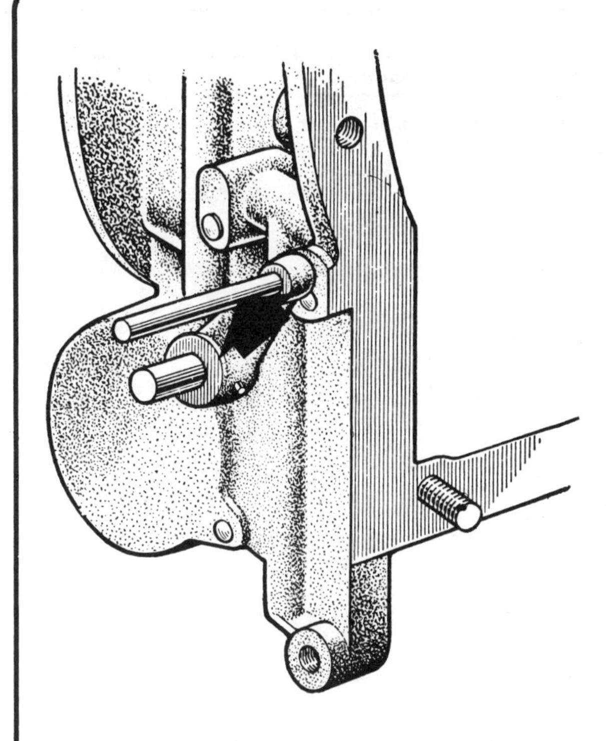

Fig 6.16 VALVE SETTING LEVER (ARROWED)

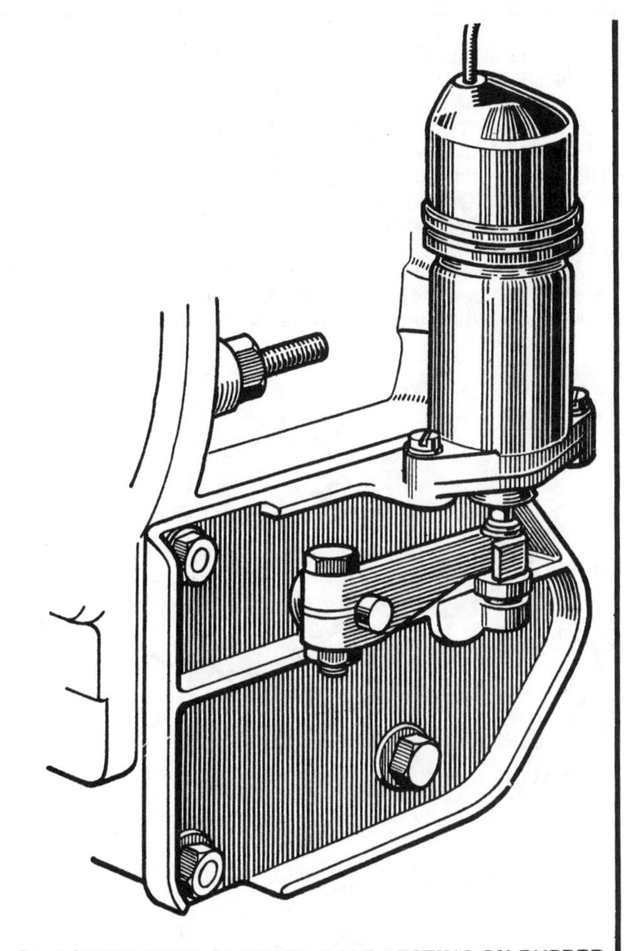

Fig 6.17 SOLENOID PLUNGER BOLT RESTING ON RUBBER STOP

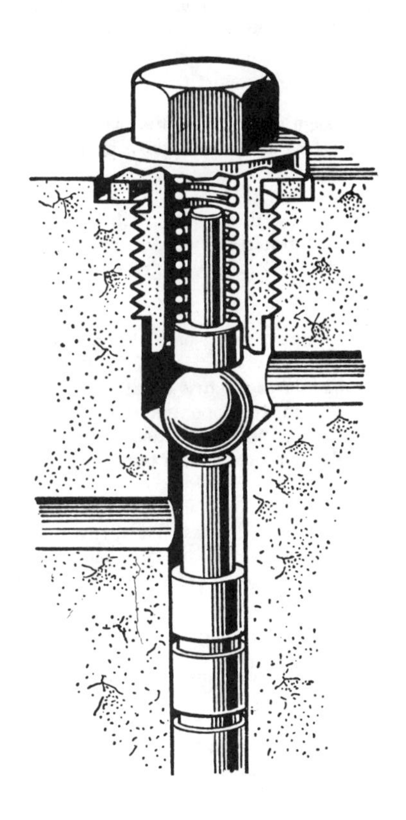

Fig 6.18 CROSS SECTIONAL VIEW THROUGH OPERATING VALVE

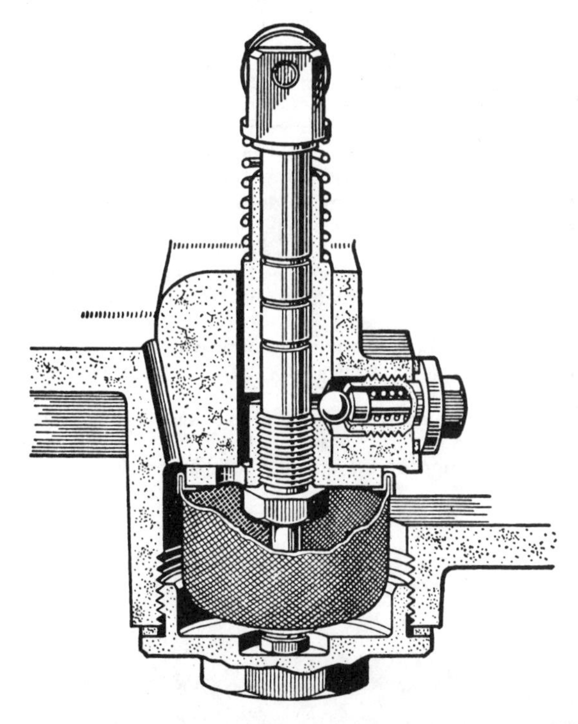

Fig 6.19 CROSS SECTIONAL VIEW THROUGH PUMP AND NON-RETURN VALVE

4 Slowly open the throttle by depressing the accelerator pedal until the test light bulb goes out. At this point check the position of the throttle opening. It should be open approximately 1/5 of its full movement.
5 To test this position it should be just possible to pass a 3/16 inch diameter rod between the throttle stop screw and the stop lever (HD type carburettor) or a 0.048 inch thick feeler gauge between the throttle screw stop and the stop on the H4 type carburettor as found on the earlier engines.
6 To adjust the switch slacken the lever clamping bolt 'C' (Fig 6.20) and turn the slotted end of the switch operating shaft 'D' with a screwdriver until the required setting is obtained.

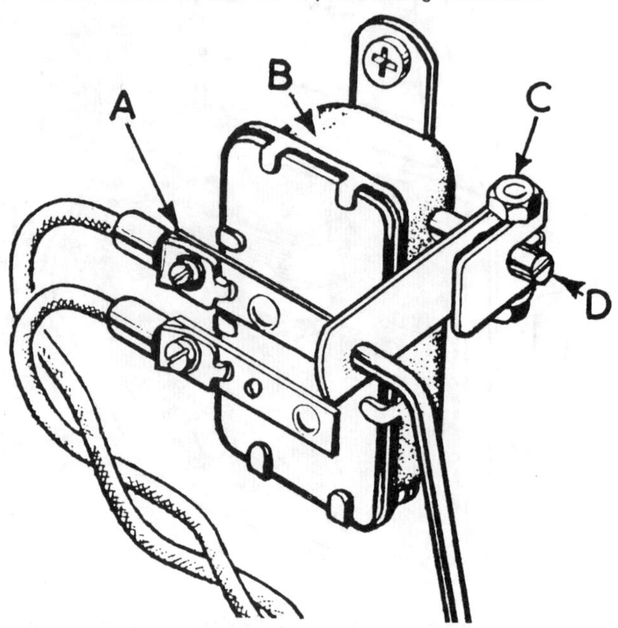

Fig 6.20 OVERDRIVE THROTTLE SWITCH

A Switch terminal *C Lever clamping bolt*
B Switch body *D Operating shaft*

20 Overdrive - modified type

The overdrive unit fitted to Mk I and II (series BN7 and BT7) and also Mk II and III (series BJ7 and BJ8) had a modified unit fitted. It is basically the same in design to the earlier type but with the below listed modifications. An illustration of this unit is shown in Fig 6.22.
1 The filter has been redesigned and is retained in the body by a boss on the inside of the drain plug. The filter is accessible through the drain plug hole.
2 New design operating pistons are fitted with synthetic rubber sealing rings and the accumulator piston with a three-piece cast iron rings.
3 The pinions are now fitted with needle roller bearings instead of 'Clevite' bushes.
4 The outer ring of the uni-directional clutch is no longer riveted.
5 A redesigned solenoid bracket and adaptor plate have been fitted.
6 An additional selective washer of 0.160 $\pm$ .0005 inch has been added to the range.

21 Overdrive - fault finding

When applicable refer to the wiring diagram (Fig 6.21). Switch ignition on.

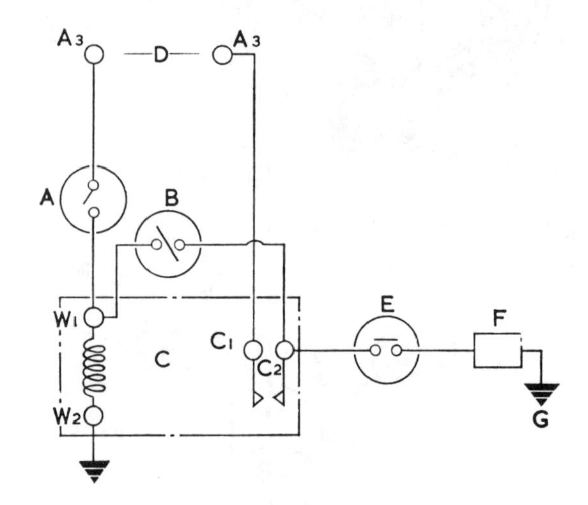

Fig 6.21 OVERDRIVE ELECTRICAL CIRCUIT

1 Relay coil
Quickly bridge relay terminal W1 and A3 and the relay should be heard to operate. If no sound occurs the relay 'C' is defective.

2 Toggle switch
Operate the toggle switch and the relay should be heard to operate. If no sound is heard the switch is defective.

3 Relay contacts
Engage top gear, close the toggle switch and open the throttle switch. The solenoid should be heard to operate. If no sound is heard test solenoid. If satisfactory the relay is defective.

4 Solenoid
With engine stationary select neutral and switch on ignition, disconnect the solenoid connection (F). Using a jumper lead quickly connect the solenoid to fuse unit supply terminal A3. The solenoid should be heard to operate. If no sound is heard the solenoid is defective or incorrectly adjusted to the operating linkage. Remove the electrical connections.

5 Gear switch
Engage top gear and depress the throttle pedal, quickly connect the relay terminal C2 to terminal A3. The solenoid should be heard to operate. If no sound is heard the switch is probably defective.

6 Throttle switch
Engage top gear and close the toggle switch. Open the toggle switch and slowly depress the accelerator pedal. The solenoid should be energised from zero to one fifth of the throttle. If the solenoid is heard to release under one fifth of the throttle the switch setting must be checked.

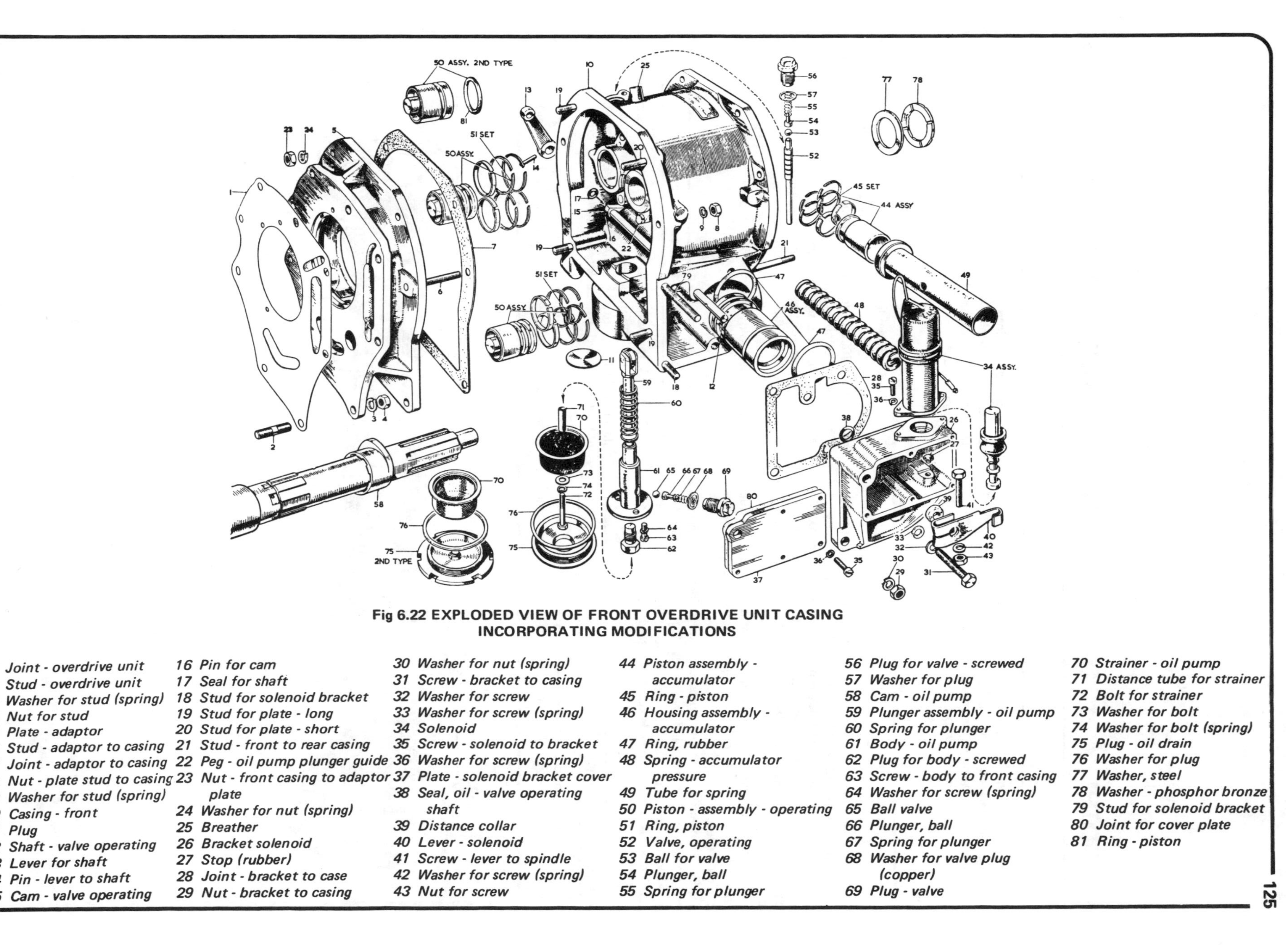

Fig 6.22 EXPLODED VIEW OF FRONT OVERDRIVE UNIT CASING INCORPORATING MODIFICATIONS

1 Joint - overdrive unit
2 Stud - overdrive unit
3 Washer for stud (spring)
4 Nut for stud
5 Plate - adaptor
6 Stud - adaptor to casing
7 Joint - adaptor to casing
8 Nut - plate stud to casing
9 Washer for stud (spring)
10 Casing - front
11 Plug
12 Shaft - valve operating
13 Lever for shaft
14 Pin - lever to shaft
15 Cam - valve operating
16 Pin for cam
17 Seal for shaft
18 Stud for solenoid bracket
19 Stud for plate - long
20 Stud for plate - short
21 Stud - front to rear casing
22 Peg - oil pump plunger guide
23 Nut - front casing to adaptor plate
24 Washer for nut (spring)
25 Breather
26 Bracket solenoid
27 Stop (rubber)
28 Joint - bracket to case
29 Nut - bracket to casing
30 Washer for nut (spring)
31 Screw - bracket to casing
32 Washer for screw
33 Washer for screw (spring)
34 Solenoid
35 Screw - solenoid to bracket
36 Washer for screw (spring)
37 Plate - solenoid bracket cover
38 Seal, oil - valve operating shaft
39 Distance collar
40 Lever - solenoid
41 Screw - lever to spindle
42 Washer for screw (spring)
43 Nut for screw
44 Piston assembly - accumulator
45 Ring - piston
46 Housing assembly - accumulator
47 Ring, rubber
48 Spring - accumulator pressure
49 Tube for spring
50 Piston - assembly - operating
51 Ring, piston
52 Valve, operating
53 Ball for valve
54 Plunger, ball
55 Spring for plunger
56 Plug for valve - screwed
57 Washer for plug
58 Cam - oil pump
59 Plunger assembly - oil pump
60 Spring for plunger
61 Body - oil pump
62 Plug for body - screwed
63 Screw - body to front casing
64 Washer for screw (spring)
65 Ball valve
66 Plunger, ball
67 Spring for plunger
68 Washer for valve plug (copper)
69 Plug - valve
70 Strainer - oil pump
71 Distance tube for strainer
72 Bolt for strainer
73 Washer for bolt
74 Washer for bolt (spring)
75 Plug - oil drain
76 Washer for plug
77 Washer, steel
78 Washer - phosphor bronze
79 Stud for solenoid bracket
80 Joint for cover plate
81 Ring - piston

Chapter 7 Propeller shaft and universal joints

Contents

1 General description

The drive from the gearbox to the rear axle is via the propeller shaft which is in fact a tube. Due to the variety of angles caused by the up and down movement of the rear axle in relation to the gearbox, universal joints are fitted to each end of the shaft to convey the drive through the constantly varying angles.

To accommodate fore and aft movement of the rear axle due to road spring deflection, a sliding joint of the telescopic yoke design is used.

Each universal joint comprises a centre spider, four needle roller bearing assemblies and two yokes. A grease nipple is fitted to each universal joint so that the needle roller bearings can be lubricated. A further nipple is also fitted to the telescopic yoke.

2 Routine maintenance

1 A lubrication nipple is fitted to each front and rear universal joint journal. There is a further nipple fitted to the telescopic yoke.

2 It is recommended that these nipples be lubricated with Castrol LM Grease or a similar recommended multi-purpose grease every 3000 miles. Four strokes of the grease gun to each nipple should be sufficient.

3 If it is obvious that the grease is being passed out from the seals, the joints must be dismantled and new seals fitted.

4 At the same time as lubricating the nipples, or at every major car service, a check should be made for signs of slackness in the universal joint bearings or at the flange bolts.

3 Propeller shaft - testing for wear whilst on car

1 To check for wear grasp each side of the universal joint and with a twisting action determine whether there is any play or slackness in the joint. This will indicate wear in the bearings. Do not be confused by backlash between the crownwheel and pinion.

2 Try an up and down rocking movement which will indicate wear of the thrust faces of the spiders and those inside the cups.

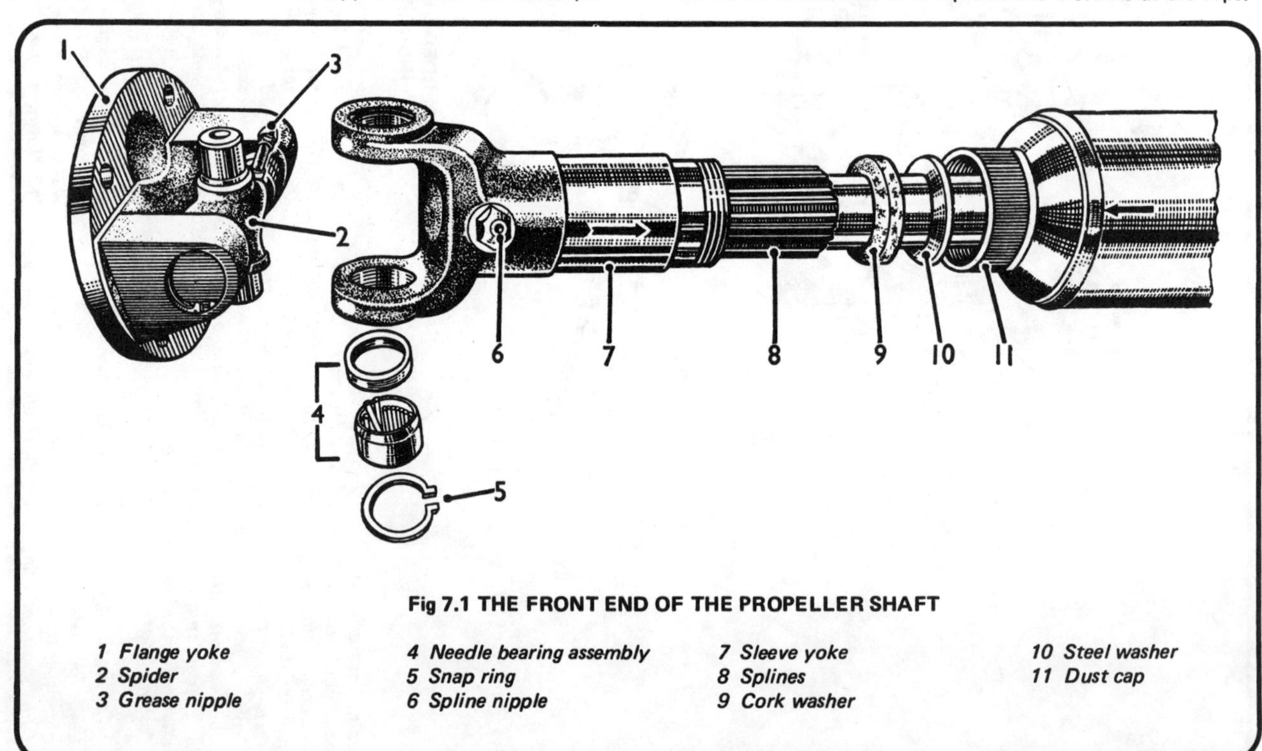

Fig 7.1 THE FRONT END OF THE PROPELLER SHAFT

1 Flange yoke
2 Spider
3 Grease nipple
4 Needle bearing assembly
5 Snap ring
6 Spline nipple
7 Sleeve yoke
8 Splines
9 Cork washer
10 Steel washer
11 Dust cap

Fig 7.2 TAP THE YOKE LIGHTLY AS SHOWN AFTER REMOVING THE RETAINING CIRCLIP TO FREE THE BEARING

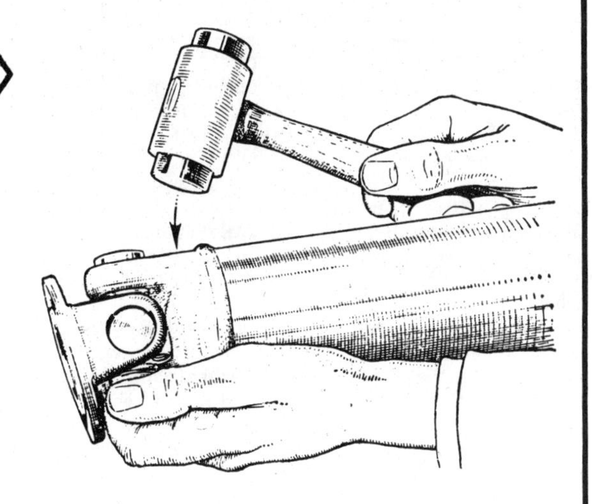

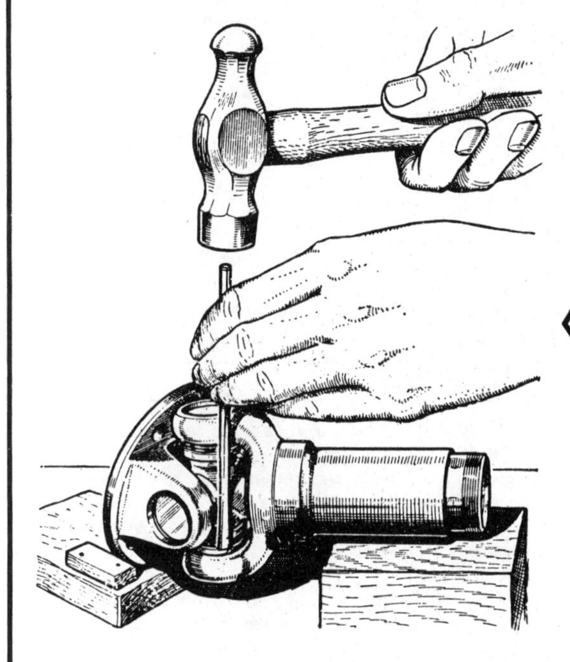

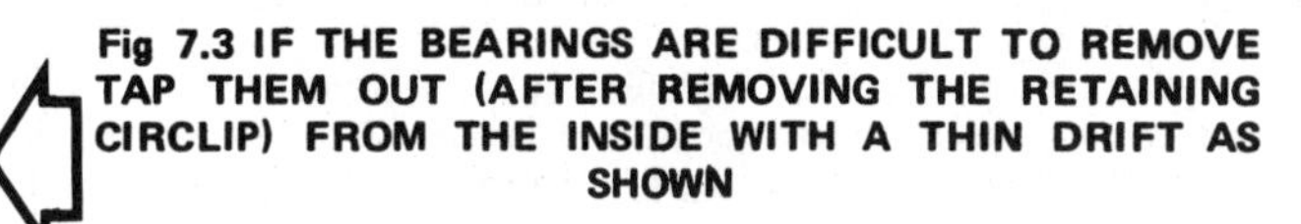

Fig 7.3 IF THE BEARINGS ARE DIFFICULT TO REMOVE TAP THEM OUT (AFTER REMOVING THE RETAINING CIRCLIP) FROM THE INSIDE WITH A THIN DRIFT AS SHOWN

Fig 7.4 BEARING CUP REMOVAL

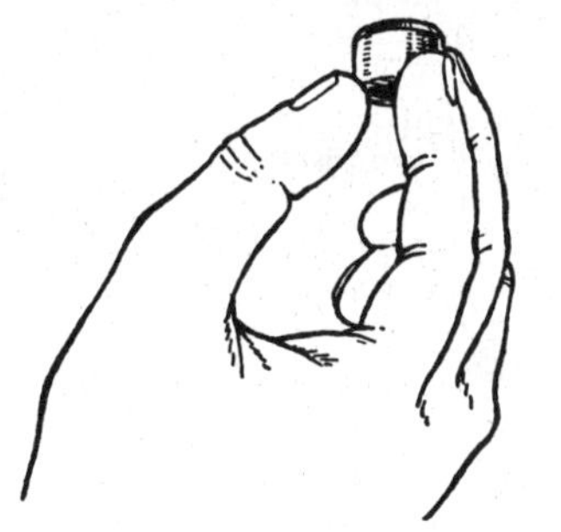

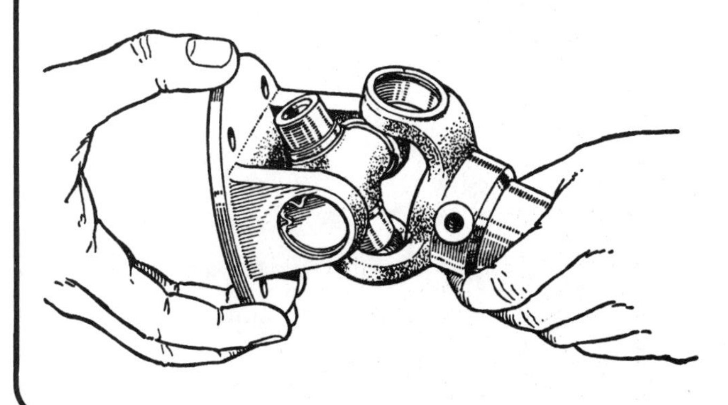

Fig 7.5 UNIVERSAL JOINT SEPARATION

4 Propeller shaft - removal, inspection and replacement

1 With a scriber or file mark the propeller shaft coupling flanges front and rear so that they may be refitted in their original position. Note that the sliding yoke is fitted to the gearbox.
2 Undo and remove the nuts and bolts that secure the propeller shaft to the rear axle flange. Carefully lower the rear of the propeller shaft to the ground.
3 Undo and remove the nuts and bolts that secure the propeller shaft to the rear of the gearbox/overdrive.
4 Carefully remove the propeller shaft from the underside of the car.
5 If the propeller shaft is removed for inspection, first examine the bore and the counter-bore of the two flanges which mate at the front and the rear of the propeller shaft and corresponding flanges at the rear of the gearbox and forward end of the rear axle. If they are damaged in any way or a slack fit, it would mean that the propeller shaft is running off centre at the flange and causing vibration in the drive.
6 If there is nothing obviously wrong and the universal joints are in good order, it is permissible to reconnect the flanges with one turn through 180° relative to the other. This may stop any vibration previously experienced.
7 To refit the propeller shaft, carefully wipe the faces of the flanges clean and position the propeller shaft under the car. Make sure that the flange registers engage correctly and that the components are refitted in exactly the same position as were found on removal except in the condition described in the previous paragraph.
8 Insert the four bolts, refit the four nuts and tighten securely, to secure each universal joint flange in position. The four bolts should be fitted with the heads towards the universal joint.

5 Universal joints - inspection and repair

1 Wear in the needle roller bearings is characterised by vibration in the transmission, 'clonks' on taking up the drive and, in extreme cases lack of lubrication, metallic squeaking, and ultimately grating and shrieking sounds as the bearings break up.
2 It is easy to check if the needle roller bearings are worn with the propeller shaft in position, by trying to turn the shaft with one hand, the other hand holding the rear axle flange when the rear universal joint is being checked and the front gearbox/overdrive unit coupling when the front universal is being checked. Any movement between the propeller shaft and the front and rear half couplings is indicative of considerable wear.
3 If worn, the old bearings and spiders will have to be discarded and a repair kit, comprising new universal joint spiders, bearings, oil seals and retainers, purchased. Check also by trying to lift the shaft and noticing any vertical movement in the joints.
4 Examine the propeller shaft splines for wear. If worn it will be necessary to purchase a new front half coupling or, if the yokes are badly worn, an exchange propeller shaft.
5 It is not possible to fit oversize bearings and journals to the trunnion bearing holes.
6 Examine the propeller shaft telescopic yoke splines for wear. To do this unscrew the dust cap from the sleeve, and then slide the sleeve from the shaft. Take off the steel washer and cork washer. With the sleeve separated from the shaft assembly, the splines can be inspected. If worn it will be necessary to purchase a new front sleeve assembly, or if the yokes are badly worn, an exchange propeller shaft.

6 Universal joints - dismantling

1 Clean away all traces of dirt and grease from the circlips located on the end of the spiders and remove the clips by pressing their open ends together with a pair of pliers and levering them out with a screwdriver. Note: If they are difficult to remove, tap the bearing face resting on the top of the spider with a soft faced hammer which will ease the pressure on the circlip.
2 Hold the propeller shaft in one hand and remove the bearing cups and needle rollers by tapping the yoke at each bearing with a soft faced hammer. As soon as the bearings start to emerge they can be drawn out with the fingers. If the bearing cup refuses to move them, place a thin bar against the inside of the bearing and tap it gently until the cup starts to emerge.
3 With the bearings removed it is relatively easy to extract the spiders from their yokes. If the bearings and spider journals are thought to be badly worn this can be easily ascertained by visual inspection once the universal joints have been dismantled.

7 Universal joints - reassembly

1 Thoroughly clean out the yokes and journal. Make certain that the grease passages are quite clear.
2 Fit new cork oil seals and retainers on the spider journals, place the spider on the propeller shaft yoke and assemble the needle rollers in the bearing races with the assistance of some thin grease. Note: It is essential to fit the spiders in the yoke flanges so the lubricating nipples are facing the propeller shaft and not the yoke flanges. If fitted the wrong way round it will be impossible to lubricate the universal joints.
3 Refit the bearing cups on the spider and tap the bearing home so that they lie squarely in position.
4 Replace the circlips and lubricate the bearings well with Castrol LM Grease.

8 Telescopic yoke - removal and refitting

1 With the propeller shaft away from the car and the exterior cleaned, unscrew the dust cap at the rear of the sleeve yoke, pull the yoke complete with the universal joint and flange yoke from the splined stub of the propeller shaft.
2 The dust cap may be released by removing the cork washer and steel washer from the side of the cap.
3 When refitting make sure that the yoke is refitted with the lugs of the front and rear universal joint in line with each other.
4 It will be observed that there is an arrow stamped on both the sliding yoke and end of the propeller shaft. These must line up on reassembly so as to ensure that the joined yokes are in their correct relative position.

Chapter 8 Rear axle

Contents

Specifications

Models fitted with 2639 cc engines	Three quarter floating hypoid gears
Ratio	
Standard	3.91 : 1
With overdrive	4.1 : 1
Final drive	Hypoid bevel
Teeth on crownwheel	
Standard	43
With overdrive	41
Teeth on pinion	
Standard	11
With overdrive	10
Crownwheel/pinion backlash	Marked on crownwheel
Adjustment	Shims
Oil capacity	3 pints

Models fitted with 2912 cc engines

As above but with the following exceptions:

Ratio	
Standard	3.545 : 1
With overdrive	3.909 : 1
Teeth on crownwheel	
Standard	39
With overdrive	43
Teeth on pinion	
Standard	11
With overdrive	11

Torque wrench settings

Differential bearing cap nuts	67 lb ft (8.99 kg m)
Crownwheel bolts	57 lb ft (7.83 kg m)
Pinion bearing nut	140 lb ft (19.4 kg m)
Wheel nuts (pressed steel wheels)	60-63 lb ft (8.3 - 8.64 kg m)

1 General description

The rear axle is of the three quarter floating type and is held in place by semi-elliptic springs which are constructed from a number of individual leaves of different lengths and are held together by a long bolt and clip. The semi-elliptic springs provide all the necessary lateral and longitudinal location of the rear axle. The rear axle incorporates a hypoid crownwheel and pinion and a two pinion differential. All repairs can be carried out to the component parts of the rear axle without removing the axle casing from the car.

The crownwheel and pinion, together with the differential gears, are mounted in the differential unit which is bolted to the front face of the banjo type axle casing.

Adjustments are provided for the crownwheel and pinion

backlash, pinion depth of mesh, pinion shaft bearing pre-load and backlash between the differential gears. All these adjustments may be made by varying the thickness of the various shims and thrust washers.

The axle or half shafts are easily withdrawn and are splined at their inner ends to fit into the splines of the differential wheel. The inner wheel bearing races are mounted on the outer ends of the axle casing and are secured by large nuts and lockwashers. The rear wheel bearing outer races are located in the hub.

2 Routine maintenance

1 Every 3000 miles remove the filler plug in the rear axle and top up the oil level with Castrol Hypoy. After topping up, do not replace the plug for approximately five minutes to allow any excess oil to run out. If the rear axle is accidentally overfilled it is likely that oil will leak out of the ends of the axle casing and contaminate the rear brake linings.
2 Every 6000 miles drain the oil when hot. Clean the drain plug and surrounding area and refill the axle with 3 pints of Castrol Hypoy.
3 Every 3000 miles lubricate the rear spring shackle pins with Castrol LM Grease. There are two grease nipples for this purpose. At the same time it is recommended that the spring seat bolts are checked to make sure that they are completely tight.

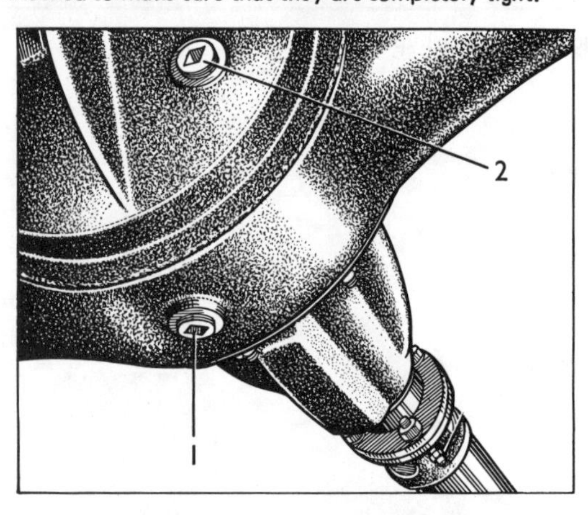

Fig 8.1 REAR AXLE DRAIN AND FILLER PLUGS

1 Drain plug *2 Filler plug*

3 Rear axle - removal and replacement

1 Remove the rear wheel trims and slacken the road wheel nuts or hub caps (wire wheels). Chock the front wheels, jack up the rear of the car and place on axle stands located under the frame members just forward of the rear spring front anchorage. Remove the two rear wheels.
2 Support the weight of the rear axle by placing the saddle of a garage hydraulic jack under the centre of the rear axle.
3 With a scriber or file, mark the relative positions of the propeller shaft and pinion driving flanges so that they may be refitted together in their original positions.
4 Undo and remove the four self-locking nuts and bolts securing the propeller shaft and pinion driving flanges. Carefully lower the propeller shaft to the ground.
5 Detach the handbrake cable from the rear axle by unscrewing it from its link to the brake balance lever. Also unscrew the nut holding the outer casing to the rear axle.
6 Wipe the union free of dust and disconnect the brake hydraulic pipe from the flexible pipe at the union just forward of the right hand shock absorber. Plug the end to stop hydraulic fluid syphoning out, or dirt ingress.
7 Unscrew the nuts that secure the shock absorber links to the axle mounting bracket. At this stage do not attempt to remove the links as this operation will be easier when the axle has been freed.
8 Undo and remove the self-locking nuts from the rear axle U bolt securing the rear axle to the spring. Note that there is a fibre pad situated between the axle and the spring.
9 Unscrew the tie bar securing nuts at the rear axle anchorage.
10 Detach the shock absorber connecting links from the axle mounting bracket.
11 Remove the rubber block that is fixed between the axle and the left hand chassis frame. It should be noted that it is not necessary to detach the corresponding block on the right hand side chassis frame.
12 The rear axle may now be removed from the right hand side of the car. Take care that the surrounding components are not damaged, particularly the petrol pump.
13 To refit the rear axle is the reverse sequence to removal. It is recommended that the springs are jacked up to meet the axle so as to locate the spring centre bolt correctly. Do not forget to refit the fibre pads.
14 The two marks previously made on the propeller shaft pinion flanges should be correctly aligned.
15 It will be necessary to bleed the brake hydraulic system as detailed in Section 15 of Chapter 9.

4 Axle shaft - removal and refitting

1 Chock the front wheels, remove the rear wheel trim and slacken the road wheel nuts or hub cap (wire wheels). Jack up the rear of the car and place on stands located between the road spring as near as possible to the axle casing. Remove the road wheel.
2 Release the handbrake, and slacken off the brake adjustment on the side on which the axle shaft is to be removed.
3 If wire wheels are fitted, remove the five self-locking nuts which secure the rear hub extension, so as to gain access to the two drum securing screws.
4 Undo and remove the two countersunk screws that secure the drum to the axle shaft and hub. If these are a little tight they may be lightly tapped with a hammer and impact screwdriver.
5 With a soft faced hammer, tap the circumference of the brake drum so releasing it from the brake shoes, and lift away the brake drum.
6 Unscrew the one countersunk screw in the axle shaft driving flange, located as shown in Fig 8.2.
7 The axle shaft may be separated from the hub either by gripping the flange and pulling over the studs or alternatively, by easing the blade of a screwdriver between the flange and the hub. If the latter method is used, the paper washer will be damaged and a new one will have to be fitted during reassembly.
8 Refitting the axle shaft is the reverse sequence to removal but the two following additional points should be noted if work has been carried out on the hub:
9 Make sure that the rubber oil seal located in the groove of the machined face of the bearing housing is in good condition. If there have been signs of oil seepage the seal must be renewed.
10 The bearing spacer must be in position when the actual shaft is being offered up into its final position.

5 Rear hub - removal and refitting

1 Refer to Section 4 and remove the brake drum and axle shaft.
2 If the bearing spacer has not been removed, it should now be done.
3 With a flat chisel, carefully knock back the tab of the locking washer and with a large box spanner undo and remove the rear hub nut.
4 Should neither box spanner be available, remove the rear brake shoes as described in Chapter 9 and undo the nut with a large adjustable spanner. DO NOT undo the nut by drifting it

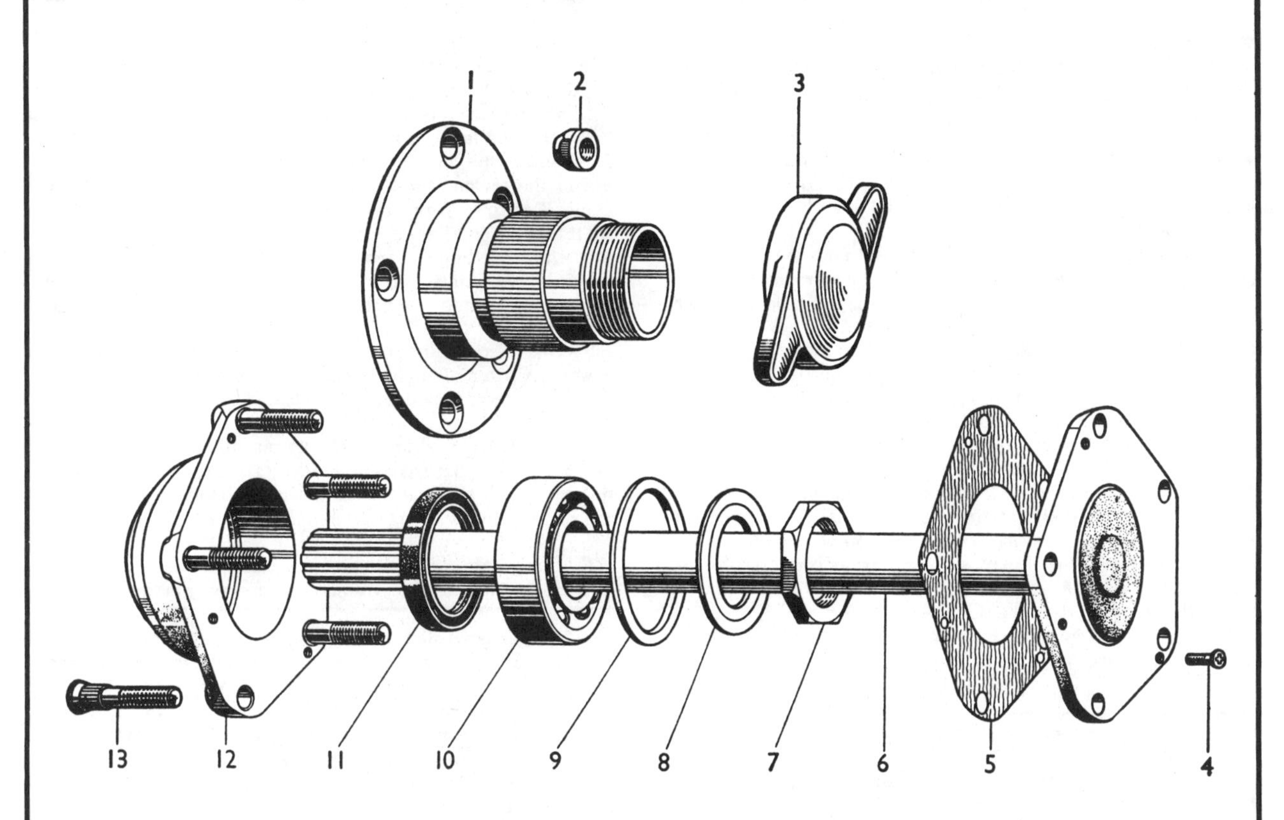

Fig 8.2 REAR HUB AND AXLE SHAFT (WITH WIRE WHEEL HUB EXTENSION)

1 Hub extension
2 Securing nut
3 Hub cap
4 Securing screw
5 Joint washer
6 Half shaft
7 Hub locknut
8 Hub lockwasher
9 Bearing spacer
10 Hub bearing
11 Oil seal
12 Hub casing
13 Hub extension stud

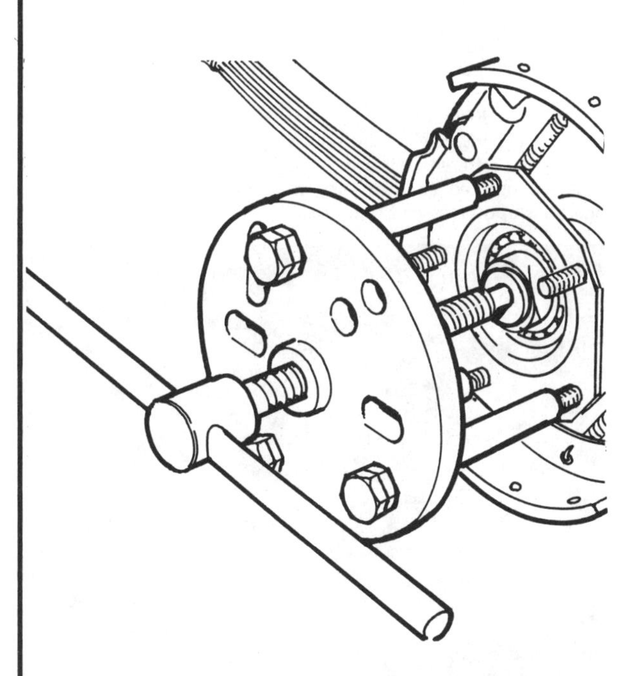

Fig 8.3 USE OF A PULLER TO REMOVE REAR HUB FROM AXLE CASING

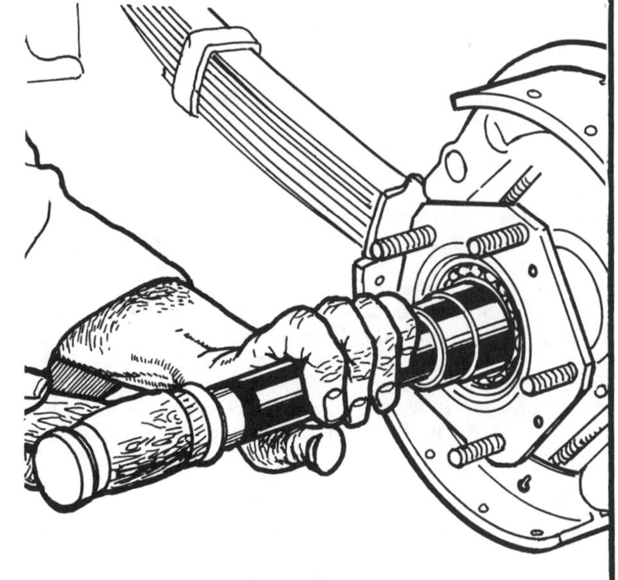

Fig 8.4 REFITTING REAR HUB USING A TUBULAR DRIFT LOCATED IN THE INNER RACE TRACK AND A SOFT FACED HAMMER

round with a chisel so as to release it.
5 Next tilt the lockwasher so as to disengage the key from the slot in the threaded section in the axle casing and lift away the washer.
6 It will now be desirable to use special BLMC tool having a part number of 18G220 with adaptors 'A', 'D' and 'E'. If this tool is not available a universal three legged puller and large diameter bolt, to be used as a thrust pad, may be used as this special tool (Fig 8.3).
7 The bearing and oil seal will be removed with the hub and if necessary may be drifted from the hub using a soft metal drift.
8 Inspect the bearing for signs of wear, by holding the inner track and rocking the outer track from side to side. Slowly turn the outer track relative to the inner track to ascertain if there are signs of stiffness or uneven movement. Finally inspect the bearing race for signs of overheating. If the race is suspect it is recommended that a new one be obtained and fitted.
9 Whenever the oil seal is removed, a new seal must be refitted, as during removal it will have become distorted.
10 Reassembly and refitting is the reverse sequence to removal, but it should be noted that the bearing spacer must protrude between 0.001 and 0.004 inch beyond the outer face of the hub and paper washer. This is to make sure that the bearing is firmly gripped between the abutment shoulder in the hub and the driving flange of the rear axle (Fig 8.4).
11 Top up the oil level in the rear axle, as some oil may have been lost by seepage once the hub was removed.

6 Pinion oil seal - removal and replacement

If oil is leaking from the front of the differential nose piece it will be necessary to renew the pinion oil seal. If a pit is not available, chock the front wheels, jack up the rear of the car and place on firmly based axle stands.
1 With a scriber, or file, mark the propeller shaft and pinion drive flanges to ensure that they are refitted in their original position.
2 Unscrew the nuts from the four bolts holding the flanges together, remove the bolts and separate the flanges. Note that the heads of the bolts face towards the universal joint.
3 If the oil seal is being renewed with the differential nose piece in position, drain the oil and check that the handbrake is firmly applied so as to prevent the pinion flange from rotating.
4 Unscrew the nut in the centre of the pinion drive flange. Although it is tightened down to a torque wrench setting of 140 lb ft it can be removed fairly easily using the long extension arm fitted to the appropriate sized spanner. Undo and remove the nut and spring washer.
5 Carefully withdraw the splined drive flange which may be a little stubborn, in which case it should be tapped with a soft faced hammer from the rear. The pressed steel end cover should next be removed and then finally the seal eased out with a screwdriver, taking care not to damage the lip of its seating.
6 Replacement is a reversal of the above procedure. Note: The new oil seal must be pushed into the differential nose piece with the edge of the sealing ring facing inwards. Great care must be taken not to damage the edge of the oil seal when replacing the end cover and drive flange. Smear the face of the flange, which bears against the oil seal, lightly with oil before fitting the flange onto its splines. Tighten the nut to a torque wrench setting of 140 lb ft.
7 Reconnect the propeller shaft, making sure that the two flanges are correctly fitted with the location marks opposite to each other. Also make sure that the joint faces are clean and that the flange registers engage correctly.

7 Differential assembly - removal and replacement

If it is wished to renew the differential carrier assembly, to exchange it for a factory reconditioned unit, or to overhaul it, first remove the axle shaft as described in Section 4.
1 Mark the propeller shaft and pinion flanges, to ensure that they are refitted in the same relative position, before removal.
2 Unscrew the nuts from the four bolts holding the two flanges together, remove the bolts and separate the flanges. Note that the bolt heads are facing towards the universal joint.
3 Remove the ring of twelve nuts and spring washers which attach the differential bevel pinion and gear carrier to the axle banjo. Carefully withdraw the carrier complete with the pinion shaft and differential assembly.
4 Before refitting the exchange or rebuilt differential assembly, carefully clean down the inside of the axle casing and wipe clean the mating faces of the axle casing and the differential assembly. Fit a new paper gasket and carefully replace the differential assembly. Secure in place with the twelve nuts and spring washers, tightening these in a diagonal manner.
5 Refill the rear axle with 3 pints of Castrol Hypoy Rear Axle Oil.
6 If the differential unit has been reconditioned or overhauled the car should be driven at moderate speeds for the first five hundred miles to allow the new parts to bed in correctly. Change the oil once the first five hundred miles has been covered.

8 Differential assembly - dismantling and examination

Most garages will prefer to renew the complete differential carrier assembly as a unit if it is worn, rather than to dismantle the unit and renew any damaged or worn parts. To do the job 'according to the book', requires the use of special and expensive tools and skilled mechanics who know how to use them, neither of which the majority of garages have.

The primary object of the special tools is to ensure that the mesh of the crownwheel to the pinion is very accurately set and thus reduce noise to a minimum. If any increase in noise cannot be tolerated (assuming the rear axle is already noisy due to a defective part), then it is best to purchase an exchange, built up differential unit.

If the possibility of a slight increase in noise can be tolerated then it is quite possible to successfully recondition the rear axle without the special tools. The differential assembly should be stripped and examined in the following manner:
1 Remove the differential assembly from the rear axle as detailed in Section 7. Note: All numbers in brackets refer to Fig 8.6.
2 With the differential assembly on the bench begin dismantling the unit by unscrewing and removing the nuts and washers (24) holding the differential bearing caps (15) in place. Ensure that the caps are marked so that they may be fitted in their original positions upon reassembly (photo).

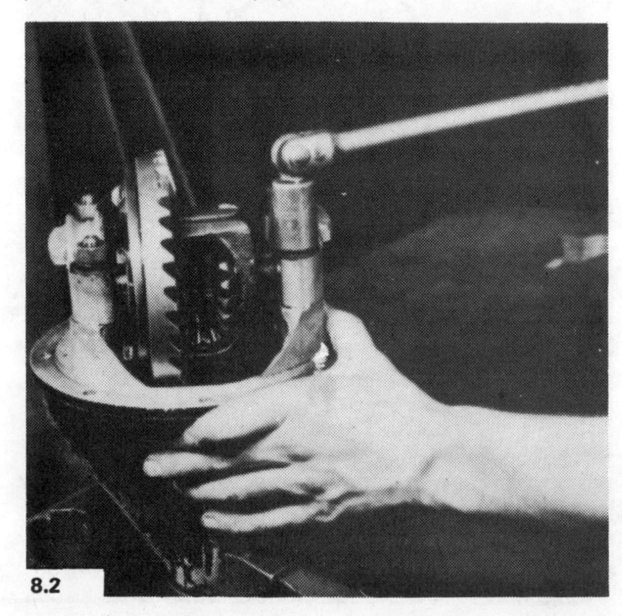
8.2

3 Pull off the caps (15) and then lever out the differential unit complete with crownwheel and differential gears (photo).

4 Check the differential bearings (17) for side play and if present draw them off from the differential cage (18) together with any shims (16) that might be placed between the bearing inner track and the differential cage (18).

5 With a chisel or screwdriver bend back the tabs on the locking washers (13) and then undo and remove the high tensile steel bolts (14) that hold the crownwheel (19) to the differential cage (18) (photo).

6 Professional fitters at BLMC garages use a special tool for holding the pinion flange (30) stationary whilst the nut (31) in the centre of the flange is unscrewed. As it is tightened to a torque wrench setting of 140 lb ft it will require considerable force to move it. The average owner will not normally have the use of the special tool so as an alternative method, clamp the pinion flange (30) in a vice and then undo the nut. Any damage caused to the edge of the flange by the vice jaws should be carefully filed smooth (photo).

7 With the nut and spring washer (31) removed, pull off the splined pinion flange (30) (tap the end of the pinion shaft (20) if the flange appears to be stuck) and remove the oil seal housing (29) and the oil seal (28) (photo).

8 Drift the pinion shaft rearwards out of the nose piece. With it will come the inner race and rollers of the rear bearing (22), the bearing spacer (23) and the shims (32). The outer race and front bearing (27) will be left in the nose piece. With the pinion shaft removed the rear outer race can be quite easily removed using a soft metal drift (photo).

9 The inner race of the front bearing (27) can now be tapped out and then the outer race extracted.

10 The inner race of the rear bearing (22) is a press fit onto the pinion shaft (20) and must be drifted off carefully. If the BLMC special tool 18G285 is available this will help the removal of the inner race considerably. Remove the thrust washer (21) located under the pinion gear head and retain it for future use.

11 Check the rollers and races for general wear, score marks and pitting and renew these components as necessary.

12 Examine the teeth of the crownwheel and pinion for pitting, score marks, chipping and general wear. If a crownwheel and pinion are required, a mated crownwheel and pinion must be obtained, and under no circumstances may only one part of the two be replaced (photo).

13 Tap out the pinion peg (12) from the crownwheel side of the differential cage (18) to free the pinion shaft (4) which is then driven out with a suitable diameter parallel pin punch. Note: The hole into which the peg fits is slightly tapered, the opposite end may be lightly peened over and should be cleared with a 3/16 inch drill.

14 Lift away the pinions (10), differential wheels (9) and thrust washers (8 and 11) from the differential cage. Check them for wear and obtain new as required. Replacement of the pinions or differential wheels is a reversal of the above process. Note: After the peg (12) has been inserted, the larger end of the hole should be lightly peened over to retain the pin in position.

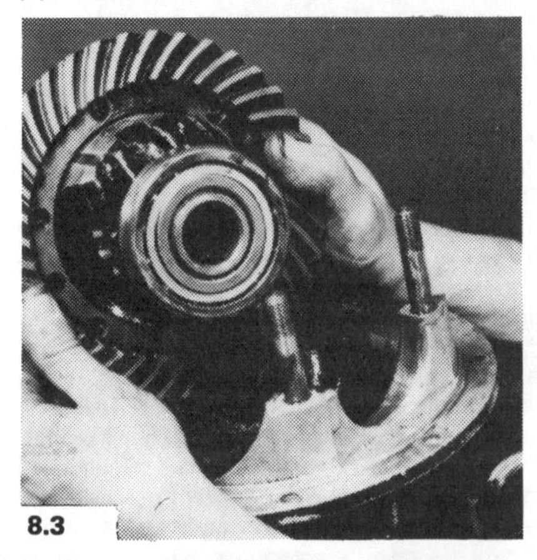
8.3

8.5

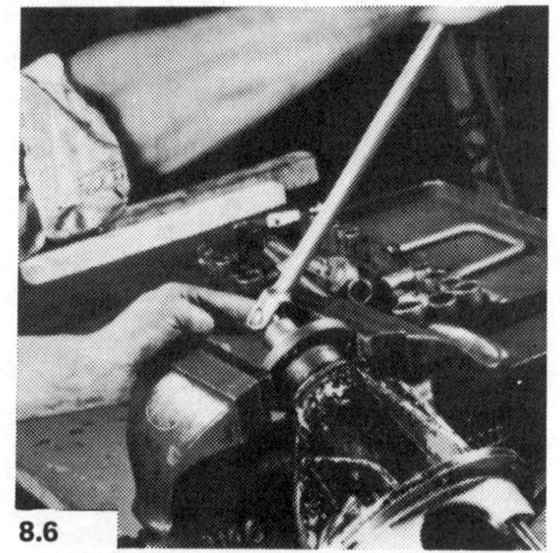
8.6

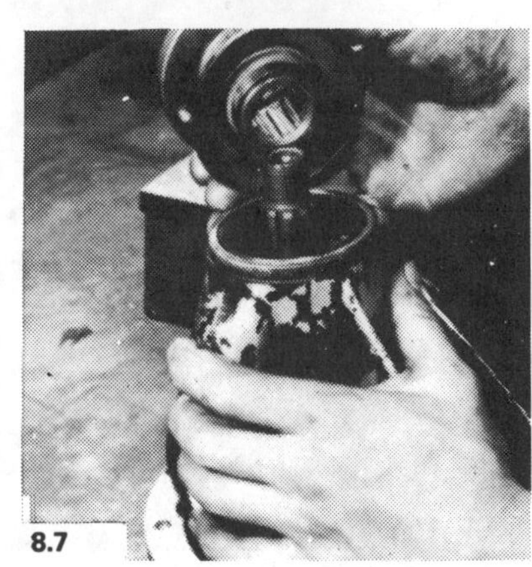
8.7

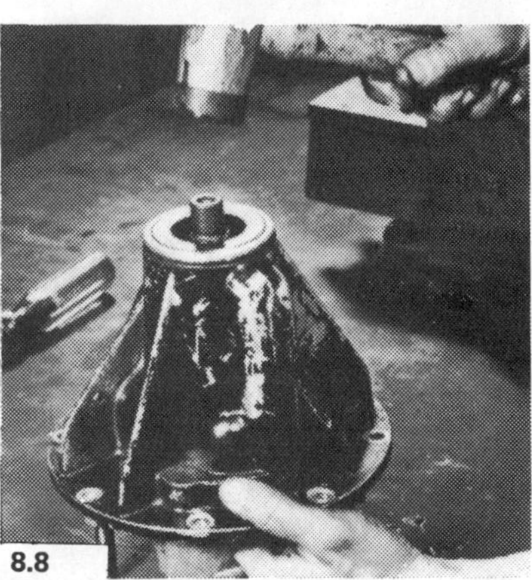
8.8

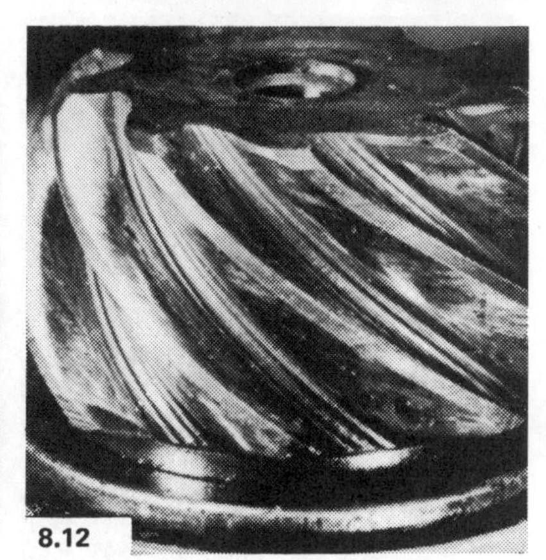
8.12

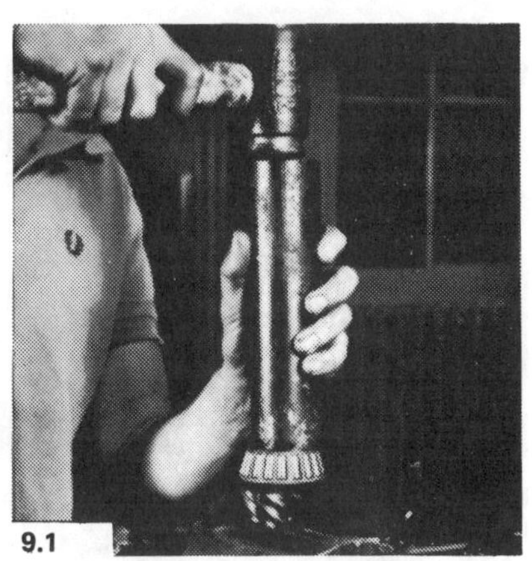
9.1

9 Differential - reassembly

1 Replace the thrust washer (21) onto the pinion shaft and then fit the inner race of the rear bearing (22). If the special BLMC removal and replacement tool 18G285 is not available it is quite satisfactory to drift the rear bearing on with a piece of steel electrical piping, 12 to 14 inches long, with sufficient internal diameter to just fit over the pinion shaft. With one end of the tube bearing against the inner track of the race, tap the top end of the tube with a hammer, so driving the bearing squarely down the shaft and hard up against the underside of the thrust washer (photo).

2 Slip the bearing spacer (23) over the pinion shaft (20) and fit the outer race of the front and rear bearings (22 and 27) onto the differential nose piece.

3 Insert the pinion shaft (20) forwards into the differential nose piece from inside the casing and then drop the front inner bearing race and rollers (27) into position (photo).

4 Lubricate the bearings with Castrol Hypoy Rear Axle Oil. Fit a new oil seal (28) with the edge of the sealing ring facing inwards. A block of wood is useful for ensuring the seal is driven on squarely (photo).

5 With the seal in position replace the oil seal housing (29), lubricate the underside of the pinion flange (20) which bears against the oil seal and drive the flange onto the splines with a soft faced hammer (photo).

6 Replace the spring washer, and with the flange held securely in a vice, tighten the flange nut (31) to a torque wrench setting of 140 lb ft (photo).

7 To obtain the correct pinion bearing pre-load. slowly tighten the nut, taking frequent readings. The correct pre-load should be 13-15 lb ins. Measure this with a spring balance hooked into one of the drive flange holes. As these holes are 1½ inches from the shaft axis a pull of 9 lb ins is the correct pre-load figure using this method. If the pre-load is too great use a thinner thrust

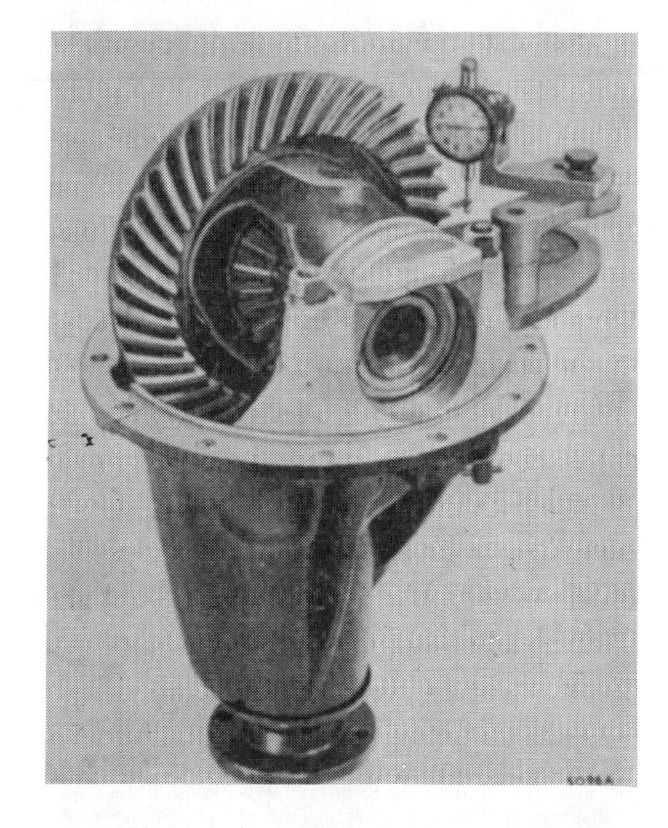

Fig 8.5 CHECKING CROWNWHEEL TO PINION BACKLASH

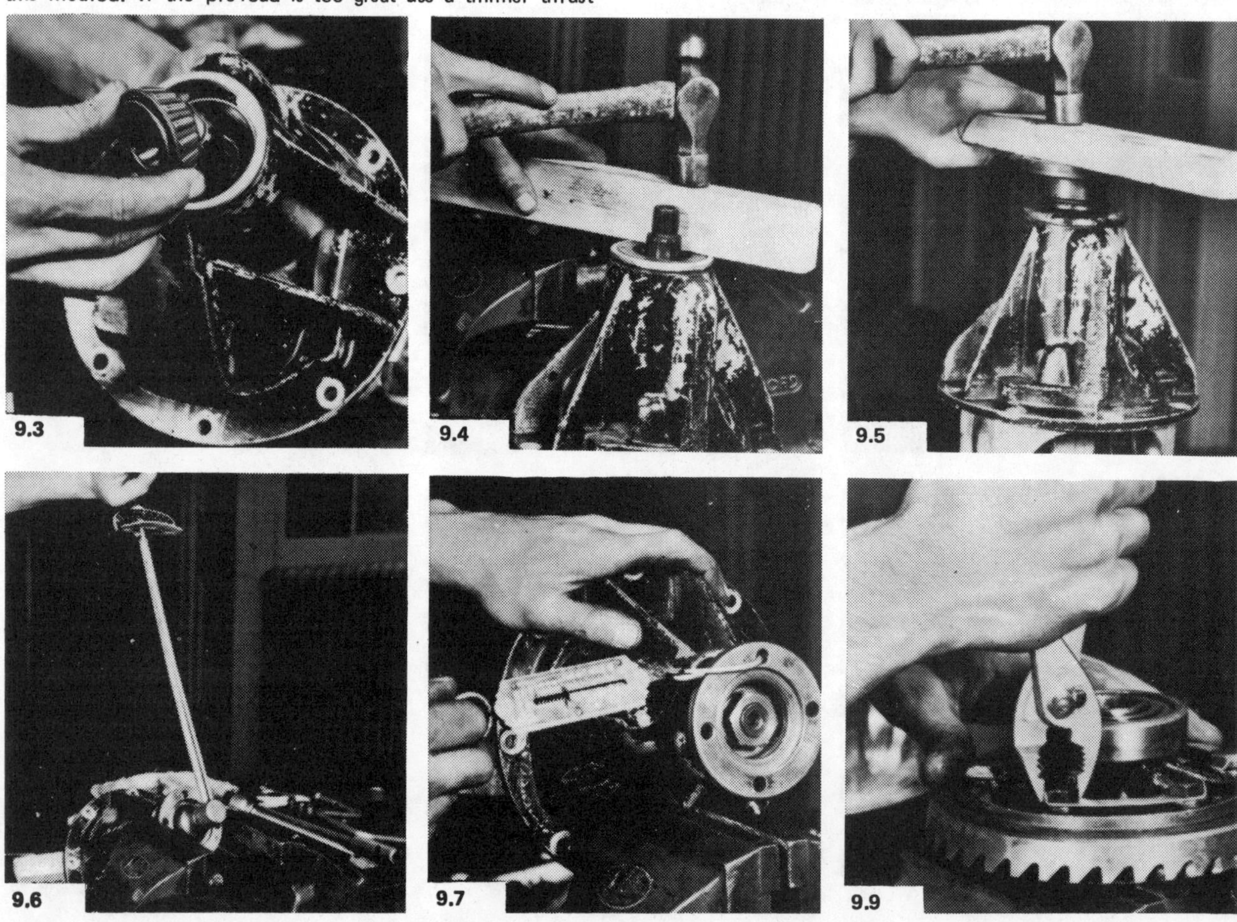

9.3 9.4 9.5 9.6 9.7 9.9

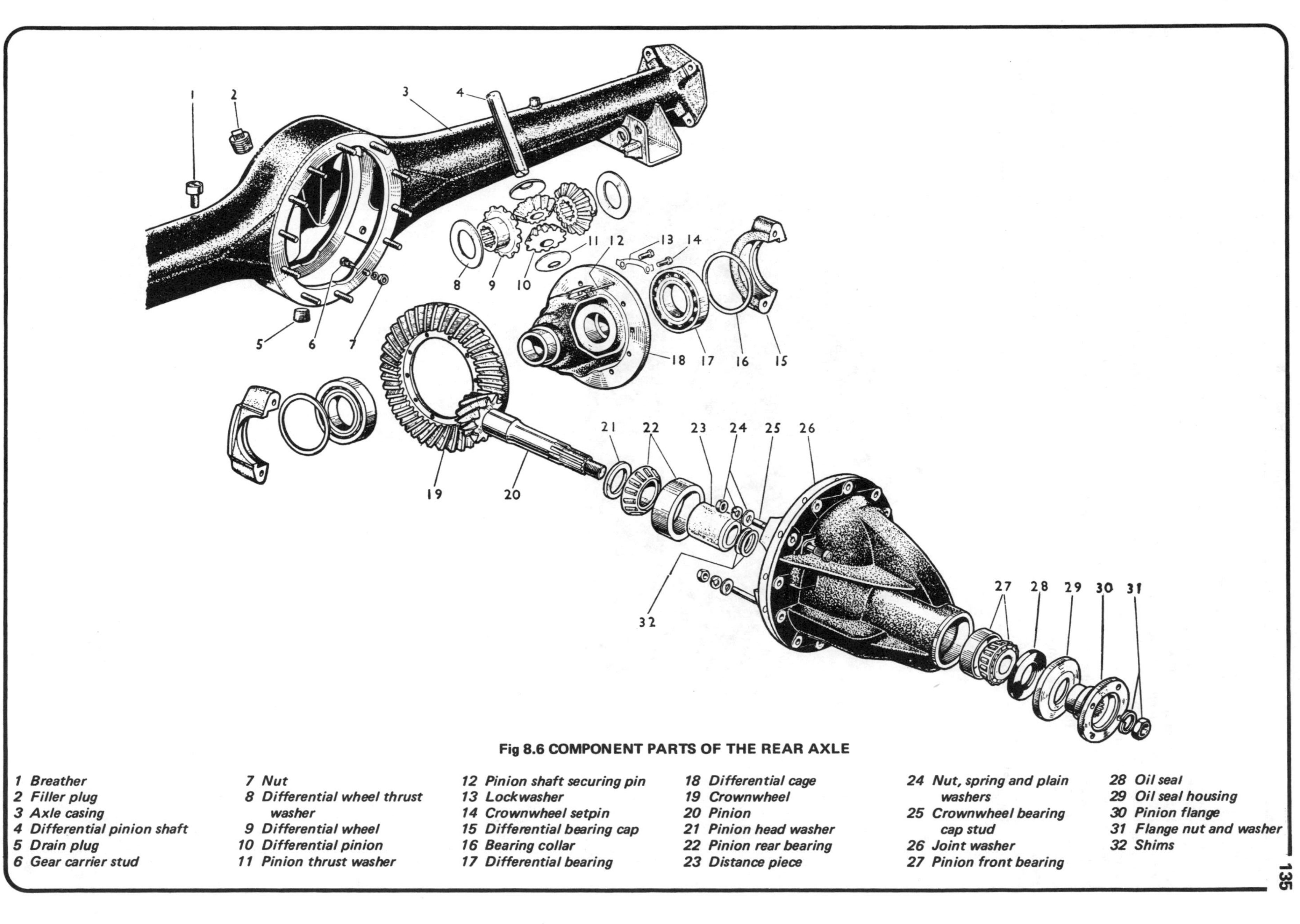

Fig 8.6 COMPONENT PARTS OF THE REAR AXLE

1 Breather
2 Filler plug
3 Axle casing
4 Differential pinion shaft
5 Drain plug
6 Gear carrier stud
7 Nut
8 Differential wheel thrust washer
9 Differential wheel
10 Differential pinion
11 Pinion thrust washer
12 Pinion shaft securing pin
13 Lockwasher
14 Crownwheel setpin
15 Differential bearing cap
16 Bearing collar
17 Differential bearing
18 Differential cage
19 Crownwheel
20 Pinion
21 Pinion head washer
22 Pinion rear bearing
23 Distance piece
24 Nut, spring and plain washers
25 Crownwheel bearing cap stud
26 Joint washer
27 Pinion front bearing
28 Oil seal
29 Oil seal housing
30 Pinion flange
31 Flange nut and washer
32 Shims

washer. If too low, use a thicker thrust washer (photo).

8 Refit the differential bearings (17) to the differential cage (18) making sure that the bearings are the correct way round.

9 Ensure that the crownwheel and cage are scrupulously clean and then bolt the crownwheel (19) to the differential cage flange (18) tightening the high tensile steel bolts (14) to a torque wrench setting of 65 lb ft. Bend up the tabs on the locking washers (13) (photo).

10 Mount a dial indicator gauge as shown in Fig 8.5 and with the probe resting on one of the teeth of the crownwheel determine the backlash. The correct figure for the backlash to be used with any particular crownwheel is etched on the rear face of the crownwheel and this must be strictly adhered to. The backlash may be varied by decreasing the thickness of the bearing collar (16) (Fig 8.6) at one side and increasing the thickness of the collar at the other side by the same amount, thus moving the crownwheel into or out of mesh as required. The total thickness of the two collars must not be changed.

11 As a further check for the correct meshing of the crownwheel and pinion, smear a little engineer's blue onto the crownwheel teeth and then rotate the pinion. The contact marks should appear right in the middle of the crownwheel teeth. If the mark appears on the toe or the heel of the crownwheel teeth, then the crownwheel must be moved either nearer or further away from the pinion as described in the last paragraph. Fig 8.7 shows the various tooth patterns that may be obtained (photo).

12 When the correct meshing between the crownwheel and the pinion has been obtained, refit the differential unit bearing caps and tighten the nuts to a torque wrench setting of 65 lb ft.

13 The differential unit can now be refitted to the axle casing.

9.11

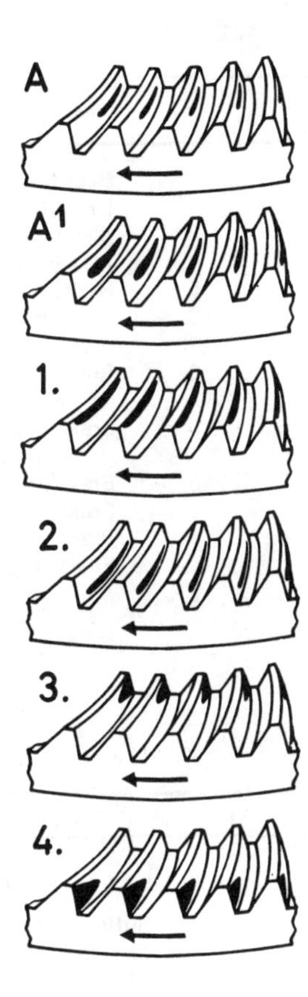

Fig 8.7 CONTACT MARKING ON CROWNWHEEL

A

Correct contact marking picture without load.

A1

When subjected to load the contact picture is displaced somewhat toward the outside.
Displacement of the crownwheel changes primarily the backlash, in addition to contact picture is displaced in the axial direction of the teeth.
Displacement of the pinion primarily moves the contact marking in the direction of the tooth height, while the backlash changes only marginally.
In addition the four fundamentally false contact markings, which usually occur in conjunction with each other, but knowledge of which simplifies the actual adjustment work.

1 High, narrow contact marking (tip contact) on crownwheel. **Correction: displace the pinion toward the crownwheel axis** and, if necessary, correct backlash by moving the crownwheel away from the pinion.
2 Deep, narrow contact marking (roof contact) on crownwheel. **Correction: move the pinion away from the crownwheel axis** and, if necessary, correct backlash by pushing the crownwheel toward the pinion.
3 Short contact marking on smallest tooth end (toe contact) of the crownwheel. **Correction: move the crownwheel away from the pinion** and, if necessary, move the pinion closer towards the crownwheel axis.
4 Short contact marking on large tooth end (heel contact) of the crownwheel. **Correction: move the crownwheel toward the pinion** and, if necessary, move the pinion away from the crownwheel axis.

Chapter 9 Braking system

Contents

Specifications

Models BN4 and BN6

Type	Girling hydraulic. Two leading shoes on front
Drum diameter	11 inch (280 mm)
Total brake lining area	188 sq in (1213 sq cm)
Shoe lining width	2.25 in (57 mm)
Shoe lining length:	
Front	10.4 in (265.6 mm)
Rear	10.4 in (265.6 mm)
Shoe lining thickness	0.167 to 0.174 inch (4.24 to 4.42 mm)
Pedal free movement	1/8 inch (3.175 mm)
Handbrake	Mechanical, rear wheels only

Models BN7 and BT7

Type	Girling hydraulic
Front	Disc
Rear	Drum. One leading, one trailing shoe
Disc diameter	11 inch (28.0 cm)
Drum diameter	11 inch (28.0 cm)
Drum width	2 7/16 inch (61.91 mm)
Total friction area (rear)	95 sq in (612.75 cm^2)
Shoe lining width	2¼ inch (57 mm)
Shoe lining length	10.57 inch (267.4 mm)
Shoe lining thickness	0.187 inch (4.76 mm)
Disc pad area	4.25 sq in (10.8 cm^2) x 4
Disc pad thickness	13/32 inch (10.32 mm)
Disc pad minimum thickness	1/8 inch (3.175 mm)
Pedal free movement	1/8 inch (3.175 mm)
Handbrake	Mechanical, rear wheels only
Brake servo unit type	Girling

Torque wrench setting

Calliper retaining bolts	45 - 50 lb ft (6.22 - 6.91 kg m)

1 General description

BN4 and BN6 models

The drum brakes fitted to all four wheels are hydraulically operated when the brake pedal is depressed. The brake pedal is connected to the brake master cylinder in which the hydraulic pressure of the hydraulic fluid is generated. A reservoir is integral with the master cylinder body and ensures that the system is kept full of fluid. The master cylinder is connected to the wheel cylinders by a system of metal pipes, flexible hoses and unions.

When the brake pedal is depressed the pressure generated within the master cylinder is transmitted as an equal pressure to all the wheel cylinders. This causes the wheel cylinder pistons to move outwards which forces the brake shoe linings onto the inner circumference of the brake drum.

Upon releasing the brake pedal the brake shoe return springs draw the brake shoes away from the drum which will in turn move the pistons back into the wheel cylinders. Fluid displaced from the wheel cylinders returns through the system to the master cylinder reservoir.

The handbrake is connected by a system of cables and levers to the rear brakes and operates mechanical expanders which are attached to the rear wheel cylinder bodies.

The front brakes are of the two leading shoe type with sliding shoes which ensure automatic centralisation of the brake shoe when initially in contact with the brake drum. Two wheel cylinders are used and are interconnected by a metal bridge pipe on the outside of the backplate.

Adjustment is by means of two serrated snail cam adjusters with square heads which should be turned clockwise to adjust the shoes.

The rear brake shoes are not mounted on rigidly fixed anchor bolts but are allowed to slide and centralise themselves in the drum. They are operated by a single acting wheel cylinder which also incorporates the handbrake mechanism.

One adjuster is fitted to each backplate and should be turned by the square end of the adjuster in a clockwise direction.

BN7 and BT7 models

Girling disc brakes are fitted to the front wheels. The calliper is mounted over the disc and houses two horizontally opposed blind cylinders and the friction pads. The friction lining material is bonded to a steel backplate and is inserted between each piston and the disc. The pads are retained in position by pins and spring clips and because of the design are self-adjusting in operation.

The rear brakes fitted to these models are basically identical to those fitted to the earlier models and the complete operation of the braking system is comparable with that previously described.

A vacuum servo unit of Girling manufacture was offered as an optional extra on 3000 Mk II cars from car number 15104 but was fitted as standard equipment on the 3000 Mk III models.

2 Maintenance

It is important that the complete braking system is at the peak of condition at all times. To safeguard against premature wear or deterioration it is suggested that the following points are adhered to:

1 The front and rear drum linings or front disc brake pads, hoses, metal pipes and unions be examined at intervals of 3000 miles.
2 The brake hydraulic fluid should be changed every 18 months or 24,000 miles, whichever is sooner.
3 All seals in the brake master cylinder, wheel cylinders and callipers (BN7, BT7 models) and the flexible hoses should be examined and preferably renewed every three years or 40,000 miles, whichever is sooner. The working surfaces of the master, wheel and calliper cylinders should be inspected for signs of wear or scoring and new parts fitted as considered necessary.
4 Only Castrol Girling Brake Fluid should be used in the hydraulic system. Never leave brake hydraulic fluid in open unsealed containers as it absorbs moisture from the atmosphere which lowers the safe operating temperature of the fluid. Also any fluid drained or used for bleeding the system should not be re-used immediately.
5 Any work performed on the hydraulic system must be done under conditions of extreme care and cleanliness.

Drum brake system

1 At intervals of 3000 miles or more frequently if pedal travel becomes excessive, adjust the brake shoes to compensate for wear of the brake linings.
2 At the same time lubricate all joints in the handbrake mechanism with an oil can filled with Castrol GTX or similar grade oil.
3 Every 3000 miles check the level of the hydraulic fluid in the master cylinder reservoir. Carefully wipe the top of the brake master cylinder reservoir, remove the cap, and inspect the level of the fluid which should be ¼ inch below the bottom of the filler neck. Ensure that the breather hole in the cap is free from dirt.
4 If the hydraulic fluid is below this level top up the reservoir with Castrol Girling Brake Fluid. It is important that no other type of fluid is used. Use of a non-standard fluid could result in brake failure caused by the perishing of the special seals used throughout the system. If topping up becomes frequent then check the metal piping, flexible hoses and unions for leaks and check for worn brake or master cylinders which will also cause loss of fluid.

Disc brakes

1 Refer to the information given for Drum brakes and follow as applicable.
2 Examine the wear in the brake disc pads and change them round if necessary as detailed in Section 5.

Brake servo unit

1 The air cleaner must be cleaned regularly. Push the air filter spring clip aside and lift away the cover.
2 Lift out the air filter element and if the element is coated with a light dust deposit this may be blown clean using a compressed air jet. Any other contamination or damage will necessitate a new filter. Do not clean it in any liquid cleaning fluid.
3 Refitting is the reverse sequence to removal.

3 Front drum brake - adjustment

1 Chock the rear wheels, apply the handbrake, jack up the front of the car and support on firmly based axle stands located under the main chassis members.
2 Upon inspection it will be seen that there are two adjusters located as shown in Fig 9.1. Soak these in penetrating oil.
3 Turn the hexagon head adjuster bolts anticlockwise as far back as possible.
4 Turn one of the two adjuster bolts in a clockwise direction until the brake shoe touches the brake drum. Back off the adjuster until the shoe is just free of the drum.
5 Repeat paragraph 4 for the second adjuster bolt.
6 Check the adjustment by rotating the wheel and applying the footbrake. Release the footbrake and turn the adjusters clockwise again until the shoes are just touching the drum and backing off slightly. This will ensure that the shoes are central relative to the drum.
7 Repeat the above procedure for the second front wheel.
8 Remove the axle stands and lower the car to the ground.

Fig 9.1 LOCATION OF THE TWO BRAKE SHOE ADJUSTERS ON THE FRONT BACKPLATE

4 Rear drum brake - adjustment

1 Chock the front wheels, jack up the front of the car and support on firmly based axle stands located under the main chassis members. Release the handbrake.
2 Upon inspection it will be seen that there is one adjuster for each brake unit and it takes the form of a square head protruding from the rear of the backplate opposite to the wheel cylinder.
3 Soak each adjuster in penetrating oil and then turn in a clockwise direction until a resistance is felt using a good fitting square ended spanner.
4 Slacken the adjuster by two clicks. The wheel should be free to rotate easily. Depress the brake pedal and check the adjustment.
5 Repeat the above procedure for the second rear wheel.
6 Remove the axle stands and lower the car to the ground.

5 Front disc brake pad - removal and replacement

1 Apply the handbrake, chock the rear wheels, jack up the front of the car and place on firmly based axle stands located under the main chassis members.
2 Remove the road wheels.
3 Extract the spring clips (1) (Fig 9.2) locking the pad retaining pins (2) in position.
4 Draw the retaining pins (2) out of the calliper and lift away the friction pad (3) assemblies.
5 Whilst the calliper is being worked upon it is recommended that it be cleaned with a stiff brush and inspected for hydraulic fluid leaks.
6 Using a piece of hardwood gently lever the pistons (6) into their respective bores as far as they will go.
7 Check the tightness of the pad lining which should be greater than 1/8 inch. If the pad is less than this obtain new pads. DO NOT use pads less than 1/16 inch.
8 If the disc brakes emitted a high pitched squeal when the brake pedal is applied this may be remedied by fitting anti-squeal shims.
9 If applicable position one shim on the back of each pad with the indicating arrow pointing towards the calliper bleed screw, in other words in the direction of the forward rotation of the wheel.
10 Slip the new pads (and shims) into position and secure with the retaining pins. Lock each pin with a spring clip.
11 Depress the brake hard several times so as to adjust the position of the pads and check the level of fluid in the reservoir.

6 Front calliper - removal and refitting

1 Chock the rear wheels, apply the handbrake, jack up the front of the car and support on firmly based axle stands located under the main chassis frame.
2 Remove the road wheel on the side which the calliper is to be removed.
3 Wipe the top of the master cylinder reservoir, remove the cap and place some thick polythene over the filler neck. Refit the cap. This is to stop hydraulic fluid syphoning out when the pipe union is removed.
4 Wipe the area around the brake pipe union nut in front of its support bracket and unscrew the union nut. Disconnect the pipe and plug the end to stop dirt ingress with a piece of tapered wood or pencil.
5 Undo and remove the two nuts and spring washers that secure the brake hose support bracket and remove the bracket.
6 Bend back the tab washers and unscrew and remove the two calliper retaining bolts and lift away the calliper from the swivel axle.
7 Refitting the calliper is the reverse sequence to removal. Make sure that the brake disc passes between the two pads. Tighten the two calliper retaining bolts to a torque wrench setting of 45 to 50 lb ft.
8 Finally bleed the brake hydraulic system as described in Section 15 of this Chapter. Apply the brake pedal firmly and check for leaks.

7 Front calliper - dismantling, overhaul and reassembly

1 With the calliper away from the car and the pads removed, first wash the exterior with petrol, methylated spirits or hydraulic fluid and wipe dry with a clean non-fluffy rag.
2 If a compressed air jet is available apply this to the hydraulic union and eject the pistons. Before commencing this operation wrap some rag around the calliper so that the pistons do not fly out, possibly causing injury.
3 As an alternative method using a small G clamp hold the piston in the mounting half of the calliper. Temporarily reconnect the calliper to the flexible hose pipe union. Do not allow the calliper to hang on the flexible hose but support it. Carefully depress the brake pedal and this will push the piston in the rim half of the calliper outwards.
4 Push back the dust cover and disengage it from the piston.
5 Using a tapered wooden rod or an old plastic knitting needle carefully extract the sealing ring from its groove in the calliper.
6 With the piston out of the calliper remove the G clamp from the mounting half calliper. Temporarily refit the rim half piston and repeat the operations in paragraphs 3, 4 and 5.
7 Thoroughly clean the internal parts of the calliper using Castrol Girling Brake Fluid or Industrial Methylated Spirits. Any other fluid cleaner will damage the internal seals between the two halves of the calliper. DO NOT SEPARATE THE TWO HALVES OF THE CALLIPER UNLESS A SEAL IS LEAKING.
8 Inspect the pistons and bores for signs of scoring which if evident, a new calliper assembly should be obtained.
9 To reassemble first fit the internal seal, previously wetted with Castrol Girling Brake Fluid, into the groove in the cylinder bore with the scraping edge (smaller diameter) innermost.
10 Position the lip of the dust cover in the outer groove.
11 Smear the piston with brake fluid and push it into the bore, closed end first.
12 Push the piston right home and then engage the outer edge of the cover with the groove in the piston body.
13 Refit the second piston into the calliper body by repeating

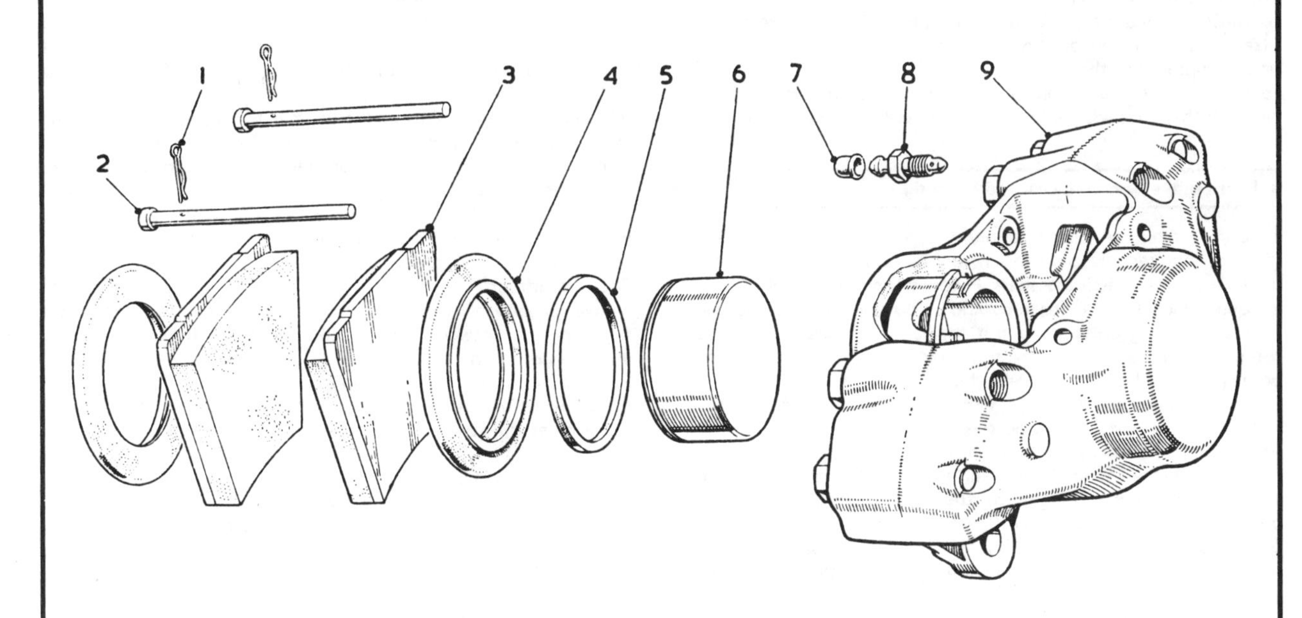

Fig 9.2 FRONT DISC BRAKE CALLIPER SHOWING INTERNAL PARTS

1 Wire clip
2 Retaining pin
3 Lining pad and steel backplate
4 Dust cover
5 Sealing ring
6 Piston
7 Bleed nipple dust cover
8 Bleed nipple
9 Calliper body

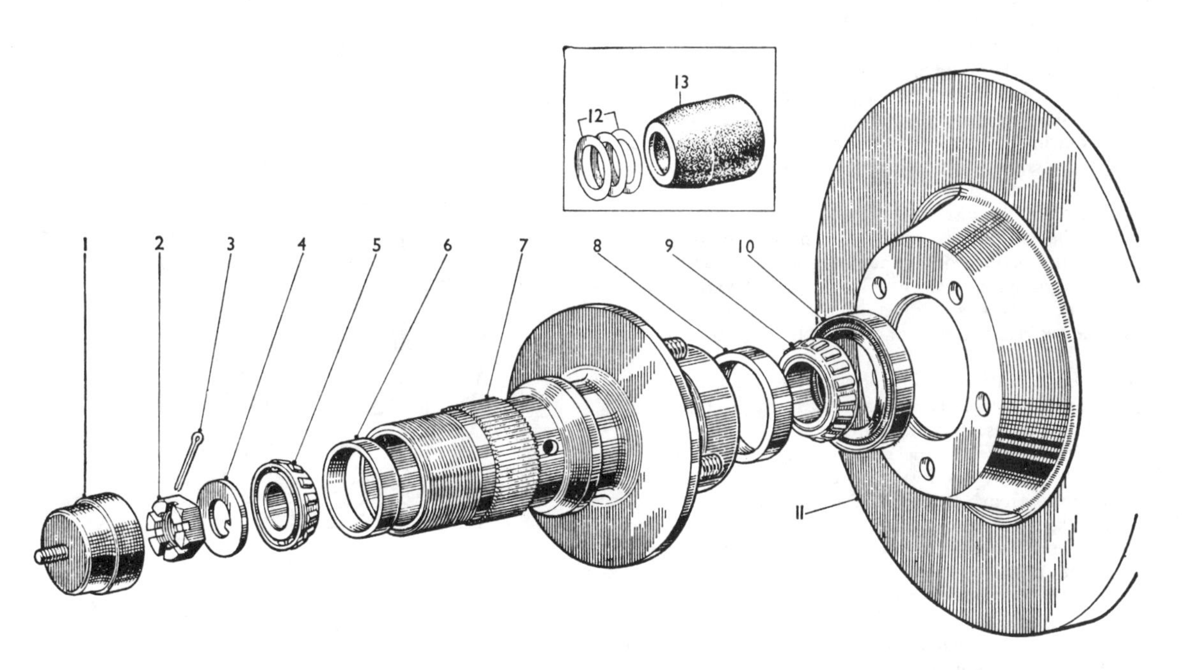

Fig 9.3 FRONT HUB AND DISC COMPONENTS

1 Grease cup
2 Axle nut
3 Split pin
4 Washer
5 Outer bearing
6 Bearing outer race
7 Hub
8 Bearing outer race
9 Bearing outer race
10 Oil seal
11 Brake disc
Inset—Distance piece and shims

operations 9 to 12 inclusive.
14 Refit the pad assemblies and anti-squeal shims, if previously fitted, and secure in position with the retaining pins. Lock the retaining pins with the wire clips.
15 Refit the calliper unit making sure that the disc passes between the two pads and proceed as described in Section 6.

8 Front brake disc - removal and refitting

1 Chock the rear wheels, apply the handbrake, jack up the front of the car and place on firmly based axle stands located under the main chassis members. Remove the road wheel.
2 Bend back the locking tabs and undo and remove the two calliper retaining bolts securing it to the swivel axle, and lift the calliper up and away from the brake disc. Note any shims placed between the calliper and swivel axle.
3 With a piece of string or wire support the weight of the calliper so that the flexible hose is not strained. By doing it this way it is not necessary to disconnect the brake hydraulic system.
4 Place a piece of wood between the two pistons so as to act as a safeguard against ejection of the pistons should the brake pedal be accidentally depressed.
5 On cars fitted with disc wheels use the special extractor supplied in the tool kit and remove the grease retaining cup from within the hub.
6 Extract the split pin and undo and remove the castellated hub nut. Withdraw the flat washer from the end of the stub axle.
7 Ease off the hub grease cup using two screwdrivers.
8 Straighten the split pin locking the hub securing nut and ease it out of the axle stub and nut on models with wire wheels. A special hole is drilled in the hub which will allow the split pin to be removed.
9 Using a three legged puller and thrust pad withdraw the hub complete with bearings and brake disc.
10 To separate the disc from the hub clamp the disc between soft faces in a vice and with a scriber or file mark the disc and hub so that the two parts may be refitted in their original positions if new parts are not to be fitted.
11 Undo and remove the five bolts and self-locking nuts so as to release the wheel hub from the disc.
12 Clean the disc and inspect for signs of hairline cracks or excessive corrosion or scoring. It should be appreciated that scoring is not detrimental provided that the scoring is concentric, even and not excessive. The braking efficiency will be impaired and the pad wear increased.
13 It is possible for the disc to be reground but only if a new disc is not available. As this work must be carried out very accurately, an engineering works should be allowed to undertake the work. The ground surface must be quite flat and parallel with the mounting face and must have a fine finish. Sharp corners must be avoided at the inner circumference of the ground area.
14 Either or both sides may be reground by no more than 0.040 inch from each disc. This means that after grinding the thickness must not be less than 0.335 inch.
15 Wipe the mating faces of the disc and hub. Refit the disc to the hub and tighten the five bolts and nuts in a diagonal manner to a torque wrench setting of 28 to 30 lb ft.
16 Refit the assembled disc and hub onto the stub axle and using a tubular drift located on the outer bearing inner track drive the hub into position.
17 Replace the plain washer and castellated nut and tighten fully.
18 Using either a dial indicator gauge or feeler gauges placed between the disc and assembled calliper body, check the run out of the disc at the outer circumference. This must not exceed 0.006 inch.
19 Should the reading obtained be in excess of the limit remove the hub and disc assembly and once again remove the disc. Reposition it on the hub and try again.
20 If the maximum limit is still exceeded the disc should be suspected for distortion and it should be reground with the limits given in paragraph 14 or a new disc be obtained.
21 When the disc run out is within the limits tighten the castellated nut and lock in position with a new split pin. Do not slacken the nut to align the slots with the split pin hole in the stub axle.
22 Remove the pads from the calliper by extracting the spring clips and withdrawing the retaining pins. Lift the pads and shims (if fitted) from the calliper taking care to note which way round they are fitted.
23 Refit the calliper to the swivel axle and secure with the two bolts and lock tab washer. Do not bend over the lockwasher tabs yet.
24 It is important that when fitted the hub disc must run centrally between the calliper cylinders. To check this insert feeler gauges between the pad abutments on the calliper body and faces of the disc.
25 It is permissible for the gap on opposite sides of the disc to differ by 0.015 inch but there must be no difference between the gaps at the two abutments on the same side of the calliper. This ensures the calliper is in correct alignment with the disc and that the pads and pistons are square with the disc.
26 Fit shims between the calliper and swivel axle mounting to counteract any misalignment. When correct bend over the lockwasher tabs.
27 Refit the pads and anti-squeal shims if fitted.
28 Operate the brake pedal firmly several times to adjust the position of the pistons within the calliper.
29 Replace the road wheel, remove the axle stands and lower the car to the ground.

9 Drum brake shoe - removal, inspection and refitting (front)

1 Chock the rear wheels, apply the handbrake, jack up the front of the car and place on firmly based axle stands located under the main chassis members. Remove the front wheels.
2 Undo and remove the four self-locking nuts and spring washers that secure the brake drum to the hub when wire wheels are fitted. Lift away the extension.
3 Soak the brake adjusters, located as shown in Fig 9.1, in penetrating oil and turn the adjusters in an anticlockwise direction to back off the brake shoe adjustment.

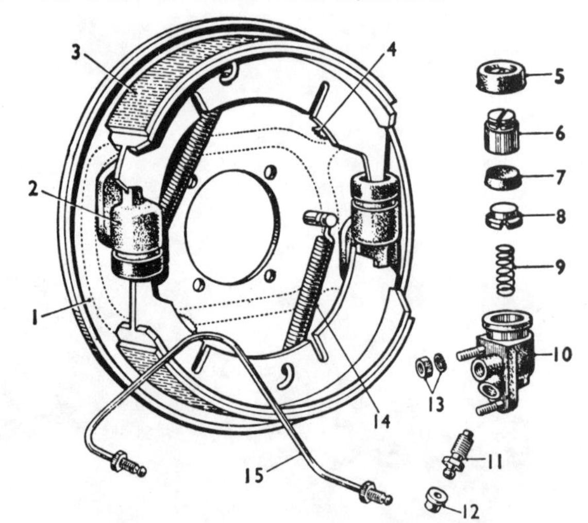

Fig 9.4 FRONT DRUM BRAKE COMPONENTS

1 Backplate
2 Wheel cylinder
3 Shoe
4 Snail cam adjuster
5 Dust cover
6 Piston
7 Seal
8 Seal support
9 Spring
10 Cylinder housing
11 Bleed screw
12 Bleed screw cover
13 Nut and washer
14 Shoe return spring
15 Cylinder connecting pipe

4 Undo and remove the two countersunk screws securing the drum to the hub and remove the drum from the front hub assembly.
5 If the drum is difficult to remove tap off with a soft faced hammer.
6 Ease one shoe out of the abutment slot of one wheel cylinder. Next release the piston slot of the other. Detach the brake shoe drom the two return springs and lift away the first shoe.
7 The second shoe may now be lifted away from the abutment slot of the second wheel cylinder and the piston slot of the first wheel cylinder.
8 To prevent the wheel cylinder piston from expanding it is advisable to place a rubber band round each cylinder to stop the pistons being accidentally ejected.
9 Clean down the backplate and check the wheel cylinders for signs of hydraulic leaks. Rectify by stripping down and overhauling as described in Section 11 of this Chapter.
10 Check the wheel cylinders for operation without signs of seizure.
11 Check the adjusters for correct operation by screwing in and out fully so ensuring that the threads are not badly rusted. Lubricate the threads with Girling (White) Brake Grease.
12 Inspect the thickness of the linings and if they are worn down to the heads of the rivets, or a minimum of 1/16 inch for bonded linings, they should be renewed.
13 Smear the tips of the brake shoe supports on the backplate as well as the operating and abutment ends of the new shoes with Girling (White) Brake Grease. The brake grease must not be allowed to contact the hydraulic wheel cylinder, pistons or rubber parts. Also keep all grease off the linings on new replacement shoes and do not handle more than necessary.
14 Fit new shoe return springs to the new shoes. Position the hooked end of the spring through the hole in the shoe web and the swan neck through the hole in the backplate near the abutment end of the same shoe.
15 Each shoe may be refitted individually to the wheel cylinders. Do not forget to remove the rubber bands.
16 Clean the inside of the brake drums and inspect them for signs of excessive scoring. If this is evident, they may be either skimmed on a lathe by the local engineering works or new drums obtained.
17 Refit the brake drum and secure with the countersunk screws.
18 Replace the hub extension if wire wheels are fitted and secure with the four self-locking nuts and spring washers.
19 Refit the road wheel.
20 Refer to Section 3 and adjust the front brakes.
21 Remove the axle stands and lower the car to the ground.
22 Check the brake adjustment after the first 500 miles covered after new linings have been fitted to allow for initial bedding in.

10 Drum brake shoe - removal, inspection and refitting (rear)

1 Refer to the previous section and follow the instructions as the procedures are basically identical but with the below described differences.
2 Paragraph 6. Lift one of the shoes out of the slots in the adjuster link and wheel cylinder piston. Both shoes may now be removed complete with springs.
3 Paragraph 14. Fit the two new shoe return springs to the new shoes and between the shoe web and the brake backplate. The shorter spring goes to the adjuster end.
4 It is important to note that the first shoe has the lining positioned towards the heel of the shoe and the second shoe towards the toe or operating end in both left hand and right hand brake units.
5 It may be found beneficial to slacken off the handbrake adjustment slightly as described in Section 17. Also when adjusting the brakes as described in Section 4 back off the adjuster one further notch on the rear brake adjuster to allow for possible lining expansion.
6 Check the brake adjustment after the first 500 miles covered after new linings have been fitted to allow for initial bedding in.

11 Front brake wheel cylinder - removal, inspection and overhaul

If hydraulic fluid is leaking from the brake wheel cylinder it will be necessary to dismantle the wheel cylinder and replace the seals. Should brake fluid be found running down the side of the wheel or it is noticed a pool of liquid forms alongside one wheel and the level in the master cylinder has dropped proceed as follows:
1 Refer to Section 9 and remove the brake shoes.
2 Wipe the top of the brake master cylinder and unscrew the cap. Place a piece of thick polythene over the top of the master cylinder and refit the cap. This will prevent loss of hydraulic fluid when the wheel cylinder hydraulic pipe is detached from the rear of the wheel cylinder.
3 Wipe the area around the fluid pressure pipe and disconnect the pressure pipe union from the rear of the wheel cylinder.
4 Unscrew and remove the two securing nuts and spring washers (13) (Fig 9.4) and lift the wheel cylinder away from the backplate.
5 Remove the dust cover (5) and lift away the piston (6) followed by the seal (7), seal support (8) and spring (9) from the wheel cylinder body (10). Note which way round the seal (7) is fitted.
6 Inspect the inside of the cylinder for score marks caused by impurities in the hydraulic fluid. If any are found the cylinder and piston will require renewal.
7 If the cylinder is sound, thoroughly clean it out with fresh hydraulic fluid.
8 The old rubber piston seal will probably be swollen and visibly worn so always fit a new seal.
9 Reassembly and refitting the wheel cylinder is the reverse sequence to dismantling. Always assemble the wheel cylinder parts well wetted with clean hydraulic fluid. It will be necessary to bleed the hydraulic system as described in Section 15.
10 Refer to Section 3 and adjust the brake shoe clearance.

12 Rear brake wheel cylinder - removal, inspection and overhaul

If hydraulic fluid is leaking from the brake wheel cylinder it will be necessary to dismantle the wheel cylinder and replace the seals. Should brake fluid be found running down the side of the wheel or it is noticed that a pool of liquid forms alongside one wheel and the level in the master cylinder has dropped, proceed as follows:
1 Refer to Section 10 and remove the brake shoes.
2 Extract the split pin, lift away the plain washer and disconnect the brake rod from the handbrake lever on the rear of the wheel cylinder.
3 Wipe the top of the brake master cylinder and unscrew the cap. Place a piece of thick polythene over the top of the master cylinder and refit the cap. This will prevent loss of hydraulic fluid when the wheel cylinder hydraulic pipe is detached from the rear of the wheel cylinder.
4 Wipe the area around the fluid pressure pipe and disconnect the pressure pipe union from the rear of the wheel cylinder and then remove the rubber dust cover from the rear of the backplate.
5 With a screwdriver prise the locking plate (2) (Fig 9.5) and spring (4) apart and carefully tap the locking plate from beneath the neck of the wheel cylinder.
6 Withdraw the handbrake lever (3) from between the backplate and wheel cylinder.
7 Remove the spring plate (4) and distance piece, lift away the wheel cylinder from the backplate and separate the rubber seal (1) from the wheel cylinder.
8 Referring to Fig 9.5 remove the dust cover (17) and lift

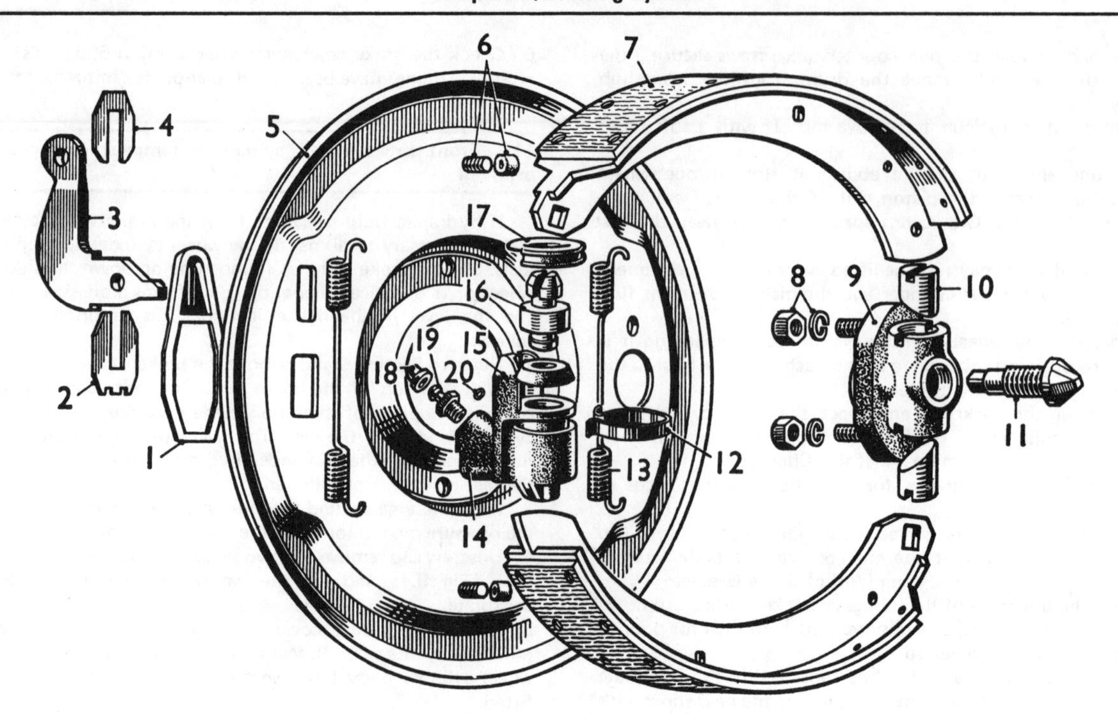

Fig 9.5 REAR DRUM BRAKE COMPONENTS

1 Rubber seal
2 Wheel cylinder locking plate
3 Handbrake lever
4 Wheel cylinder locking plate
5 Backplate
6 Steady post
7 Brake shoe
8 Nut and spring washer
9 Adjuster body
10 Adjuster tappets
11 Adjuster wedge
12 Dust cover clip
13 Shoe return spring
14 Pipe orifice
15 Cylinder body
16 Piston
17 Dust cover
18 Bleed nipple dust cover
19 Bleed nipple
20 Bleed valve ball

away the piston (16) together with its seal from the cylinder body (14).

9 Remove the piston seal using a non-metal pointed rod or the fingers. Do not use a metal screwdriver as this could scratch the piston.

10 Inspect the inside of the cylinder for score marks caused by impurities in the hydraulic fluid. If any are found the cylinder and piston will require renewal.

11 If the cylinder is sound, thoroughly clean it out with fresh hydraulic fluid.

12 The old rubber piston seals will probably be swollen and visibly worn. Smear the new rubbers with hydraulic fluid and fit the piston seal to the piston (16) taking care not to roll it otherwise it will stretch. The smallest diameter should be facing to the open end of the wheel cylinder when assembled.

13 Insert the piston (16) into the bore of the wheel cylinder with the slot to the outside of the wheel cylinder. Take care not to nick the leading edge of the seal as it is inserted.

14 Reassemble the dust cover (17) to the wheel cylinder.

15 Locate the neck of the wheel cylinder in the larger slot and replace the distance piece with cranked lips outwards.

16 Refit the handbrake lever (3) the correct way round engaging it in the piston slot and tap the locking plate into position between the distance piece and the spring plate until located by the spring plate.

17 Refitting the wheel cylinder is the reverse sequence to removal. It will be necessary to bleed the hydraulic system as described in Section 15.

18 Refer to Section 4 and adjust the brake shoe clearance.

13 Rear brake adjuster - removal and replacement

1 Refer to Section 10 and remove the brake shoes.

2 Undo and remove the two nuts and spring washers (8) (Fig 9.5) that secure the adjuster (9) to the backplate (5). Lift away the adjuster (9) from the backplate.

3 Remove the two adjuster tappets (10) from the adjuster body (9).

4 Unscrew the adjuster wedge (11) from the adjuster body (9).

5 Wash the parts in petrol or paraffin and dry thoroughly. Check that the tappets are a good fit in the adjuster body without signs of binding.

5 Inspect the squared end of the adjuster wedge (11) and if badly rounded obtain a new wedge.

6 Lubricate all moving parts with white brake grease and refit the wedge and two tappets.

7 Refitting the adjuster is the reverse sequence to removal.

8 Refer to Section 4 and adjust the brake shoe clearance.

14 Drum brake backplate - removal and replacement

1 To remove the brake backplate jack up the front or rear of the car depending on which backplate is to be removed and support on firmly based axle stands. Remove the road wheel.

2 Front backplate. Remove the front hub as described in Chapter 11.

3 Rear backplate. Refer to Chapter 8, Section 5 and remove the rear axle shaft and hub.

4 Wipe the top of the brake master cylinder and unscrew the cap. Place a piece of thick polythene over the top of the filler neck and replace the cap. This is to stop the fluid syphoning out when the brake pipe is disconnected from the rear of the wheel cylinder.

5 Disconnect the union from the rear of the wheel cylinder and plug the ends to stop dirt ingress into the hydraulic system.

6 Finally on rear brakes recover the handbrake attachment to the operating lever by extracting the split pin, lifting away the plain washer and withdrawing the clevis pin from the connection.

7 Unscrew the four bolts and self-locking nuts that secure the backplate to its mounting flange. The backplate complete with brake shoes and wheel cylinder/s may now be lifted away from the rear axle.

8 Refitting is the reverse sequence to removal. It will be necessary to bleed the hydraulic system and full details of the operation will be found in Section 15 of this Chapter.

15 Bleeding the hydraulic system

1 Removal of all the air from the hydraulic system is essential to the correct working of the braking system, but before undertaking this, examine the fluid reservoir cap to ensure that both vent holes, one on top and the second underneath but not in line, are clear; check the level of fluid and top up if required.
2 Check all brake line unions and connections for possible seepage, and at the same time check the condition of the rubber hoses which may be perished.
3 If the condition of the wheel cylinders is in doubt, check for possible signs of fluid leakage.
4 If there is any possibility of incorrect fluid having been put into the system drain all the fluid out and flush through with methylated spirits. Renew all piston seals and cups since they will be affected and could possibly fail under pressure.
5 Gather together a clean jar, a 9 inch length of tubing which fits tightly over the bleed nipple, and a tin of the correct brake fluid, ie: Castrol Girling Brake Fluid.
6 To bleed the system clean the areas around the bleed valve and start on the rear brakes first by removing the rubber cup over the bleed valve and fitting a rubber tube in position. The nipple location is shown in Fig 9.4 (front drum), Fig 9.2 (front disc) or Fig 9.5 (rear).
7 Place the end of the tube in a clean glass jar containing sufficient fluid to keep the end of the tube underneath during the operation.
8 Open the bleed valve with a spanner and quickly press down the brake pedal. After slowly releasing the pedal, pause for a moment to allow the fluid to recoup in the master cylinder and then depress again. This will force air from the system. Continue until no more air bubbles can be seen coming from the tube. At intervals make certain that the reservoir is kept topped up, otherwise air will enter at this point again.
9 Repeat this operation on all four brakes bleeding each calliper, mounting half first and then the rim half and when completed, check the level of the fluid in the reservoir. Check the feel of the brake pedal. This should be firm and free from any spongy action which is normally associated with air in the system. Tighten the bleed screws to a torque wrench setting of 5 to 7 lb ft.

16 Flexible hose - inspection, removal and replacement

Inspect the condition of the flexible hydraulic hoses leading from the chassis mounting metal pipes to the brake backplates. If they are swollen, damaged, cut or chafed they must be renewed.
1 Unscrew the metal pipe union nut from its connection to the hose, and then holding the hexagon on the base with a spanner, unscrew the attachment nut and washer.
2 The chassis end of the hose can now be pulled from the chassis mounting brackets and will be quite free.
3 Disconnect the flexible hydraulic hose at the backplate by unscrewing it from the brake cylinder or calliper. Note: When releasing the hose from the backplate, the chassis end must always be freed first.
4 Replacement is a straightforward reversal of the removal procedure.

17 Handbrake cable - adjustment

1 The lever type handbrake incorporates a thumb operated ratchet release in the handle. By pulling on the handbrake lever a system of cable operated levers operates on the rear wheel brake shoes only.
2 It is usual that when the rear brakes are adjusted, either manually or automatically, any excessive free movement of the handbrake will automatically be taken up. However, in time the cables will stretch and it will be necessary to take up the free play by shortening the cable at the point where the inner cable is connected to the handbrake lever.
3 Never try to adjust the handbrake to compensate for wear on the rear brake linings. It is usually badly worn brake linings that lead to the excessive handbrake travel.
4 If upon inspection the rear brake linings are in good condition, or recently renewed, and the handbrake tends to reach the end of its ratchet travel before the brakes operate, the cable must be shortened as follows:
5 Apply a little penetrating oil onto the threads of the sleeve nut at the front end of the longitudinal cable. Screw either in or out the brass adjusting nut until the cable slackness is removed with the rear shoes locked and the handbrake lever on the third notch of the ratchet.

18 Brake master cylinder - removal and replacement

1 Working inside the car disconnect the brake lever for the master cylinder pushrod by extracting the split pin and lifting away the plain washer and clevis pin.
2 Wipe the top of the brake master cylinder and unscrew the cap. Place a piece of thick polythene over the top of the master cylinder neck and refit the cap. This will prevent loss of hydraulic fluid when the master cylinder is removed.
3 Undo the union securing the pressure pipe to the master cylinder body and detach the pressure pipe. Plug the end of the pipe with a piece of tapered wood or a pencil to stop dirt ingress.
4 Undo and remove the two master cylinder securing bolts and spring washers.
5 Carefully lift away the master cylinder from its mounting, making sure that no hydraulic fluid comes into contact with paintwork as it acts as a solvent.
6 Refitting the brake master cylinder is the reverse sequence to removal. It will be necessary to bleed the hydraulic system as described in Section 15 of this Chapter.

19 Brake master cylinder - dismantling and reassembly

1 The brake master cylinder comprises an alloy body with a polished finish bore and reservoir with cap. The inner assembly is made up of the pushrod, dished washer circlip, plunger, plunger seal, spring thimble, plunger return spring, valve spacer, spring washer, valve stem and valve seal. The open end of the master cylinder is protected by a rubber dust cover. All these parts are shown in Fig 9.6.
2 Carefully ease the rubber boot (13) (Fig 9.6) from the master cylinder body (3) and slide up the pushrod (12).
3 Push the brake pushrod in slightly so as to relieve the pressure from the spring (7) inside the master cylinder body and, using a pair of long nosed pliers, extract the circlip (11) from the notched groove in the body of the master cylinder.
4 Lift out the pushrod assembly (10, 12) from the master cylinder together with the locating washer (10) and rubber boot (13).
5 Pull the piston and valve assembly (4 - 9) as one unit, from the master cylinder body (3).
6 Using a small screwdriver, raise the leaf of the spring thimble (8) to clear it from the shoulder on the plunger (9) and pull the thimble (8) from the plunger (9).
7 Compress the coil spring (7) and slip the valve stem (4) into the larger offset in the base of the thimble (8).
8 Slide the thimble (8) off the head of the valve stem (4). Lift away the spacer (6) and spring washer (5) from the valve stem. Note which way round the spring washer is fitted.
9 Remove the valve seat from the valve stem and the seal from the plunger using the fingers or pieces of tapered softwood or

plastic such as an old knitting needle.
10 Clean and carefully examine all parts, especially the plunger cup and rubber washer for signs of distortion, swelling, or other wear. Check the piston and cylinder for wear and scoring. Replace any parts that are faulty.
11 During the inspection of the plunger seal it has been found advisable to maintain the shape of this seal as regular as possible and for this reason do not turn it inside out as slight permanent distortion may be caused.
12 Rebuild the plunger and valve assembly in the following manner:

a) Fit the plunger seal to the plunger so that the larger circumference of the rubber lip will enter the cylinder bore first. The seal sits in the groove.
b) Fit the valve seal to the valve in the same manner.
c) Place the valve spring seal washer (5) so its convex face abuts the valve stem flange and then fit the seal spacer (6) and spring (7).
d) Fit the spring thimble (8) to the spring which must then be compressed so the valve stem (4) can be re-inserted in the thimble.
e) Replace the front of the plunger (9) in the thimble (8) and then press down the thimble lead so it is located under the shoulder at the front of the plunger.
f) Generously lubricate the assembly with hydraulic fluid and carefully replace it in the master cylinder body, taking care not to damage the rubber seal as it is inserted into the cylinder bore.

13 Fit the pushrod (12) and washer (10) in place and secure with the circlip (11). Replace the rubber boot (13).

20 Brake master cylinder - removal and replacement (servo assisted brakes)

1 Working inside the car disconnect the brake lever for the master cylinder pushrod by extracting the split pin and lifting away the plain washer and clevis pin.
2 Wipe the top of the brake master cylinder independent reservoir and unscrew the cap. Place a piece of thick polythene over the top of the master cylinder independent reservoir and refit the cap. This will prevent loss of hydraulic fluid when the master cylinder independent reservoir is removed.
3 Undo the unions securing the inlet and pressure pipes to the master cylinder independent reservoir and detach the pressure pipe. Plug the end of the pipe with a piece of tapered wood or a pencil to stop dirt ingress.
4 Undo and remove the two master cylinder independent reservoir securing bolts and spring washers.
5 Carefully lift away the master cylinder independent reservoir from its mounting making sure that no hydraulic fluid comes into contact with paintwork as it acts as a solvent.
6 Refitting the brake master cylinder independent reservoir is the reverse sequence to removal. It will be necessary to bleed the hydraulic system as described in Section 15 of this Chapter.

21 Brake master cylinder - dismantling and reassembly (servo assisted brakes)

1 The brake master cylinder comprises an alloy body with a polished finish bore. An independent reservoir is connected to the master cylinder by a metal pipe. The inner assembly is made up of the pushrod, dished washer circlip, plunger, plunger seal, spring thimble, plunger return spring, valve spacer, spring washer, valve stem and valve seal. The open end of the master cylinder is protected by a rubber dust cover. All these parts are shown in Fig 9.7.
2 Carefully ease the rubber boot (6) (Fig 9.7) from the master cylinder body (3) and slide up the pushrod (7).
3 Push the brake pushrod (7) in slightly so as to relieve the pressure from the spring (12) inside the master cylinder body (3) and using a pair of long nosed pliers extract the circlip (5) from the notched groove in the body of the master cylinder.
4 Lift out the pushrod assembly (4, 7) from the master cylinder body (3).
5 Pull the piston and valve assembly (8 - 16) as one unit from the master cylinder body (3).
6 Using a small screwdriver raise the leaf of the spring thimble (11) to clear it from the shoulder on the plunger (9) and pull the thimble (11) from the plunger (9).
7 Compress the coil spring (12) and slip the valve stem (15) into the larger offset in the base of the thimble (11).
8 Slide the thimble (11) off the head of the valve stem (15). Lift away the spacer (13) and spring washer (14) from the valve stem. Note which way round the spring washer is fitted.
9 Remove the valve seal (16) from the valve stem (15) and the seals (8, 10) from the plunger (9) using the fingers or a piece of tapered softwood or plastic such as an old knitting needle.
10 Clean and carefully examine all parts especially the plunger seal (10), end seal (8) and valve seal (16) for signs of distortion, swelling or other wear. Check the piston and cylinder for wear and scoring. Replace any parts that are faulty.
11 During the inspection of the plunger seals (8, 10) it has been found advisable to maintain the shape of this seal as regular as possible and for this reason do not turn it inside out as slight permanent distortion may be caused.
12 Rebuild the plunger and valve assembly in the following manner:

a) Fit the plunger seals (8, 10) to the plunger so that the larger circumference of the rubber lip will enter the cylinder bore first. The seals sit in the grooves.
b) Fit the valve seal (16) to the valve stem (15) in the same manner.
c) Place the valve spring seal washer (14) so that its convex face abuts the valve stem flange (15) and then fit the seal spacer (13) and spring (12).
d) Fit the spring thimble (11) to the spring (12) which must then be compressed so the valve stem (15) can be re-inserted in the thimble.
e) Replace the front of the plunger (9) in the thimble (11) and then press down the thimble leaf so it is located under the shoulder at the front of the plunger.
f) Generously lubricate the assembly with hydraulic fluid and carefully replace it in the master cylinder body taking care not to damage the rubber seal as it is inserted into the cylinder bore.

13 Fit the pushrod (7) and washer (4) in place and secure with the circlip (5). Replace the rubber boot (6).

22 Brake pedal - removal and replacement

1 Upon inspection it will be seen that the brake and clutch pedal linkages are mounted in a common bracket and therefore have to be removed as one unit before separating the brake pedal from the bracket.
2 Working inside the car, withdraw the split pin and remove the clevis pin so as to release the brake pedal from the brake master cylinder pushrod.
3 Withdraw the split pin and remove the clevis pin so as to release the clutch pedal from the clutch master cylinder pushrod.
4 Open the bonnet and locate the six bolts securing the pedal bracket to the engine bulkhead. Partially unscrew the six bolts until there is sufficient room to allow the brake and clutch pedal linkage bracket to be removed from the inside of the car. It is not necessary to completely remove the bolts.
5 Disconnect the pedal return springs from the brake and clutch pedals.
6 Undo and remove the nut that secures the brake and clutch pedal pivot shaft and carefully withdraw the pivot shaft.
7 Lift away the brake and clutch pedal levers and recover the distance piece.
8 If excessive pedal movement on the shaft other than in the normal direction is evident inspect the lever bush for wear and

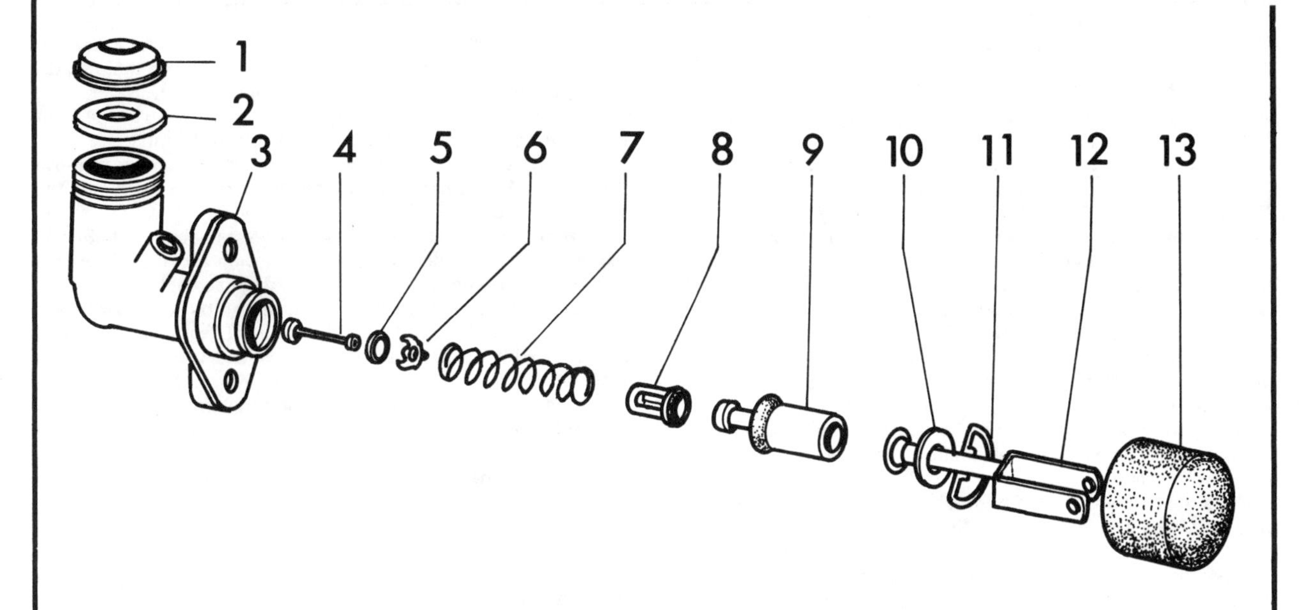

Fig 9.6 BRAKE MASTER CYLINDER COMPONENTS

1 Filler cap
2 Washer
3 Master cylinder
4 Valve stem
5 Spring washer
6 Valve spacer
7 Return spring
8 Thimble
9 Plunger
10 Dished washer
11 Circlip
12 Fork
13 Dust cover

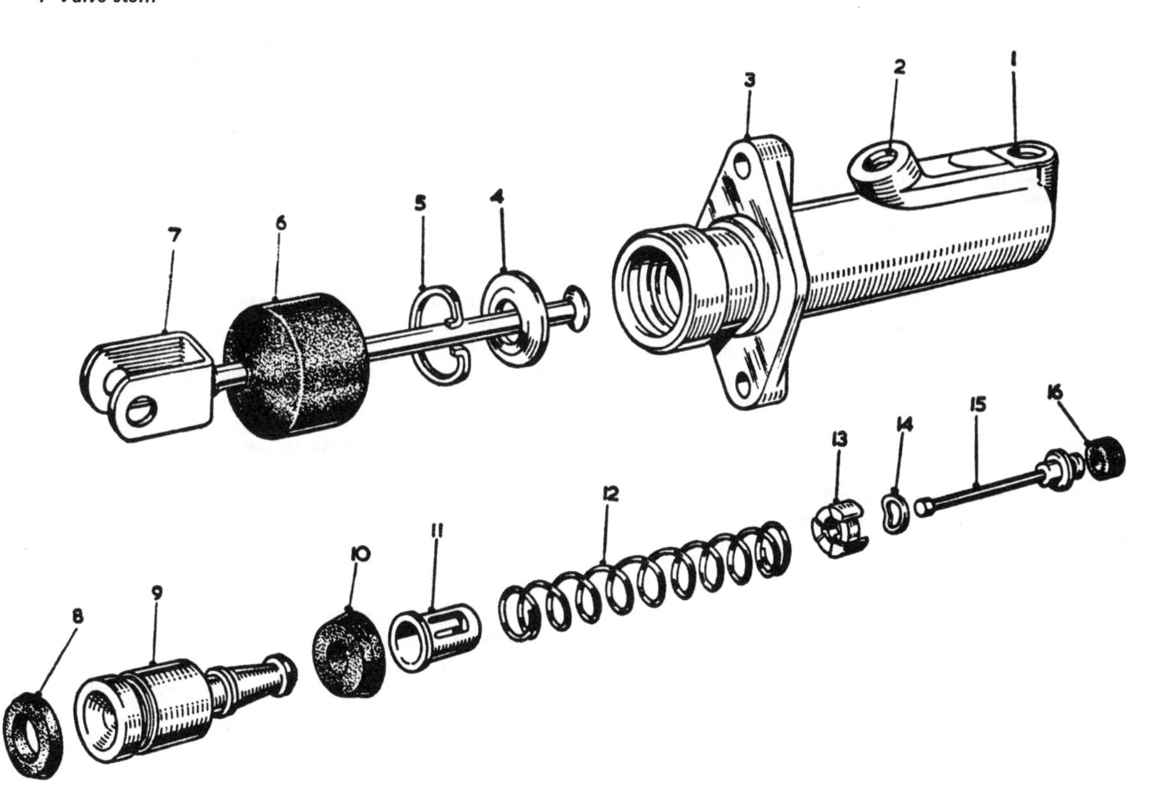

Fig 9.7 BRAKE MASTER CYLINDER COMPONENTS (SERVO ASSISTED)

1 Fluid inlet
2 Fluid outlet
3 Master cylinder
4 Dished washer
5 Circlip
6 Dust cover
7 Pushrod
8 End seal
9 Plunger
10 Plunger seal
11 Thimble
12 Return spring
13 Valve spacer
14 Spring washer
15 Valve stem
16 Valve seal

if evident it is possible to drift out the old bush and fit a new one.
9 Refitting is the reverse sequence to removal. Lubricate the pedal pivot bushes with Castrol GTX.

23 Brake servo unit - removal and refitting

1 The brake servo unit is located under the right hand front wing behind the road wheel.
2 Working under the bonnet slacken the clip and detach the top end of the rubber vacuum hose from the vacuum pipe adjacent to the brake and clutch master cylinders.
3 Unscrew the servo unit hydraulic inlet pipe union from the three-way connection which is situated next to the horn on the wing valance.
4 Jack up the front of the car and support on firmly based axle stands located under the main chassis members. Remove the right hand wheel.
5 Wipe the area around the hydraulic outlet pipe union on the servo unit body and then unscrew the union.
6 Wipe the top of the brake master cylinder reservoir and unscrew the cap. Place a piece of thick polythene over the top of the master cylinder reservoir neck and refit the cap. This will prevent loss of hydraulic fluid when the servo unit is removed.
7 Using a piece of tapered wood, such as a pencil, plug the ends of the pipes to prevent dirt ingress into the hydraulic system.
8 Undo and remove the three set bolts securing the servo unit to the mounting bracket under the front wing.
9 Remove the servo unit with vacuum hose and inlet pipe which may now be pulled downwards and lifted away from under the wing.
10 Refitting the servo unit is the reverse sequence to removal. It will be necessary to bleed the hydraulic system as described in Section 15 of this Chapter.

24 Brake servo unit - dismantling and reassembly

1 Before commencing to dismantle the servo unit thoroughly clean it with a stiff brush and wipe with a non-fluffy rag. It cannot be too strongly emphasised that cleanliness is important when working on the unit. To dismantle the unit proceed as follows, making reference to Fig 9.8.
2 Firmly hold the servo unit in a vice at the mounting lugs on the body.
3 Undo and remove the seven nuts and bolts from the vacuum cylinder flange. This will release the end cover but take care as it will be under the influence of the strong vacuum cylinder piston return spring (65).
4 Gently allow the spring to extend and then lift away the end cover, gasket (67), piston and seal and finally the return spring.
5 Inspect the piston rod for signs of scoring and if this is evident the complete servo unit must be replaced.
6 Unscrew and remove the three set bolts from inside the closed end of the cylinder. It will be noticed that there are three copper washers (17) under the set bolt heads and these must be renewed upon reassembly.
7 Lift away the clamp plate from inside the cylinder.
8 Carefully pull the body away from the cylinder (1) easing the vacuum pipe from the rubber grommet in the cylinder flange.
9 Push the air filter cover spring clip aside and lift away the cover and air filter element (71). Recover the filter to body rubber sealing washer (72).
10 Undo and remove the four screws which retain the valve chest cover and take off the cover with vacuum pipe and gasket (69).
11 Undo and remove the two screws from inside the valve chest and lift out the valve retaining plate and the valves together with their rocking lever.
12 It will be noticed that the valve plates are attached to the rocking lever by two separate wire clips.
13 Carefully pull the gasket (16) off the face of the body and then by tapping the face on a wooden block remove the plug (9) sealing the valve operating piston bore (upper bore).
14 The control piston assembly (2 - 8) will now be pushed out by its spring (6). Lift out the complete piston assembly.
15 Should it be necessary to dismantle this assembly, remove the circlip (8) from the large diameter end so as to release the washer (7) and spring (6).
16 Lift off the square seating spring washer abutment (5) and remove the two tapered seals (2, 4) using a piece of tapered softwood or plastic such as an old knitting needle.
17 To remove the components from the output cylinder (lower bore) pull out the end guide bush (15), ease up the gland seal (14) with a thin blade screwdriver and lift out the nylon spacer.
18 Using a pair of circlip pliers extract the circlip (13) from its groove in the bore.
19 With the circlip removed the output piston (11) complete with stop washers will now be ejected from the bore by the piston return spring (10).
20 Upon inspection the piston assembly will be found to have two seals, a tapered seal which is located in an annular groove in the outside of the piston and a second washer like seal is held in place by a metal cup which is pressed into the end of the piston. This second seal also seals the end of the piston rod when the brakes are applied. This piston assembly cannot be dismantled and if necessary a new assembly must be fitted.
21 Thoroughly wash all parts in methylated spirits or clean Castrol Girling Brake Fluid.
22 Examine all metal parts for signs of corrosion, pitting or scoring. The piston rod, pistons and bores must be free from scoring or steps. Any parts that are suspect must be renewed.
23 Before reassembly thoroughly lubricate all hydraulic parts such as pistons, seals and bore with Castrol Girling Brake Fluid.
24 Carefully fit a new taper seal (12) to a new output piston assembly with the taper facing the smaller end of the piston.
25 Assemble the return spring (10) piston and washers (11, 12) to the output cylinder (lower bore).
26 Press these components down into the bore against the tension of the spring and, using suitable circlip pliers, insert the circlip (13) into its groove in the bore.
27 Insert the large end of the seal spacer first into the bore and then ease in the gland seal (14) with the taper towards the output piston and then push in the guide bush (15) until the flange is level with the face of the body.
28 Fit new seals (2, 4) to the valve control piston (3) positioning the large tapered seal (4) with the taper facing the spring (6) and the small tapered seal (2) with the taper facing away from the spring (6).
29 Position the abutment washer (5), spring (6) and retaining washer (7) on the piston (3).
30 Press the spring (6) down and insert the circlip (8) into the groove. Position the piston assembly into the bore aligning the hole in the piston with the hole in the side of the bore.
31 Fit the valve plates to the valve rocking lever and secure with wire clips. Position the assembly into the valve chest, taking care to engage the ball end of the lever in the valve control piston (3).
32 Position the valve retainer over the valve assembly and secure with the two screws and washers.
33 Test the operation of the valve gear by depressing and releasing the valve operating piston. The valves must move freely and in the normal position the valve nearest the body flange should be open and the other valve closed.
34 Fit a new seal (9) into the groove in the valve control bore plug. Insert the plug into the bore until about 1/16 inch of the plug stands proud of the body face.
35 Refit the valve chest cover gasket (69) cover and vacuum pipe and secure the cover with four screws and shakeproof washers.
36 The vacuum cylinder may now be refitted to the body. Fit a new gasket (16), positioning the retaining plate in the vacuum pipe and fit the cylinder to the body whilst at the same time easing the vacuum pipe into the grommet in the cylinder flange.
37 Place the clip plate inside the cylinder and secure with the

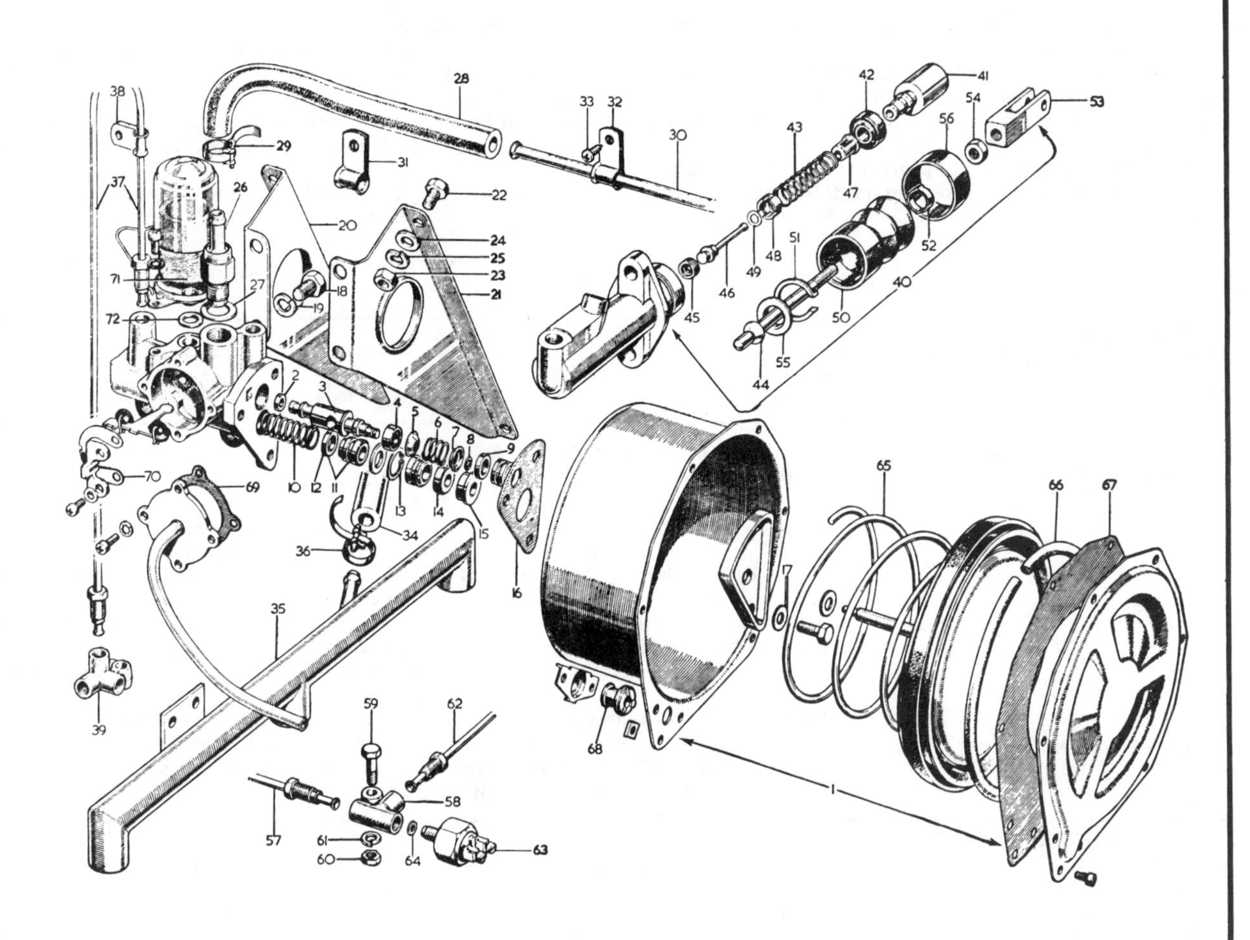

Fig 9.8 SERVO ASSISTED BRAKING SYSTEM COMPONENTS

1 Vacuum servo unit
2 Valve operating piston seal (small)
3 Valve operating piston
4 Valve operating piston seal (large)
5 Valve operating piston spring abutment
6 Valve operating piston return spring
7 Return spring retainer
8 Return spring retainer circlip
9 Cylinder plug seal
10 Hydraulic piston return spring
11 Hydraulic piston assembly
12 Hydraulic piston tappet seal
13 Sealing spacer retaining circlip
14 Vacuum cylinder piston gland seal
15 Vacuum cylinder piston guide bush
16 Vacuum cylinder to body gasket
17 Washer (copper)
18 Servo unit to mounting bracket screw
19 Washer (spring)
20 R/H servo unit mounting bracket
21 L/H servo unit mounting bracket
22 Bracket to pedal box screw
23 Nut
24 Washer (plain)
25 Washer (spring)
26 Non-return valve
27 Valve to servo unit gasket
28 Valve to vacuum pipe hose
29 Hose clip
30 Vacuum pipe
31 Pipe to pedal box top clip
32 Pipe to dash panel clip
33 Screw
34 Vacuum pipe to balance pipe hose
35 Balance pipe
36 Hose clip
37 Servo unit to 4-way connection pipe
38 Pipe clip
39 4-way connection
40 Brake master cylinder assembly
41 Plunger
42 Plunger seal
43 Spring
44 Pushrod
45 Valve seal
46 Valve stem
47 Spring retainer
48 Valve spacer
49 Washer (spring)
50 Dust cover
51 Circlip
52 Dust cover clip
53 Fork-end
54 Locknut
55 Retaining washer
56 Retaining band
57 Master cylinder to 3-way connection pipe
58 3-way connection
59 3-way connection to wheel arch bolt
60 Nut
61 Washer (spring)
62 3-way connection to servo unit pipe
63 Stop light switch
64 Switch to 3-way connection gasket
65 Vacuum cylinder piston return spring
66 Locking plate seal
67 Vacuum cylinder end cover gasket
68 Vacuum pipe to cylinder sleeve (rubber)
69 Vacuum pipe to body gasket
70 Control valve return spring
71 Air filter element
72 Filter to body sealing washer (rubber)

three bolts and copper washers. Tighten the bolts to a torque wrench setting of 10 to 12 lb ft.

38 As the piston seal and cylinder are specially processed during manufacture it is not necessary to lubricate them. The sealing sponge rubber backing ring (66) and the end cover gasket, however, must be renewed.

39 Insert the return spring (65) piston and piston seal assembly into the cylinder taking extreme care not to damage the rod or central bearing guide bush (15).

40 Push the piston fully home and fit the end cover using a new gasket (60). Secure the end cover and vacuum pipe retaining plate with the seven nuts and bolts.

41 Refit the moulded cellular air filter element (71) with a new sealing washer (72). Replace the cover and secure with the spring clip.

42 The servo unit is now ready for refitting to the car.

25 Fault diagnosis

Cause	Trouble	Remedy
SYMPTOM: PEDAL TRAVELS ALMOST TO FLOORBOARDS BEFORE BRAKES OPERATE		
Leaks and air bubbles in hydraulic system	Brake fluid level too low	Top up master cylinder reservoir. Check for leaks.
	Wheel cylinder leaking	Dismantle wheel cylinder, clean, fit new rubbers and bleed brakes.
	Master cylinder leaking (bubbles in master cylinder fluid)	Dismantle master cylinder, clean, and fit new rubbers. Bleed brakes.
	Brake flexible hose leaking	Examine and fit new hose if old hose leaking.
	Brake lining fractured	Replace with new brake pipe. Bleed brakes.
	Brake system unions loose	Check all unions in brake system and tighten as necessary. Bleed brakes.
Normal wear	Linings over 75 per cent worn	Fit replacement shoes and brake linings.
Incorrect adjustment	Brakes badly out of adjustment	Jack up car and adjust brakes.
	Master cylinder pushrod out of adjustment causing too much pedal free movement	Reset to manufacturer's specification.
SYMPTOM: BRAKE PEDAL FEELS SPRINGY		
Brake lining renewal	New linings not yet bedded-in	Use brakes gently until springy pedal feeling leaves.
Excessive wear or damage	Brake drums badly worn and weak or cracked	Fit new brake drums.
Lack of maintenance	Master cylinder securing nuts loose	Tighten master cylinder securing nuts. Ensure spring washers are fitted.
SYMPTOM: BRAKES TEND TO BIND, DRAG OR LOCK-ON		
Incorrect adjustment	Master cylinder pushrod out of adjustment giving too little brake pedal free movement	Reset to manufacturer's specifications.
Wear or dirt in hydraulic system or incorrect fluid	Reservoir vent hole in cap blocked with dirt	Clean and blow through hole.
	Master cylinder bypass port restricted brakes seize in 'on' position	Dismantle, clean and overhaul master cylinder. Bleed brakes.
	Wheel cylinder seizes in 'on' position	Dismantle, clean and overhaul wheel cylinder. Bleed brakes.
Mechanical wear	Brake shoe pull off springs broken, stretched or loose	Examine springs and replace if worn or loose.
Incorrect brake assembly	Brake shoe pull off springs fitted wrong way round, omitted or wrong type used	Examine and rectify as appropriate.
Neglect	Handbrake system rusted or seized in the 'on' position	Apply Plus Gas to free, clean and lubricate.
SYMPTOM: BRAKE PEDAL FEELS SPONGY AND SOGGY		
Leaks or bubbles in hydraulic system	Wheel cylinder leaking	Dismantle wheel cylinder, clean, fit new rubbers and bleed brakes.
	Master cylinder leaking (bubbles in master cylinder reservoir)	Dismantle master cylinder, clean and fit new rubbers and bleed brakes. Replace cylinder if internal walls scored.
	Brake pipe line or flexible hose leaking	Fit new pipeline or hose.
	Unions in brake system loose	Examine for leaks, tighten as necessary.
SYMPTOM: EXCESSIVE EFFORT REQUIRED TO BRAKE CAR		
Lining type or condition	Linings badly worn	Fit replacement brake shoes and linings.
	New linings recently fitted - not yet bedded-in	Use brakes gently until braking effort normal.
	Harder linings fitted than standard causing increase in pedal pressure	Remove linings and replace with normal units.

Cause	Trouble	Remedy
Oil or grease leaks	Linings and brake drums contaminated with oil, grease or hydraulic fluid	Rectify source of leak, clean brake drums, fit new linings.
SYMPTOM: BRAKES UNEVEN AND PULLING TO ONE SIDE		
Oil or grease leaks	Linings and brake drums contaminated with oil, grease or hydraulic fluid	Ascertain and rectify source of leak, clean brake drums, fit new linings.
Lack of maintenance	Tyre pressures unequal	Check and inflate as necessary.
	Radial ply tyres fitted at one end of car only	Fit radial ply tyres of the same make to all four wheels.
	Brake backplate loose	Tighten backplate securing nuts and bolts.
	Brake shoes fitted incorrectly	Remove and fit shoes correct way round.
	Different type of linings fitted at each wheel	Fit the linings specified by the manufacturers all round.
	Anchorages for front suspension or rear axle loose	Tighten front and rear suspension pick-up points including spring anchorage.
	Brake drums badly worn, cracked or distorted	Fit new brake drums.
SYMPTOM: BRAKES TEND TO BIND, DRAG OR LOCK-ON		
Incorrect adjustment	Brake shoes adjusted too tightly	Slacken off brake shoe adjusters two clicks.
	Handbrake cable over-tightened	Slacken off handbrake cable adjustment.

Chapter 10 Electrical system

Contents

Specifications

Battery	12 volt
Home (standard)	GTW 9A
Dry charged (export)	GTZ 9A
Capacity - 10 hour rate	51 amp hr
20 hour rate	58 amp hr
Electrolyte to fill one cell	1 pint (0.57 litre)
Initial charging current	3.5 amp
Normal recharge current	5 amp
Master switch	Lucas Type ST 330
Dynamo	Lucas C45 PV-5
Cutting-in speed	1100 to 1250 dynamo rpm
Maximum output	22 amps, 13.5 volts at 1700 to 1900 dynamo rpm
Field resistance	6 ohms
Starter motor	Lucas M418G
Lock torque	17 lb ft (1.2858 kg m) at 440 to 460 amps and 7.0 to 7.4 volts
Light running current	45 amps at 7400 to 8500 rpm
Solenoid switch	Lucas Type ST 950
Overdrive (optional extra)	
Control switch	Lucas Type 2TS
Transmission gear solenoid	Lucas Type TGS1
Relay - overdrive	Lucas Type SB 40-1
Interrupter switch	Lucas Type 5510-1
Rotary throttle switch	Lucas Type RTS1

Control box	Lucas RB 106/2
Cut-out: Cut-in voltage	12.7 to 13.3 volts
Drop off voltage	8.5 to 11 volts
Reverse current	3.5 to 5 amps

Regulator

Setting on open circuit at 68°F (20°C)	16.0 to 16.6 volts at 3000 dynamo rpm

Note: For circuit temperature other than 20°C the following allowances should be made to the above setting:

For every 10°C (18°F) above 20° subtract 0.1 volt
For every 10°C (18°F) below 20° add 0.1 volt

Windscreen wiper	Lucas DR2
Normal running current	2.3 to 3.1 amp at 12 volts
Stall current (motor hot)	8 amp
Stall current (motor cold)	14 amp
Armature resistance (adjacent commutator segments)	0.34 to 0.41 ohms
Field resistance	12.8 to 14.00 ohms

NB – On some high output motors, usually identified by a red insulating piece above the terminals, the field resistance is 8.0 to 11.5 ohms

Austin Healey BN6

The following information is applicable to the Austin Healey BN6 and should be used in conjunction with the preceeding specifications

Battery

Home (standard)	SLG 11E
Export (dry charged)	SLGZ 11E
Voltage	(2) 6 volt
Capacity: 10 hour rate	50 amp hr
20 hour rate	58 amp hr
Electrolyte to fill one cell	1 pint (0.57 litre)

Austin Healey Mk I and II, Series BN7 and BT7, Series BJ7, Mk III Series BJ8

Battery (Series BN7 - two seater):

Standard	SLG 11E (two)
Dry charged (export only)	SLGZ 11E (two)

(Series BT7, BJ7 and BJ8 - four seater)

Standard	BT9A
Dry charged (export only)	BTZ9A
Voltage	12 volt positive earth
Capacity: 20 hour rate	58 amp hr
Electrolyte to fill one cell	1 pint (0.57 litre)
Initial charging current	3.5 amp
Normal recharge current	5 amp
Master switch	Lucas Type ST 330

Dynamo

Mk I and II	Lucas C45 PV-6
Cutting-in speed	1100 to 1250 dynamo rpm
Maximum output	25 amps, 13.5 volts at 1700 to 1900 dynamo rpm
Field resistance	6 ohms
Brush length (minimum)	7/16 inch (11.11 mm)
Brush spring tension (maximum)	34 - 44 oz (965 - 1248 gm)
Mk III	Lucas C42
Cutting-in speed	1250 dynamo rpm
Maximum output	30 amps, 13.5 volts at 1250 dynamo rpm
Field resistance	4.5 ohms

Starting motor	Lucas M418G
Lock torque	17 lb ft (1.2858 kg m) at 440 to 460 amps at 7.0 to 7.4 volts
Lock current draw	430 to 450 amps at 7.0 to 7.4 volts
Light running current	45 amps at 7400 to 8500 rpm
Solenoid switch	Lucas type ST 950
Brush length (minimum)	5/16 inch (7.94 mm)
Brush spring tension (maximum)	30 - 40 oz (850 - 1133 gm)

Overdrive (optional extra)

Control switch	Lucas Type 2TS
Transmission gear solenoid	Lucas Type 11S
Relay - overdrive	Lucas Type SB 40-1
Interrupter switch	Lucas Type SS10
Rotary throttle switch	Lucas Type RTS1

Control box
Mk I and II

Type	Lucas RB 106/2
Cut-out relay: Cut-in voltage	12.7 to 13.3 volts
Drop off voltage	8.5 to 11 volts
Reverse current	3.5 to 5 amps
Setting on open circuit at 20°C (68°F)	16.0 to 16.6 volts at 3000 dynamo rpm

Note: For circuit temperature other than 20°C the following allowances should be made to the above setting:

For every 10°C (18°F) above 20° subtract 0.1 volt
For every 10°C (18°F) below 20° add 0.1 volt

Mk III	Lucas RB 340
Cut-out relay: Cut-in voltage	12.7 to 13.3 volts
Drop off voltage	9.5 to 11.0 volts
Voltage setting at 4500 rpm:	
10°C (50°F)	14.9 to 15.5 volts
20°C (68°F)	14.7 to 15.3 volts
30°C (86°F)	14.5 to 15.1 volts
40°C (104°F)	14.3 to 14.9 volts
Reverse current	3.0 to 5 amps
Current regulator	30 ± 1 amp

Windscreen wiper motor

Mk I and II	Lucas DR2
Normal running current	2.3 to 3.1 amp at 12 volts
Stall current (motor hot)	8 amp
Stall current (motor cold)	14 amp
Armature resistance (adjacent commutator segments)	0.34 to 0.41 ohms
Field resistance	8.0 to 9.5 ohms

NB – On some high output motors, usually identified by a red insulating piece above the terminals, the field resistance is 8.0 to 11.5 ohms

Mk II and III (from body No 60792 Mk II)	Lucas DR3A
Drive to wheelboxes	Rack and cable
Armature end float	0.008 to 0.012 inch (0.20 to 0.30 mm)
Running current	2.7 to 3.4 amps
Wiping speed	45 to 50 cycles per minute

Fuses

A1 - A2	50 amp
A3 - A4	35 amp

Replacement bulbs	Watts	BLMC Part No
Headlamp Rhd	50/40	BFS 414
Headlamp Lhd (not Europe)	50/40	BFS 415
Pilot lamps (combined flashing indicators)	6/21	BFS 380
Stop/tail lamp	6/21	BFS 380
Number plate lamp	6	BFS 989
Warning and panel lights	2.2	BFS 987

Series BN7, BT7, BJ7 and BJ8 as above but with the following exceptions:

Pilot lamps	6	BFS 207
Stop and tail lamps	21/6	BFS 380
Direction indicator lamps	21	BFS 382
Number plate illumination lamp	6	BFS 207

1 General description

The electrical system fitted to all models covered by this manual is of the conventional 12 volt type. The major components comprise a 12 volt battery (except Series BN6 which has two 6 volt batteries) with the positive terminal earthed, a control box and fuse box, a Lucas dynamo which is belt driven from the crankshaft pulley wheel and a starter motor.

The ignition system is also part of the electrical system but because of its importance and complexity is covered separately in Chapter 4.

The battery is located at the rear of the car either in the rear luggage compartment but on BN6 models the two batteries are located under the floor.

The dynamo is mounted on the right hand side of the cylinder block and its position is adjustable so that the drive belt can be kept at the correct tension.

The control box normally does not require attention but should this be necessary an accurate moving coil voltmeter and ammeter will be required.

The starter motor is mounted on the flywheel housing on the right hand side of the engine and operates on the starter ring gear located on the circumference of the flywheel through a sliding pinion drive.

The headlamps are of the double filament replaceable bulb design.

In cases of serious mechanical or electrical failures in the major units, such as the dynamo, starter motor or windscreen wiper motor, the best course of action to be taken after confirmation is to obtain and fit a new unit on a factory exchange basis. When considered necessary the author indicates this step in the relevant section.

It is important that the battery positive earth lead is always disconnected if the battery is to be boost charged or if body repairs are to be carried out using electric arc welding equipment as otherwise serious damage could be caused to the more delicate instruments.

2 Battery - removal and replacement

1 One battery is fitted to all models except the BN6 which has two 6 volt batteries. The single battery is accessible through the rear luggage compartment whereas the twin battery installation is accessible from the inside once the hinged lid has been released.

2 The earthed battery should always be removed first by slackening the retaining nuts and bolts or unscrewing the retaining screws if these are fitted.

3 Remove the battery securing clamp if one is fitted and carefully lift the battery out of its compartment. Hold it vertically to ensure that none of the electrolyte is spilled.

4 Replacement is a direct reversal of this procedure. Note: Replace the negative lead before the earth (positive) lead and smear the terminals with petroleum jelly (Vaseline) to prevent corrosion. NEVER use ordinary grease as applied to other parts of the car.

3 Battery - maintenance and inspection

1 Normal weekly battery maintenance consists of checking the electrolyte level of each cell to ensure that the separators are covered by ¼ inch of electrolyte. If the level has fallen, top up the battery using distilled water only. Do not overfill. If a battery is overfilled or any electrolyte spilled, immediately wipe away the excess as electrolyte attacks and corrodes any metal it comes into contact with, very rapidly.

2 As well as keeping the terminals clean and covered with petroleum jelly, the top of the battery, and especially the top of the cells, should be kept clean and dry. This helps prevent corrosion and ensures that the battery does not become partially discharged by leakage through dampness and dirt.

3 Once every three months remove the battery and inspect the battery securing bolts, the battery clamp plate, tray and battery leads for corrosion (ie: white fluffy deposits on the metal which are brittle to touch). If any corrosion is found, clean off the deposits with ammonia and paint over the clean metal with an anti-rust/anti-acid paint.

4 At the same time inspect the battery case for cracks. If a crack is found, clean and plug it with one of the proprietary compounds marketed by firms such as Holts for this purpose. If leakage through the crack has been excessive then it will be necessary to refill the appropriate cell with fresh electrolyte as detailed later. Cracks are frequently caused to the top of the battery cases by pouring in distilled water, in the middle of winter, AFTER instead of BEFORE a run. This gives the water no chance to mix with the electrolyte and so the former freezes and splits the battery case.

5 If topping up the battery becomes excessive and the case has been inspected for cracks that could cause leakage, but none are found, the battery is being overcharged and the voltage regulator will have to be checked and reset.

6 With the battery on the bench at the three monthly interval check, measure its specific gravity with a hydrometer to determine the state of charge and condition of the electrolyte. There should be very little variation between the different cells and if a variation in excess of 0.025 is present it will be due to either:

a) Loss of electrolyte from the battery sometimes caused by spillage or a leak, resulting in a drop in the specific gravity of the electrolyte when the deficiency was replaced with distilled water instead of fresh electrolyte.

b) An internal short circuit caused by buckling of the plates or a similar malady pointing to the likelihood of total battery failure in the near future.

7 The specific gravity of the electrolyte for fully charged conditions at the electrolyte temperature indicated is listed in Table A. The specific gravity of a fully discharged battery at different temperatures of the electrolyte is given in Table B.

8 Specific gravity is measured by drawing up into the body of a hydrometer sufficient electrolyte to allow the indicator to float freely (see Fig 10.1). The level at which the indicator floats indicates the specific gravity.

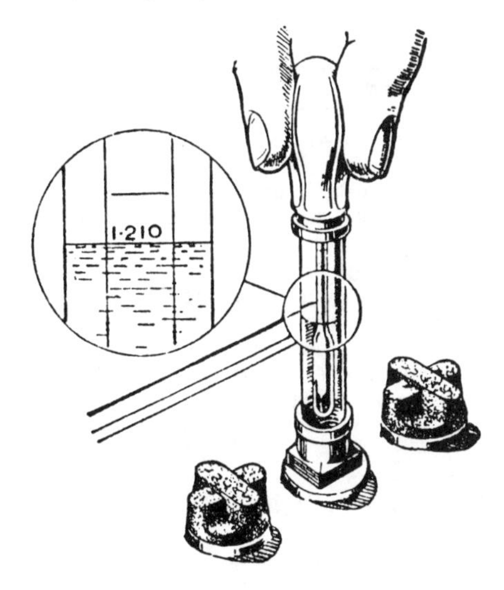

Fig 10.1 USING HYDROMETER TO TEST BATTERY SPECIFIC GRAVITY

Table A

Specific gravity - battery fully charged

1.268 at	100°F or 38°C	electrolyte temperature
1.272 at	90°F or 32°C	" "
1.276 at	80°F or 27°C	" "
1.280 at	70°F or 21°C	" "
1.284 at	60°F or 16°C	" "
1.288 at	50°F or 10°C	" "
1.292 at	40°F or 4°C	" "
1.296 at	30°F or -1.5°C	" "

Table B

Specific gravity - battery fully discharged

1.098 at	100°F or 38°C	electrolyte temperature
1.102 at	90°F or 32°C	" "
1.106 at	80°F or 27°C	" "
1.110 at	70°F or 21°C	" "
1.114 at	60°F or 16°C	" "
1.118 at	50°F or 10°C	" "
1.122 at	40°F or 4°C	" "
1.126 at	30°F or -1.5°C	" "

4 Battery - electrolyte replenishment

1 If the battery is in a fully charged state and one of the cells maintains a specific gravity reading which is 0.025 or more lower than the other, and a check of each cell has been made with a voltage meter to check for short circuits (a four to seven second test should give a steady reading of between 1.2 and 1.8 volts), then it is likely that electrolyte has been lost from the cell with the low reading at some time.
2 Top the cell up with a solution of 1 part sulphuric acid to 2.5 parts of water. If the cell is already fully topped up draw some electrolyte out of it with a pipette. The total capacity of each cell is ¾ pint.
3 When mixing the sulphuric acid and water NEVER ADD WATER TO SULPHURIC ACID - always pour the acid slowly onto the water in a glass container. IF WATER IS ADDED TO SULPHURIC ACID IT WILL EXPLODE.
4 Continue to top up the cell with the freshly made electrolyte and then recharge the battery and check the hydrometer readings.

5 Battery - charging

1 In winter time when heavy demand is placed upon the battery such as when starting from cold, and much electrical equipment is continually in use, it is a good idea to occasionally have the battery fully charged from an external source at the rate of 3.5 to 4 amps.
2 Continue to charge the battery at this rate until no further rise in specific gravity is noted over a four hour period.
3 Alternatively, a trickle charger, charging at the rate of 1.5 amps can be safely used overnight.
4 Specially rapid boost chargers which are claimed to restore the power of the battery in 1 to 2 hours are most dangerous as they can cause serious damage to the battery plates through overheating.
5 While charging the battery note that the temperature of the electrolyte should never exceed 100°F.

6 Dynamo - maintenance

1 Routine maintenance consists of checking the tension of the fan belt, and lubricating the dynamo rear bearing once every 6000 miles.
2 The fan belt should be tight enough to ensure no slip between the belt and the dynamo pulley. If a shrieking noise comes from the engine when the unit is accelerated rapidly, it is likely that it is the fan belt slipping. On the other hand, the belt must not be too taut or the bearings will wear rapidly and cause dynamo failure or bearing seizure. Ideally ½ inch of total free movement should be available at the fan belt midway between the fan and the dynamo.
3 To adjust the fan belt tension, slightly slacken the three dynamo retaining bolts, and swing the dynamo on the upper two bolts outwards to increase the tension, and inwards to lower it.
4 It is best to leave the bolts fairly tight so that considerable effort has to be used to move the dynamo, otherwise it is difficult to get the correct setting. If the dynamo is being moved outwards to increase the tension and the bolts have only been slackened a little, a long spanner acting as a lever placed behind the dynamo with the lower end resting against the block works very well in moving the dynamo outwards. Retighten the dynamo bolts and check that the dynamo pulley is correctly aligned with the fan belt.
5 Lubrication on the dynamo consists of inserting three drops of SAE 30 Engine Oil in the small oil hole in the centre of the commutator end bracket. This lubricates the rear bearing. The front bearing is pre-packed with grease and requires no attention.

7 Dynamo - testing in position

1 If, with the engine running, no charge comes from the dynamo, or the charge is very low, first check that the fan belt is in place and is not slipping. Then check that the leads from the control box to the dynamo are firmly attached and that one has not come loose from its terminal.
2 The lead from the D terminal on the dynamo should be connected to the D terminal on the control box, and similarly the F terminals on the dynamo and control box should also be connected together. Check that this is so and that the leads have not been incorrectly fitted.
3 Make sure none of the electrical equipment such as the lights or radio is on, and then pull the leads off the dynamo terminals marked D and F. Join the terminals together with a short length of wire.
4 Attach to the centre of this length of wire the negative clip of a 0-20 volts voltmeter and run the other clip to earth on the dynamo yoke. Start the engine and allow it to idle at approximately 750 rpm. At this speed the dynamo should give a reading of about 15 volts on the voltmeter. There is no point increasing the engine speed above a fast idle as the reading will then be inaccurate.
5 If no reading is recorded then check the brushes and brush connections. If a very low reading of approximately 1 volt is observed then the field winding may be suspect.
6 If a reading of between 4 to 6 amps is recorded it is likely that the armature winding is at fault.
7 On early dynamos it was possible to remove the dynamo cover band and check the dynamo and brushes in position. With the Lucas C40-1 windowless yoke dynamo it must be removed and dismantled before the brushes and commutator can be attended to.
8 If the voltmeter shows a good reading, then with the temporary link still in position connect both leads from the control box to D and F on the dynamo (D to D and F to F). Release the lead from the D terminal at the control box end and clip one lead from the voltmeter to the end of the cable, and the other lead to a good earth. With the engine running at the same speed as previously, an identical voltage to that recorded at the dynamo should be noted on the voltmeter. If no voltage is recorded there is a break in the wire. If the voltage is the same as recorded at the dynamo then check the F lead in similar fashion. If both readings are the same as at the dynamo then it will be necessary to test the control box.

8 Dynamo - removal and replacement

1 Slacken the three dynamo attachment bolts and also the nut that secures the adjustment link to the cylinder block. Move the dynamo towards the cylinder block so that the fan belt can be lifted over the dynamo pulley.
2 Disconnect the two leads from the dynamo terminals at the rear of the back end plate.
3 Remove the three dynamo attachment bolts and lift away the dynamo.
4 Replacement is the reverse sequence to removal. Do not finally tighten the retaining bolt and nut on the adjustment link until the fan belt has been tensioned correctly.

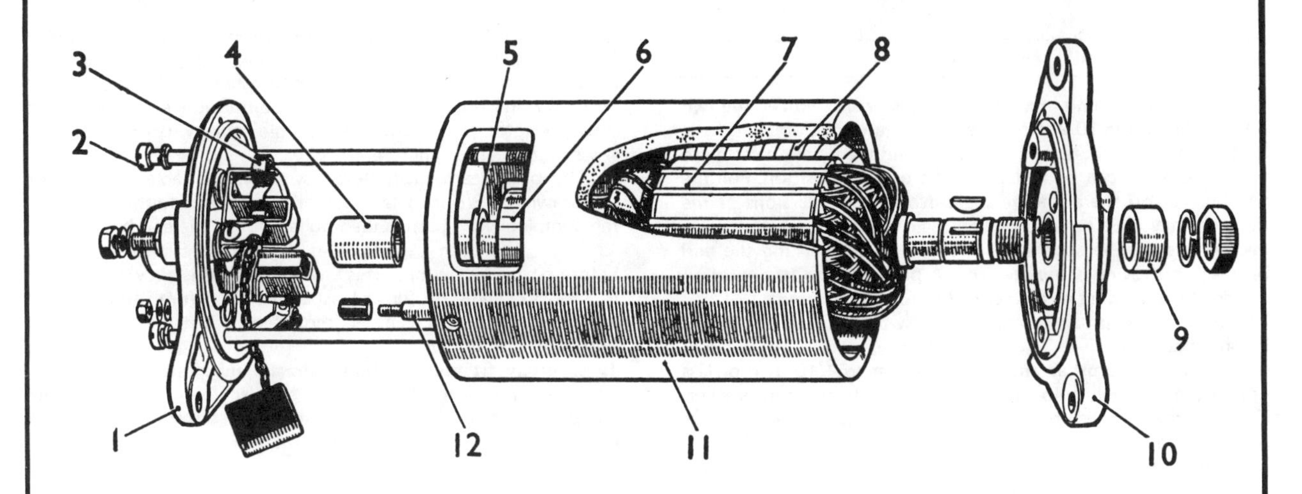

Fig 10.2 EXPLODED VIEW OF DYNAMO TYPE C45 PV-5

1 Commutator end bracket
2 Through bolt
3 Brush spring
4 Brush
5 Thrust collar
6 Commutator
7 Armature
8 Field coil
9 Distance collar
10 Drive end bracket
11 Yoke
12 Field terminal post

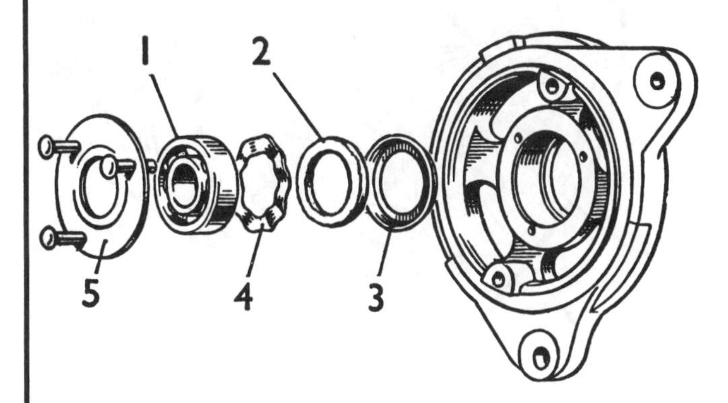

Fig 10.3 DYNAMO DRIVE END BRACKET

1 Bearing
2 Felt washer
3 Oil retaining washer
4 Corrugated washer
5 Bearing retaining plate

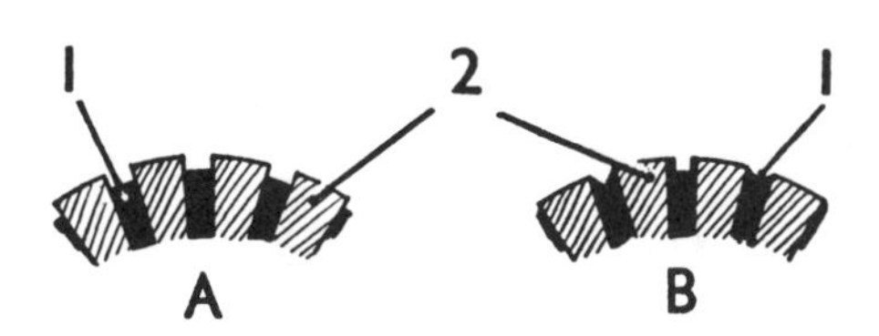

Fig 10.4 COMMUTATOR UNDERCUTTING

'A' is the correct and 'B' is the incorrect method
1 Insulation *2 Segments*

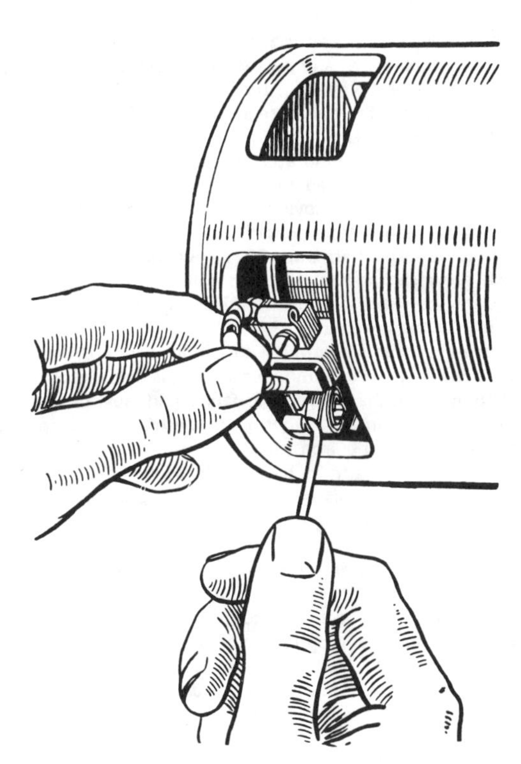

Fig 10.5 CORRECT METHOD OF REMOVING BRUSHES

9 Dynamo - dismantling and inspection

1 Mount the dynamo in a vice and unscrew and remove the two through bolts (Fig 10.2) from the commutator end bracket.
2 Mark the commutator end bracket and the dynamo casing so the end bracket can be replaced in its original position. Pull the end bracket off the armature shaft. Note: Some versions of the dynamo may have a raised pip on the end bracket which locates in a recess on the edge of the casing. If so, marking the end bracket and casing is unnecessary. A pip may also be found on the drive end bracket at the opposite end of the casing.
3 Lift the two brush springs and draw the brushes out of the brush holders.
4 Measure the brushes and if worn down to 7/16 inch or less unscrew the screws holding the brush leads to the end bracket. Take off the brushes complete with leads. Old and new brushes should be compared.
5 If no locating pip can be found, mark the drive end bracket and the dynamo casing so the drive end bracket can be replaced in its original position. Then pull the drive end bracket complete with armature out of the casing.
6 Check the condition of the ball bearing in the drive end plate by firmly holding the plate and noting if there is visible side movement of the armature shaft in relation to the end plate. If play is present, the armature assembly must be separated from the end plate. If the bearing is sound there is no need to carry out the work described in the following two paragraphs.
7 Hold the armature in one hand (mount it carefully in a vice if preferred) and undo the nut holding the pulley wheel and fan in place. Pull off the pulley wheel and fan.
8 Next remove the Woodruff key from its slot in the armature shaft and also the bearing locating ring.
9 Place the drive end bracket across the open jaws of a vice with the armature downwards and gently tap the armature shaft from the bearing in the end plate with the aid of a suitable drift.
10 Carefully inspect the armature and check it for open or short circuited windings. It is a good indication of an open circuited armature when the commutator segments are burnt. If the armature has short circuited the commutator segments will be very badly burnt, and the overheated armature windings badly discoloured. If open or short circuits are suspected then test by substituting the suspect armature for a new one.
11 Check the resistance of the field coils. To do this, connect an ohmmeter between the field terminal and the yoke and note the reading on the ohmmeter which should be about 6 ohms. If the ohmmeter reading is infinity this indicates an open circuit in the field winding. If the ohmmeter reading is below 6 ohms this indicates that one of the field coils is faulty and must be replaced.
12 Field coil replacement involves the use of a wheel operated screwdriver, a soldering iron, caulking and riveting and this operation is considered to be beyond the scope of most owners. Therefore if the field coils are at fault either purchase a rebuilt dynamo, or take the casing to an auto electrician for new field coils to be fitted.
13 Next check the condition of the commutator. If it is dirty and blackened as shown in the photo clean it with a petrol dampened rag. If the commutator is in good condition the surface will be smooth and quite free from pits or burnt areas and the insualted segments clearly defined.
14 If, after the commutator has been cleaned, pits and burnt spots are still present wrap a strip of glass paper round the commutator taking great care to move the commutator ¼ of a turn every ten rubs till it is thoroughly clean.
15 In extreme cases of wear the commutator can be mounted in a lathe and with the lathe turning at high speed, a very fine cut may be taken off the commutator. Then polish the commutator with glass paper. If the commutator has worn so that the insulators between the segments are level with the top of the segments, then undercut the insulators to a depth of 1/32 inch (see Fig 10.4). The best tool to use for this purpose is half a hacksaw blade ground to a thickness of the insulator, and with the handle end of the blade covered in insulating tape to make it comfortable to hold. The photo shows the sort of finish the surface of the commutator should have when finished.
16 Check the brush bearing in the commutator end bracket for wear by noting if the armature spindle rocks when placed in it. If worn it must be renewed.
17 The bush bearing can be removed by a suitable extractor or by screwing a 5/8 inch tap four or five times into the bush. The tap complete with bush is then pulled out of the end bracket.
18 Note: The bush bearing is of the porous bronze type and, before fitting a new one, it is essential that it is allowed to stand in SAE 30 engine oil for at least 24 hours before fitment. In an emergency the bush can be immersed in hot oil (100°C) for two hours.
19 Carefully fit the new bush into the end plate, pressing it in until the end of the bearing is flush with the inner side of the end plate. If available press the bush in with a smooth shouldered mandrel the same diameter as the armature shaft.

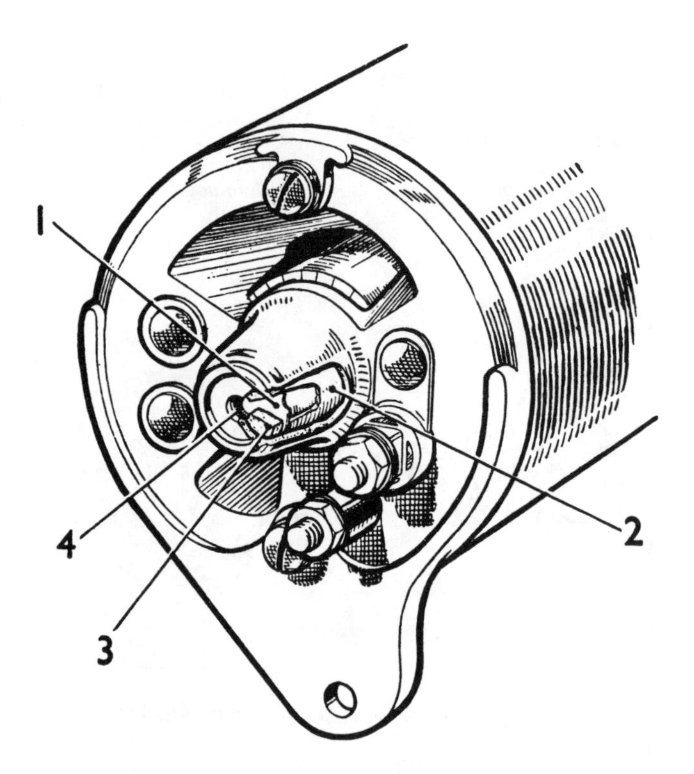

Fig 10.6 DYNAMO REAR ENDPLATE

1 Aluminium disc *3 Felt ring*
2 Porous bronze bush *4 Oil hole*

10 Dynamo - repair and reassembly

1 To renew the ball bearing fitted to the drive end bracket, drill out the rivets which hold the bearing retainer plate to the end bracket and lift off the plate.
2 Press out the bearing from the end bracket and remove the corrugated and felt washers as well as the oil retaining washer from the bearing housing.
3 Thoroughly clean the bearing housing and the new bearing, and pack with high melting point grease.

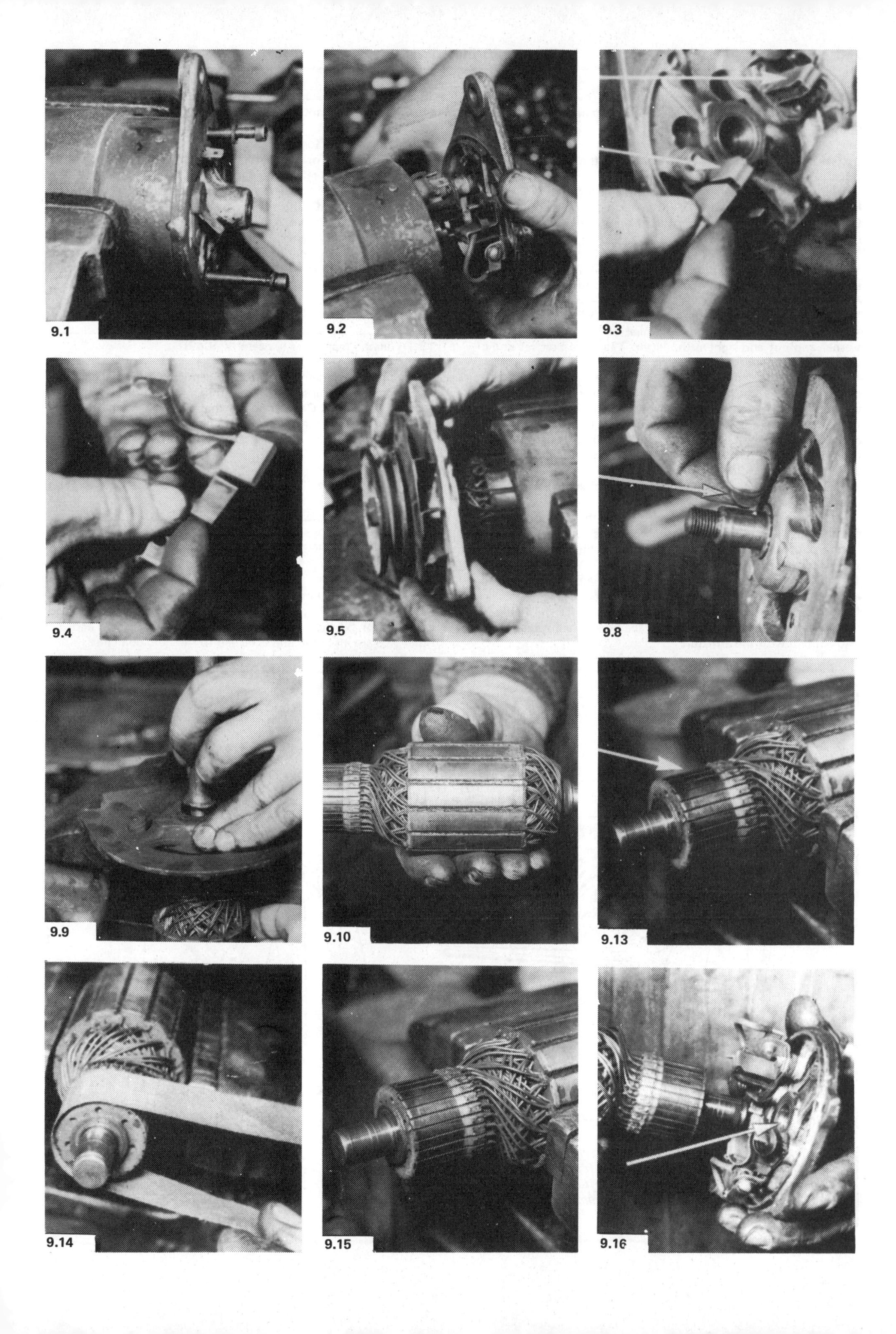
9.1
9.2
9.3
9.4
9.5
9.8
9.9
9.10
9.13
9.14
9.15
9.16

4 Place the oil retaining washer followed by the felt washer and corrugated washer in that order, in the end bracket bearing housing.
5 Then fit the new bearing as shown.
6 Gently tap the bearing into place with the aid of a suitable drift.
7 Replace the bearing plate and fit three new rivets.
8 Open up the rivets with the aid of a suitable cold chisel.
9 Finally peen over the open end of the rivets with the aid of a ball hammer as illustrated.
10 Refit the drive bracket to the armature shaft. Do not try and force the bracket on but with the aid of a suitable socket abutting the bearing, tap the bearing on gently, so pulling the end bracket down with it.
11 Slide the spacer up the shaft and refit the Woodruff key.
12 Replace the fan and pulley wheel and then fit the spring washer and nut and tighten the latter. The drive bracket end of the dynamo is now fully assembled as shown.
13 If the brushes are little worn and are to be used again then ensure that they are placed in the same holders from which they were removed. When refitting brushes either new or old, check that they move freely in their holders. If either brush sticks, clean with a petrol moistened rag and if still stiff, lightly polish the sides of the brush with a very fine file until the brush moves quite freely in its holder.
14 Tighten the two retaining screws and washers which hold the wire leads to the brushes in place.
15 It is far easier to slip the end piece with brushes over the commutator, if the brushes are raised in their holders as shown in the photo and held in this position by the pressure of the springs resting against the flanks.
16 Refit the armature to the casing and then the commutator end plate, and screw up the two through bolts.
17 Finally hook the ends of the two springs off the flanks of the brushes and onto their heads so the brushes are forced down into contact with the armature.

11 Dynamo - dismantling and inspection (C42)

The procedure for dismantling the C42 dynamo is basically identical to that for the earlier dynamo. Fig 10.7 shows the components of this later type dynamo.

If the dynamo brushes are worn, new brushes should be obtained and fitted. The replacement brushes must be bedded to the commutator and for this fine sand paper must be used. DO NOT use emery paper.

When checking the brush spring pressures note that the maximum pressure is 44 oz when testing with a new brush fitted or 34 oz on a brush worn to ¼ inch.

12 Dynamo - repair and reassembly (C42)

As the C42 dynamo is basically identical to the earlier type, the repair and reassembly is similar to that described in Section 10 of this Chapter. Note that the earlier type dynamo driving end bracket has an oil retaining washer in addition to the felt washer whereas the C42 dynamo just has the felt washer.

13 Starter motor - general description

The starter motor is mounted on the right hand lower side of the engine end plate, and is held in position by two bolts which also clamp the bellhousing flange. The motor is of the four field coil, four pole piece type, and utilises four spring loaded commutator brushes. Two of these brushes are earthed and the other two are insulated and attached to the field coil ends.

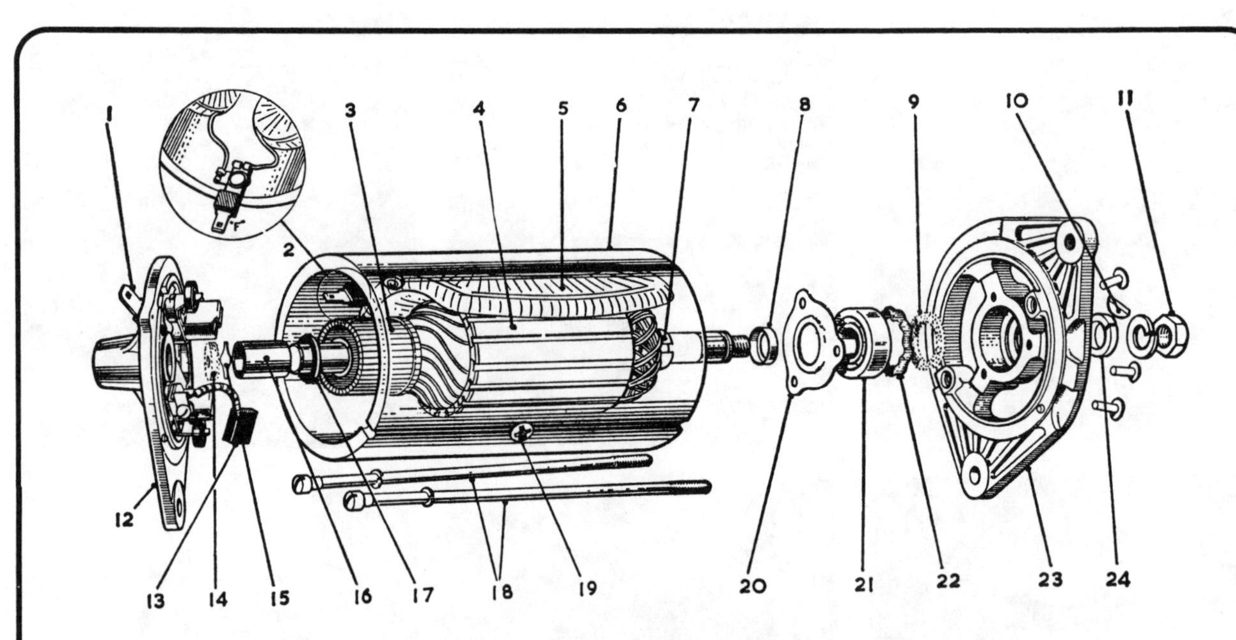

Fig 10.7 EXPLODED VIEW OF DYNAMO TYPE C42

1 Output terminal 'D'
2 Field terminal connections
3 Commutator
4 Armature
5 Field coils
6 Yoke
7 Shaft collar
8 Shaft collar retaining cup
9 Felt ring
10 Shaft key
11 Shaft nut
12 Commutator end bracket
13 Brushes
14 Felt ring
15 Felt ring retainer
16 Porous bronze bush
17 Fibre thrust washer
18 Through bolts
19 Pole-shoe securing screws
20 Bearing retaining plate
21 Ball bearing
22 Corrugated washer
23 Drive end bracket
24 Pulley spacer

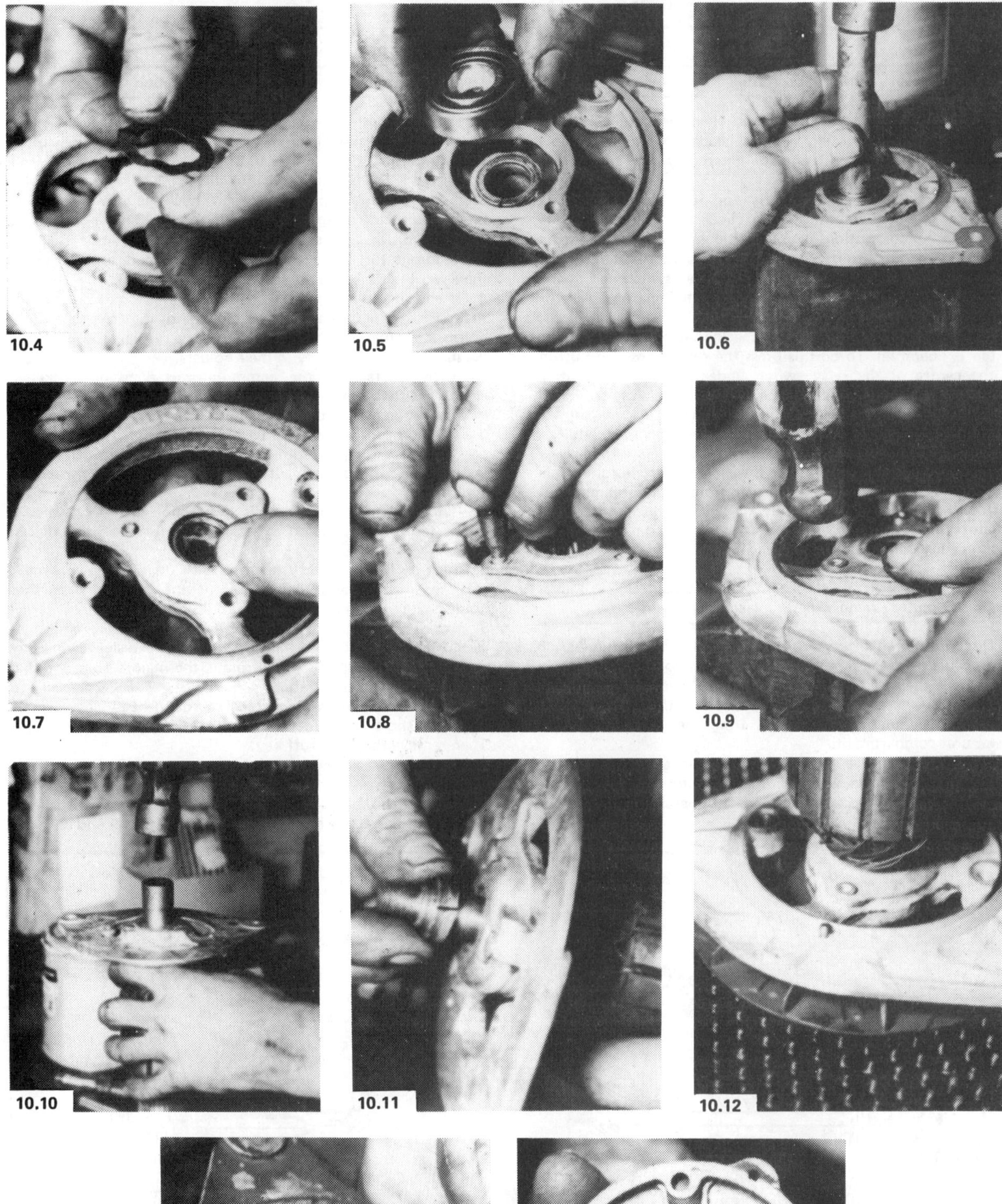

10.4 10.5 10.6

10.7 10.8 10.9

10.10 10.11 10.12

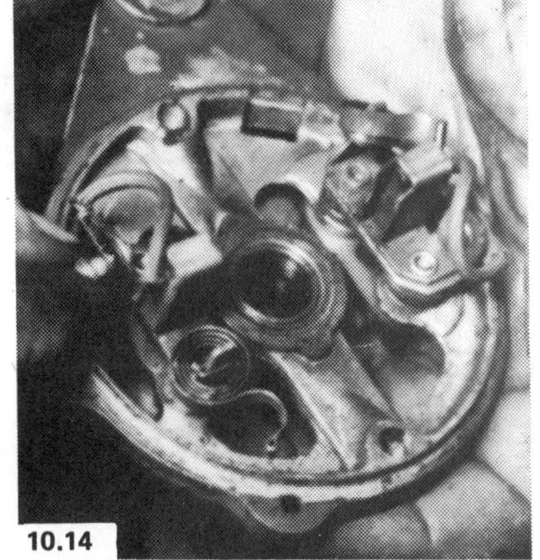

10.14

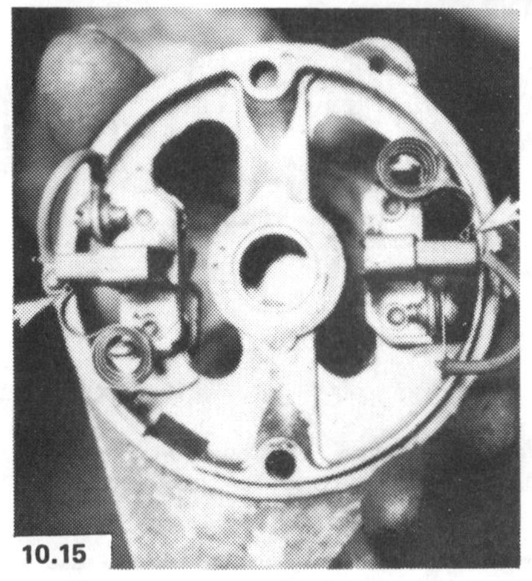

10.15

14 Starter motor - testing in position

1 If the starter motor fails to operate then check the condition of the battery by turning on the headlamps. If they glow brightly for several seconds and then gradually dim, the battery is in an uncharged condition.
2 If the headlamps glow brightly and it is obvious that the battery is in good condition then check the tightness of the battery wiring connections (and in particular the earth lead from the battery terminal to its connection on the bodyframe). Check the tightness of the connections at the relay switch and at the starter motor. Check the wiring with a voltmeter for breaks or shorts.
3 If the wiring is in order then check that the starter motor switch is operating. To do this press the rubber covered button in the centre of the relay switch under the bonnet. If it is working the starter motor will be heard to click as it tries to rotate. Alternatively, check it with a voltmeter.
4 If the battery is fully charged, the wiring in order, and the switch working and the starter motor fails to operate then itwill have to be removed from the car for examination. Before this is done, however, ensure that the starter pinion has not jammed in mesh with the flywheel. Check by turning the square end of the armature shaft with a spanner. This will free the pinion if it is stuck in engagement with the flywheel teeth.

15 Starter motor - removal and replacement

1 Disconnect the battery positive terminal and also the starter motor cable from the terminal on the starter motor end plate.
2 Remove the two bolts which secure the starter motor to the flywheel housing and engine rear plate. Lift the starter motor away by manipulating downwards from the underside of the power unit compartment.
3 Refitting is the reverse procedure to removal. Make sure that the starter motor cable, when secured in position by its terminal, does not touch any part of the body or power unit which could damage the insulation.

16 Starter motor - dismantling and reassembly

1 With the starter motor on the bench, loosen the screw on the cover band and slip the cover band off. An exploded view of the starter motor is shown in Fig 10.8.
2 With a piece of wire bent into the shape of a hook, lift back each of the brush springs in turn and check the movement of the brushes in their holders by pulling on the flexible connectors.
3 If the brushes are so worn that their faces do not rest against the commutator, or if the ends of the brush leads are exposed on their working face, they must be renewed.
4 If any of the brushes tend to stick in their holders then wash them with a petrol moistened cloth and, if necessary, lightly polish the sides of the brush with a very fine file, until the brushes move quite freely in their holders.
5 If the surface of the commutator is dirty or blackened, clean it with a petrol dampened rag. Secure the starter motor in a vice and check it by connecting a heavy gauge cable between the starter motor terminal and a 12 volt battery.
6 Connect the cable from the other battery terminal to earth in the starter motor body. If the motor turns at high speed it is in good order.
7 If the starter motor still fails to function or if it is wished to renew the brushes, then it is necessary to further dismantle the motor.
8 Start by lifting the brush spring with the aid of a wire hook, off the brushes, and then take out the brushes from their holders one at a time.
9 Working from the drive end of the starter motor, remove the circlip from off the outer end of the drive head sleeve, and then remove the front spring anchor plate, the main spring, and then the rear spring anchor plate.
10 Pull out the pin which holds the drive head sleeve to the armature shaft and slide the sleeve assembly down the shaft. Then remove the Woodruff key.
11 The complete drive assembly can now be pushed off the armature shaft.
12 Extract the barrel retaining ring from the inside of the pinion and barrel assembly and pull off the barrel and anti-drift

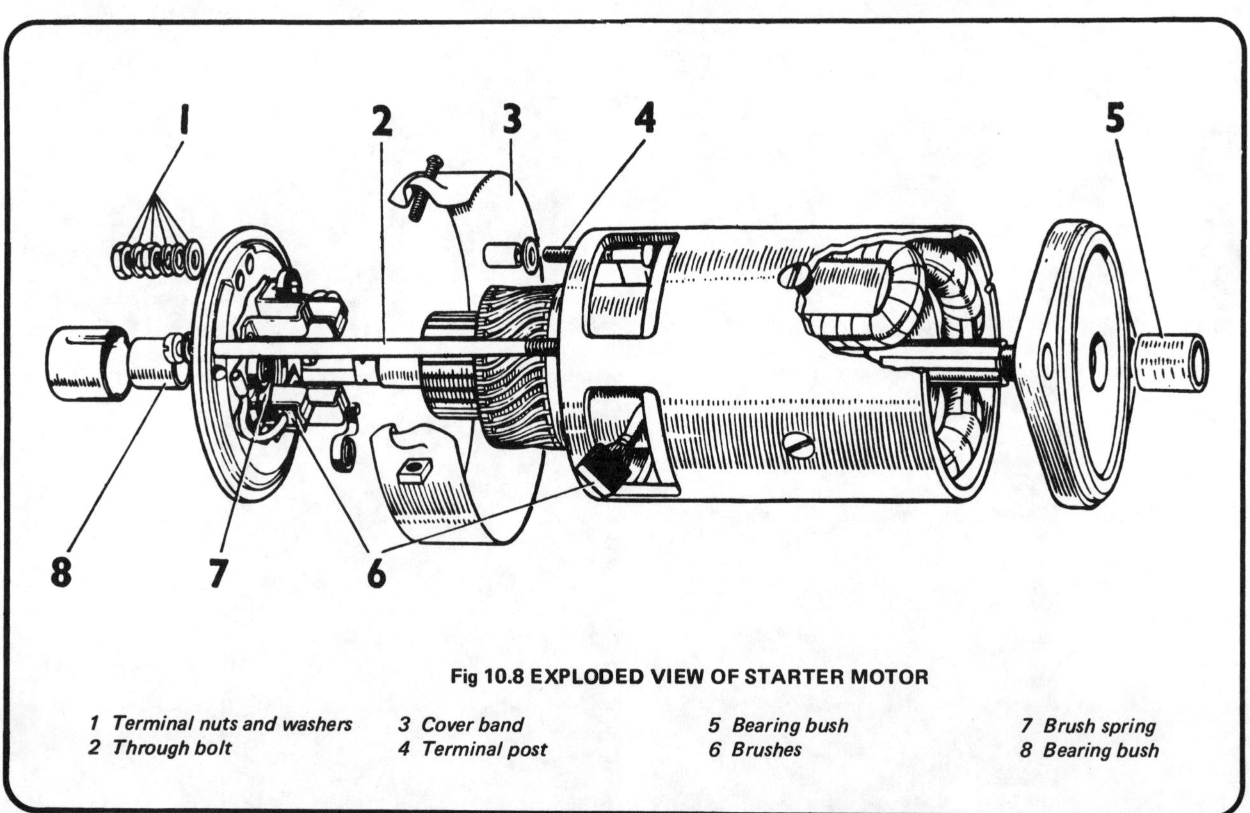

Fig 10.8 EXPLODED VIEW OF STARTER MOTOR

1 Terminal nuts and washers *2 Through bolt* *3 Cover band* *4 Terminal post* *5 Bearing bush* *6 Brushes* *7 Brush spring* *8 Bearing bush*

spring from the screwed sleeve.
13 From the inner end of the drive head sleeve take off the circlip, locating collar, control nut thrust washer, cushioning spring, control nut, screwed sleeve and the drive head thrust washer.
14 Undo the terminal nuts and washers from the terminal post and unscrew and remove the two through bolts and spring washers.
15 The commutator end bracket, the drive end bracket and the armature can now be removed.
16 At this stage if the brushes are to be renewed their flexible connectors must be unsoldered and the connectors of new brushes soldered in their place. Check that the new brushes move freely in their holders. If cleaning the commutator with petrol fails to remove all the burnt areas and spots then wrap a piece of glass paper round the commutator and rotate the armature.
17 If the commutator is very badly worn mount the armature in a lathe and with the lathe turning at high speed, take a very fine cut out of the commutator and finish the surface by polishing with glass paper. DO NOT UNDERCUT THE MICA INSULATORS BETWEEN THE COMMUTATOR SEGMENTS.
18 With the starter motor dismantled, test the four field coils for an open circuit. Connect a 12 volt battery with a 12 volt bulb in one of the leads between the field terminal post and the tapping point of the field coils to which the brushes are connected. An open circuit is proved by the bulb not lighting.
19 If the bulb lights, it does not necessarily mean that the field coils are in order, as there is a possibility that one of the coils will be earthing to the starter yoke or pole shoes. To check this, remove the lead from the brush connector and place it against a clean portion of the starter yoke. If the bulb lights the field coils are earthing.
20 Replacement of the field coils calls for the use of a wheel operated screwdriver, a soldering iron, caulking and riveting operations and is beyond the scope of the majority of owners. The starter yoke should be taken to an automobile electrical engineering works for new field coils to be fitted. Alternatively, purchase an exchange Lucas starter motor.
21 If the armature is damaged this will be evident after visual inspection. Look for signs of burning, discoloration and for conductors that have lifted away from the commutator. Ensure that if any parts of the drive gear are worn or damaged they are renewed.
22 Reassembly is a straight reversal of the dismantling procedure.

17 Starter motor drive - general description

The starter motor drive is of the outboard type. When the starter motor is operated the pinion moves forwards into contact with the flywheel ring gear by moving forwards away from the starter motor.

If the engine kicks back, or the pinion fails to engage with the flywheel ring gear when the starter motor is actuated no undue strain is placed on the armature shaft as the pinion sleeve disengages from the pinion and turns independently.

18 Starter motor drive - removal and replacement

1 With the starter motor away from the car, first remove the cotter pin from the end of the armature shaft securing the nut at the starter drive.
2 Holding the squared end of the armature shaft opposite to the starter motor drive and with a spanner, unscrew the shaft nut.
3 Lift away the main spring, washer, screwed sleeve with pinion, collar, pinion restraining spring and spring retaining sleeve.
4 It is most important that the drive gear is completely free from oil, grease and dirt. With the drive gear removed clean all parts thoroughly in paraffin. UNDER NO CIRCUMSTANCES OIL THE DRIVE COMPONENTS. Lubrication of the drive components could easily cause the pinion to stick.
5 Reassembly is the reverse sequence to removal. When the nut has been replaced, secure with a new split pin.

19 Starter motor bushes - inspection, removal and replacement

1 With the starter motor stripped down check the condition of the bushes. They should be renewed when they are sufficiently worn to allow visible side movement of the armature shaft.
2 The old bushes are simply driven out with a suitable drift and the new bushes inserted by the same method. As the bearings are of the phosphor bronze type it is essential that they are allowed to stand in SAE 30 engine oil for at least 24 hours before fitment.

20 Control box - general description (Lucas RB 106/2)

The control box, located on the right hand side of the bulkhead, comprises the voltage regulator and the cut-out. The voltage regulator controls the output from the dynamo depending on the state of the battery and the demands of the electrical equipment and ensures that the battery is not overcharged. The cut-out is really an automatic switch and connects the dynamo to the battery when the dynamo is turning fast enough to produce a charge. Similarly, it disconnects the battery from the dynamo when the engine is idling or stationary so that the battery does not discharge through the dynamo.

21 Voltage regulator adjustment (Lucas RB 106/2)

1 If the battery is in sound condition, but is not holding its charge, or is being continually overcharged and the dynamo is in sound condition, then the voltage regulator in the control box must be adjusted.
2 Check the regulator setting by removing and joining together the cables from the control box terminals A1 and A. Then connect the negative lead of a 20 volt voltmeter to the D terminal on the dynamo and the positive lead to a good earth. Start the engine and increase its speed until the voltmeter needle flicks and then steadies. This should occur at about 2000 rpm. If the voltage at which the needle steadies is outside the limits listed below, then remove the control box cover and turn the adjusting screw (1) (Fig 10.11) clockwise a quarter of a turn at a time to raise the setting and a similar amount, anticlockwise to lower it.

Air temperature	Type RB 106/2 Open circuit voltage
10°C or 50°F	16.1 to 16.7
20°C or 68°F	16.0 to 16.6
30°C or 86°F	15.9 to 16.5
40°C or 104°F	15.8 to 16.4

3 It is vital that the adjustments be completed within 30 seconds of starting the engine as otherwise the heat from the shunt coil will affect the readings.

22 Cut-out adjustment (Lucas RB 106/2)

1 Check the voltage required to operate the cut-out by connecting a voltmeter between the control box terminals D and E.
2 Remove the control box cover, start the engine and gradually increase its speed until the cut-outs close. This should occur when the reading is between 12.7 and 13.3 volts.
3 If the reading is outside these limits turn the cut-out adjusting screw (2) in Fig 10.11 a fraction at a time clockwise to raise the voltage, and anticlockwise to lower it. To adjust the drop off voltage, bend the fixed contact blade carefully. The adjustment

to the cut-out should be completed within 30 seconds of starting the engine as otherwise heat build up from the shunt coil will affect the readings.

4 If the cut-out fails to work, clean the contacts, and, if there is still no response, renew the cut-out and regulator unit.

23 Control box - general description (Lucas RB 340)

1 The control box is positioned on the right hand side of the bulkhead and comprises three units; two separate vibrating armature type single contact regulators and a cut-out relay. One of the regulators is sensitive to change in current and the other to changes in voltage.

2 Adjustments can be made only with a special tool which resembles a screwdriver with a multi-toothed blade. These can be obtained through Lucas agents.

3 The regulators control the output from the dynamo depending on the state of the battery and the demands of the electrical equipment, and ensure that the battery is not overcharged. The cut-out is really an automatic switch and connects the dynamo to the battery when the dynamo is turning fast enough to produce a charge. Similarly it disconnects the battery from the dynamo when the engine is idling or stationary so that the battery does not discharge through the dynamo.

24 Voltage regulator adjustment (Lucas RB 340)

1 The regulator requires very little attention during its service life, and if there should be any reason to suspect its correct functioning, tests of all circuits should be made to ensure that they are not the reason for the trouble.

2 These checks include the tension of the fan belt to make sure that it is not slipping and so providing only a very low charge rate. The battery should be carefully checked for possible low charge rate due to a faulty cell, or corroded battery connections.

3 The leads from the dynamo may have been crossed during replacement, and if this is the case, then the regulator points will have stuck together as soon as the dynamo starts to charge. Check for loose or broken leads from the dynamo to the regulator.

4 If, after a thorough check, it is considered advisable to test the regulator this should be carried out only by an electrician who is well acquainted with the correct method, using the test bench equipment.

5 Pull off the Lucar connections from the two adjacent control box terminals B. To start the engine it will now be necessary to join together the ignition and battery leads with a suitable wire.

6 Connect a 0-20 volt voltmeter between terminal D on the control box and terminal WL. Start the engine and run it at 3000 rpm. The reading on the voltmeter should be steady and lie between the limits detailed in the Specifications.

7 If the reading is unsteady this may be due to dirty contacts. If the reading is outside the specified limits stop the engine and adjust the voltage regulator in the following manner:

8 Take off the control box cover and start and run the engine at 3000 rpm. Using the correct tool turn the voltage adjustment cam anticlockwise to raise the setting and clockwise to lower it. To check that the setting is correct, stop the engine and then start it and run it at 3000 rpm noting the reading. Refit the cover and the connections to the WL and D terminals.

25 Current regulator adjustment (Lucas RB 340)

1 The output from the current regulator should equal the maximum output from the dynamo which is 22 amps. To test this it is necessary to bypass the cut-out by holding the contacts together.

2 Remove the cover from the control box and with a bulldog clip hold the cut-out contacts together.

3 Pull off the wires from the adjacent terminals B and connect a 0.40 moving coil ammeter to one of the terminals and to the leads.

4 All the other lead connections including the ignition must be made to the battery.

5 Turn on all the lights and other electrical accessories and run the engine at 3000 rpm. The ammeter should give a steady reading between 19 and 22 amps. If the needle flickers it is likely that the points are dirty. If the reading is too low turn the special Lucas tool clockwise to raise the setting and anticlockwise to lower it.

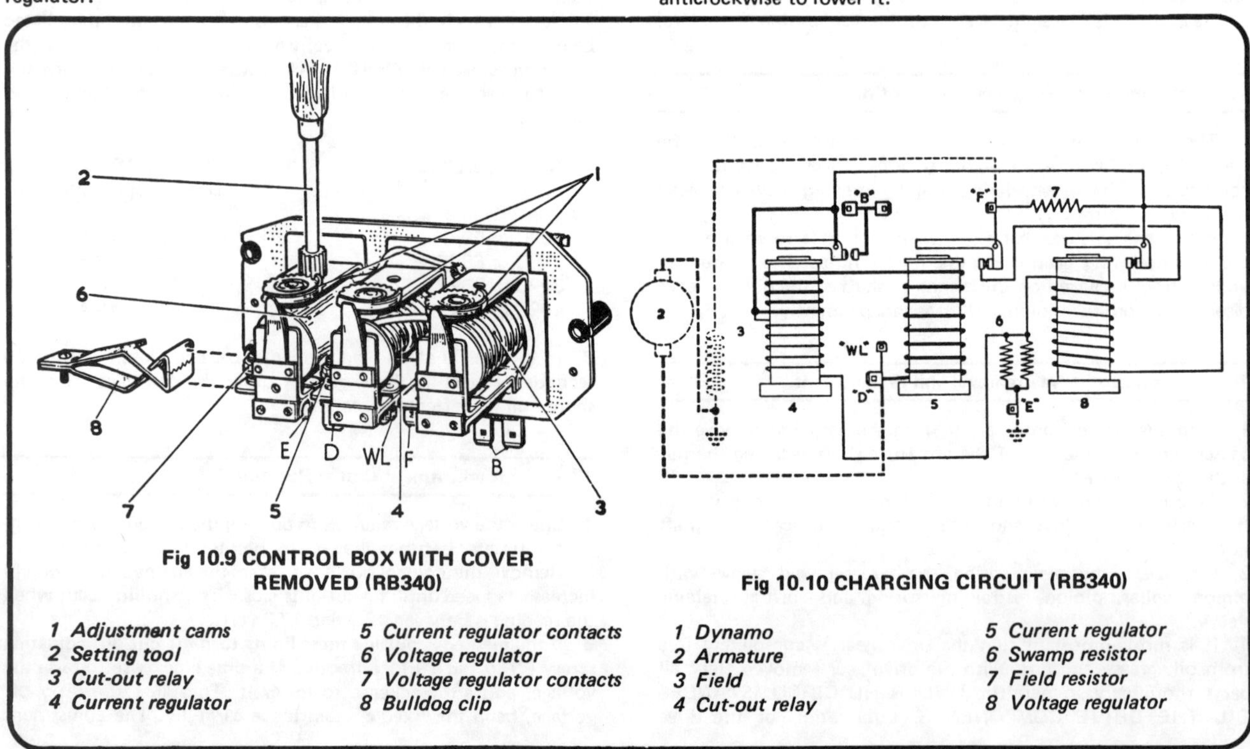

Fig 10.9 CONTROL BOX WITH COVER REMOVED (RB340)

1 Adjustment cams
2 Setting tool
3 Cut-out relay
4 Current regulator
5 Current regulator contacts
6 Voltage regulator
7 Voltage regulator contacts
8 Bulldog clip

Fig 10.10 CHARGING CIRCUIT (RB340)

1 Dynamo
2 Armature
3 Field
4 Cut-out relay
5 Current regulator
6 Swamp resistor
7 Field resistor
8 Voltage regulator

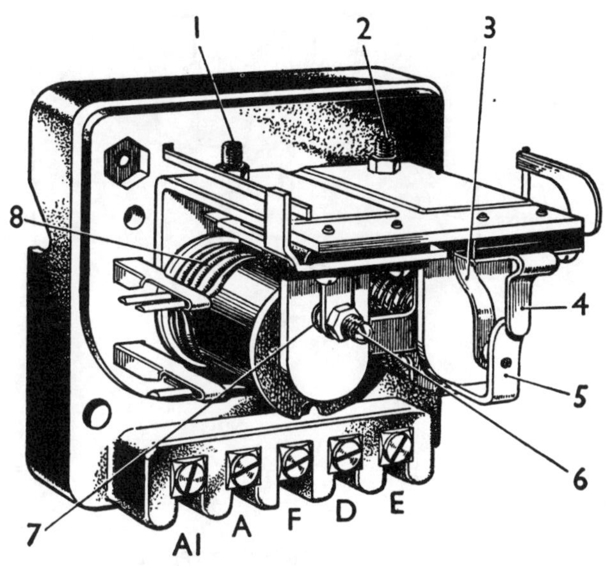

Fig 10.11 CONTROL BOX TYPE RB 106/2

1 Regulator adjusting screw
2 Cut-out adjusting screw
3 Fixed contact blade
4 Stop arm
5 Armature tongue and moving contact
6 Regulator moving contact
7 Fixed contact
8 Regulator series windings

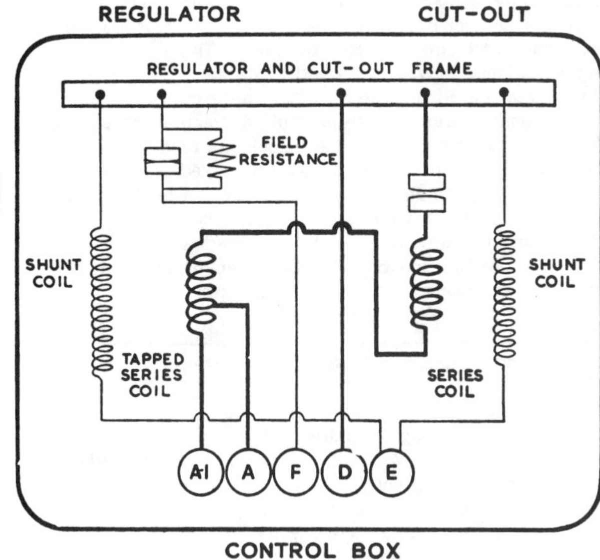

Fig 10.12 INTERNAL CONNECTIONS OF CONTROL BOX

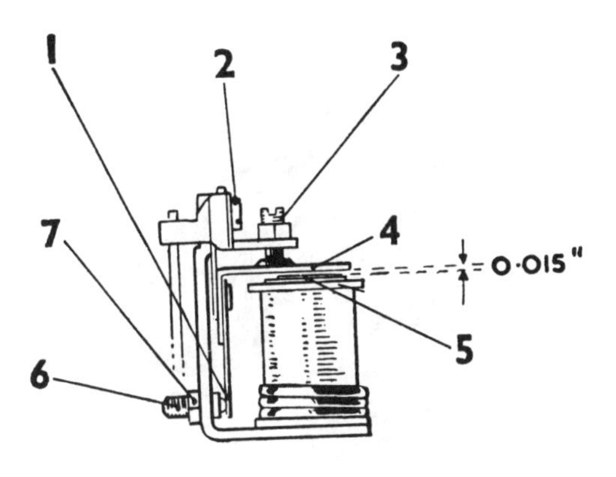

Fig 10.14 CUT-OUT MECHANICAL SETTINGS

1 Stop arm
2 Armature tongue and moving contact
3 Armature screw
4 Fixed contact blade
5 Cut-out adjusting screw
6 Armature tension spring

Fig 10.13 REGULATOR MECHANICAL SETTING

1 Armature tension spring
2 Armature securing screws
3 Fixed contact adjustment screw
4 Armature
5 Core face and shim
6 Voltage adjusting screw
7 Locknut

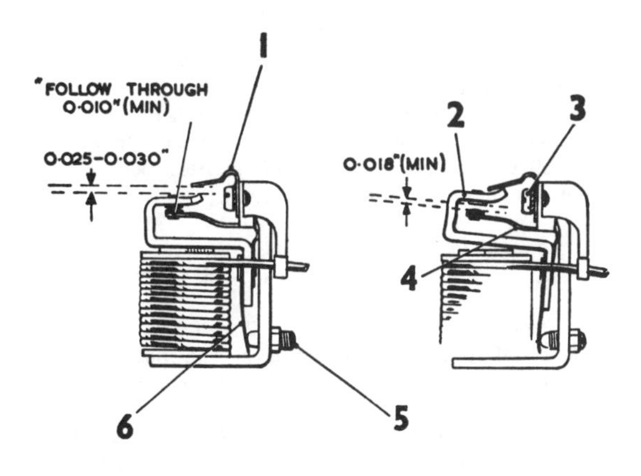

26 Cut-out adjustment (Lucas RB 340)

1 Check the voltage required to operate the cut-out by connecting a voltmeter between the control box terminals D and WL. Remove the control box cover, start the engine and gradually increase its speed until the cut-out closes. This should occur when the reading is between 12.7 and 13.3 volts.
2 If the reading is outside these limits turn the cut-out adjusting cam by means of the adjusting tool, a fraction at a time clockwise to raise the voltage and anticlockwise to lower it.
3 To adjust the drop off voltage bend the fixed contact blade carefully. The adjustment to the cut-out should be completed within 30 seconds of starting the engine as otherwise heat build up from the shunt coil will affect the readings.
4 If the cut-out fails to work, clean the contacts, and if there is still no response, renew the cut-out and regulator unit.

27 Cut-out and regulator contacts - maintenance

1 Every 12,000 miles check the cut-out and regulator contacts. If they are dirty or rough or burnt, place a piece of fine glass paper (DO NOT USE EMERY PAPER OR CARBORUNDUM PAPER) between the cut-out contacts, close them manually and draw the glass paper through several times.
2 Clean the regulator contacts in exactly the same way, but use emery paper or carborundum paper and not glass paper. Carefully clean both sets of contacts from all traces of dust with a rag moistened in methylated spirits.

28 Fuses

The fuse unit is mounted on the bulkhead under the engine cowl and is an open insulated moulding carrying two single pole cartridge fuses which are held by spring clips between grub screw type terminal blocks.

Two spare fuses are carried in recesses in the fuse unit base and are held in position by a common retaining spring.

The fuse which bridges the terminal blocks A1-A2 is to protect general auxiliary circuits which are independent of the ignition switch. The second fuse which bridges terminal blocks A3-A4 is to protect ignition auxiliary circuits which only operate when the ignition is switched on.

On BN7, BT7, BJ7 and BJ8 models an additional 10 amp fuse protects the number plate lamp circuit. It is located in a nylon tube situated in the boot wiring loom to the right of the boot floor catch. To renew the fuse, twist and release the end of the tube and withdraw the fuse.

29 Flasher circuit - fault tracing and rectification

1 The flasher unit is located in the engine compartment and is actuated by a self-cancelling steering column direction switch. A warning light is provided in the centre of the fascia panel.
2 If the flasher unit fails to operate, or works very slowly or very rapidly, check out the flasher circuit as described below before assuming there is a fault in the unit itself.

a) Examine the bulbs for broken filaments.
b) With the aid of the wiring diagrams check all flasher circuit connections.
c) Switch on the ignition and with a 12 volt test lamp or voltmeter check that battery voltage is at the flasher unit terminal B by putting one lead on the terminal and the other to earth.
d) Connect the flasher unit terminals B (or X) and L together and operate the direction indicator switch. If the flasher lamps now light the flasher unit is defective and must be renewed.
e) If in test (d) the lamps do not light the brake switch overriding relay should be tested as described in the next section.

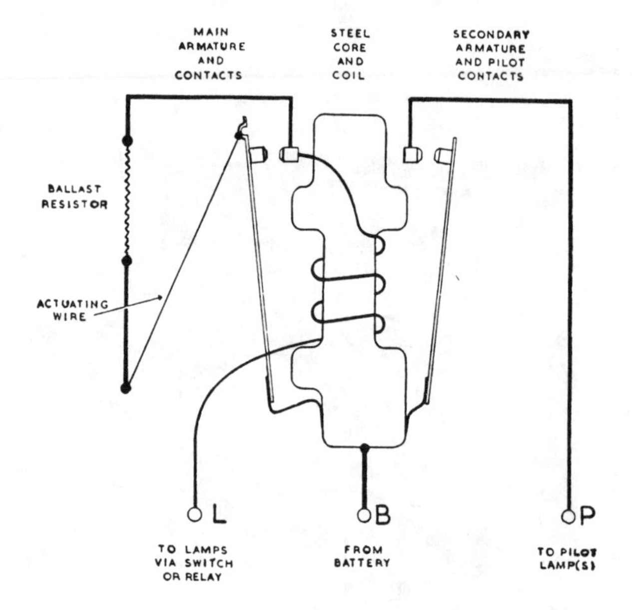

Fig 10.15 FLASHER UNIT TYPE FL5

f) In the event of total failure of the direction indicator circuit check the A3-A4 fuse.

30 Brake switch overriding relay

On the Healey models covered by this manual where stop light filaments are used also as direction indicator lights it is essential that responses to the flasher unit should override simultaneous applications of the brake switch.

In the event of simultaneous applications being made the relay which is shown in Fig 10.16 allows the appropriate stop light filament to flash and the other to remain steadily illuminated as long as the brake pedal is depressed.

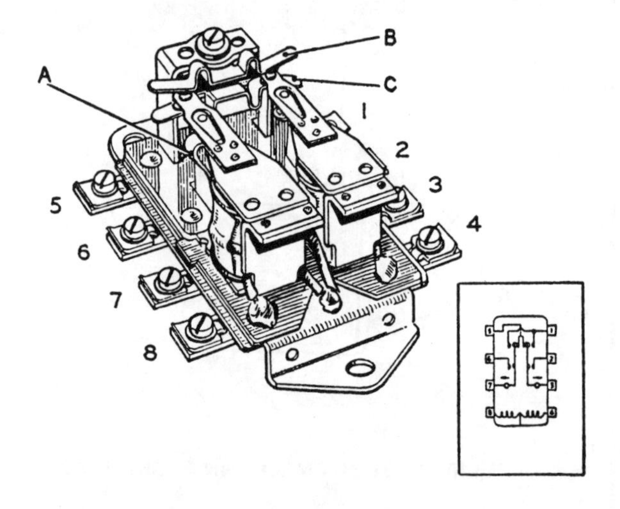

Fig 10.16 BRAKE SWITCH OVERRIDING RELAY TYPE DB10

A Inner contacts
B Outer upper contacts
C Outer lower contacts

When the direction indicator switch is turned to the left or right the appropriate relay operating coil is energised and this effects movement of its associated armature in the direction shown by the arrow in the inset. By this means the flasher unit terminal 2 is connected to relay terminals 2 and 3 or 6 and 7 and so to the indicating lamps. As long as the relay coil remains energised connection to the brake switch on the corresponding side is interrupted. To check the brake switch overriding relay is as follows:

a) Temporarily link relay 1 (Fig 10.16) to terminal 2 and 3. The left hand lights should now flash.
b) Temporarily link the relay terminal 1 to terminals 6 and 7. The right hand lights should now flash.
c) If the lamps do not flash in the above tests the relay is defective and requires resetting as described below.
d) Check the direction indicator switch by substitution.

Relay air gap - checking and resetting

1 With a screwdriver prise off the relay cover making a note of the non-reversible locating slot between terminals 6 and 7 (Fig 10.16).
2 Upon inspection it will be seen that each armature controls three pairs of contacts, two pairs being normally open and one pair normally closed. Identification of the three contacts may be made as follows:

a) Inner pairs, adjacent to bobbins, normally open
b) Outer lower pairs, normally open
c) Outer upper pairs, normally closed

3 When an inner pair of contacts is just touching, a relay in correct adjustment will have an armature to bobbin core gap of 0.010 to 0.015 inch.
4 In addition when these contacts are separated by a 0.007 to 0.013 inch gap, the outer lower contacts must be separated by a 0.012 to 0.018 inch gap.
5 If the gaps are not within these limits the relay must be reset and adjustments are made by bending the fixed contact carrier with a suitably slotted bending tool such as a spark plug electrode resetting tool.
6 Insert a 0.010 inch feeler gauge between one of the armatures and its bobbin core. Press down the armature and adjust the height of the inner contact carrier until the inner pair of contacts is just touching. Remove the feeler gauge.
8 With the outer lower contacts just touching adjust the upper contact carrier until a 0.015 inch feeler gauge is a sliding fit between the outer upper contacts. Remove the feeler gauge and refit the cover.

31 Windscreen wiper mechanism - maintenance

1 Renew the windscreen wiper blades at intervals of 12,000 miles, or more frequently if the screen is not efficiently cleaned of water.
2 The cable which drives the wiper blades from the gearbox attached to the windscreen wiper motor is prepacked with grease and requires no maintenance.
3 The washer around the wheelbox spindle can be lubricated with several drops of glycerine every 6000 miles.

32 Windscreen wiper blades - removal and replacement

1 Lift the wiper arm away from the windscreen and remove the old blade by turning it towards the arm and then disengaging the arm from the slot in the blade.
2 To fit a new blade slide the end of the wiper arm into the slotted spring fastening in the centre of the blade. Push the blade firmly onto the arm until the raised portion of the arm is fully home in the hole in the blade.

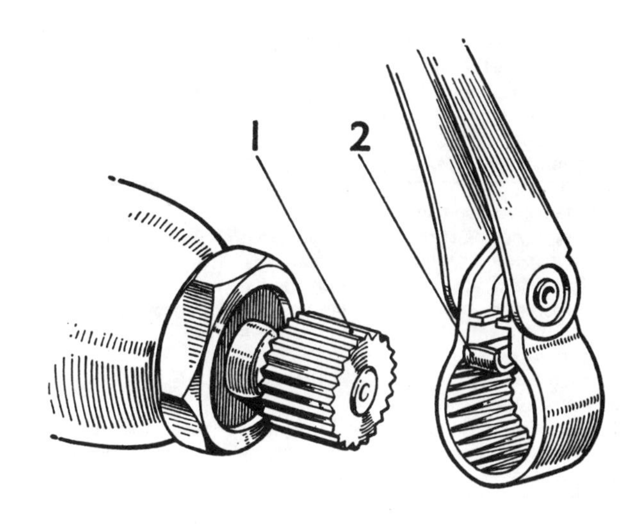

Fig 10.17 WINDSCREEN WIPER ARM ASSEMBLY

1 Splined driving drum *2 Retaining clip*

33 Windscreen wiper arms - removal and replacement

1 Before removing a wiper arm turn the windscreen wiper switch to the fully off position, with the ignition switch to ensure that the arms are in their normal parked position, parallel with the bottom of the windscreen.
2 To remove an arm, pivot the arm and using a screwdriver ease back the clip (2) (Fig 10.17) in the hub of the arm. Pull the arm from the splines of the drive spindle.
3 When replacing an arm position it so it is in the correct relative parked position and then press the arm lead onto the splined drive until the retaining clip clicks into place.

34 Windscreen wiper mechanism - fault diagnosis and rectification

Should the windscreen wiper fail to park or park badly then check the limit switch on the gearbox cover. Loosen the four screws which retain the gearbox cover and place the projection close to the rim of the limit switch in line with the groove in the gearbox cover. Rotate the limit switch anticlockwise 25° and tighten the four screws retaining the gearbox cover. If it is wished to park the windscreen wipers on the other side of the windscreen rotate the limit switch 180° clockwise.

Should the windscreen wipers fail, or work very slowly, then check the current the motor is taking by connecting up a 0-20 volt voltmeter in the circuit and turning on the wiper switch. Consumption should be between 2.3 to 3.1 amps.

If no current is passing through check the A3-A4 fuse. If the fuse has blown replace it after having checked the wiring of the motor and other electrical circuits serviced by this fuse for short circuits. If the fuse is in good condition check the wiper switch and the current operated thermostat by substitution.

If the wiper motor takes a very high current check the wiper blades for freedom of movement. If this is satisfactory check the gearbox cover and gear assembly for damage and measure the armature end float which should be between 0.009 and 0.012 inch (0.20 and 0.30 mm). The end float is set by the adjusting screw. Check that excessive friction in the cable connecting tubes caused by too small a curvature is not the cause of the high current consumption.

If the motor takes a very low current ensure that the battery is fully charged. Check the brush gear after removing the commutator end bracket and ensure that the brushes are bearing on the commutator. If not, check the brushes for freedom of movement and if necessary renew the tension spring. Check the armature by substitution if this unit is suspected.

35 Windscreen wiper motor - removal and refitting

1 The windscreen wiper motor and gearbox is located under the passenger side of the fascia panel and is mounted on a bracket which is secured to the bulkhead panel by three screws.
2 The cable rack which is connected to the crosshead in the gearbox passes through outer casings which connect the gearbox to the first wheelbox and the first wheelbox to the second wheelbox.
3 Remove the windscreen wiper arms as described in the previous section.
4 For safety reasons switch off the battery master switch. Make a note of the electrical connections to the motor and disconnect these cables.
5 Undo the cable outer casing securing nut to the motor gearbox housing so releasing the outer casing from the motor.
6 Undo and remove the three screws that secure the bracket to the bulkhead panel and withdraw the motor, bracket and cable rack from beneath the fascia panel.
7 Slacken the cover screws in each wheelbox and lift away each of the cable rack outer casings.
8 Undo and remove the nut, front bush and washer from the front of each wheelbox and remove the wheelbox together with the rear bush and spindle tube from beneath the fascia panel.
9 Refitting the wiper system is the reverse sequence to removal. Care must be taken to ensure that the wheelboxes are correctly lined up and that the cable rack engages the gear and spindle assemblies. Carefully pack the inner cable with Castrol LM Grease.

36 Windscreen wiper motor and gearbox - dismantling and reassembly (Type DR2)

1 Refer to Figs 10.18 and 10.19 and undo the four screws that secure the gearbox cover to the casting and lift away the cover.

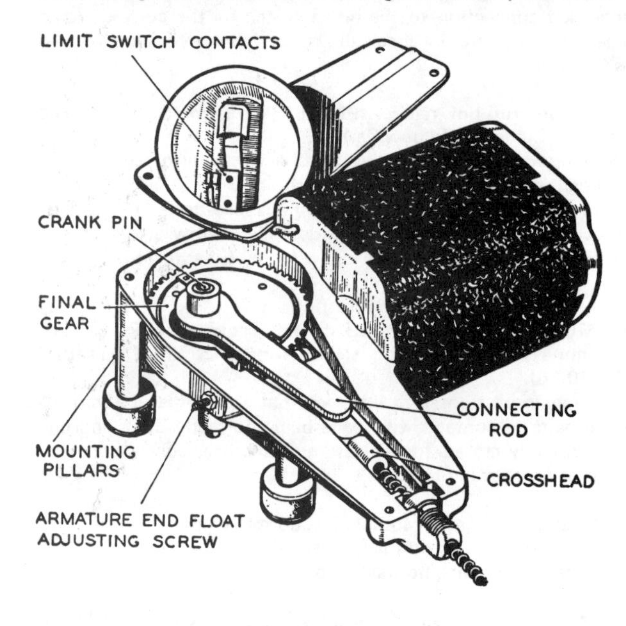

Fig 10.18 WIPER MOTOR GEARBOX WITH COVER REMOVED (TYPE DR2)

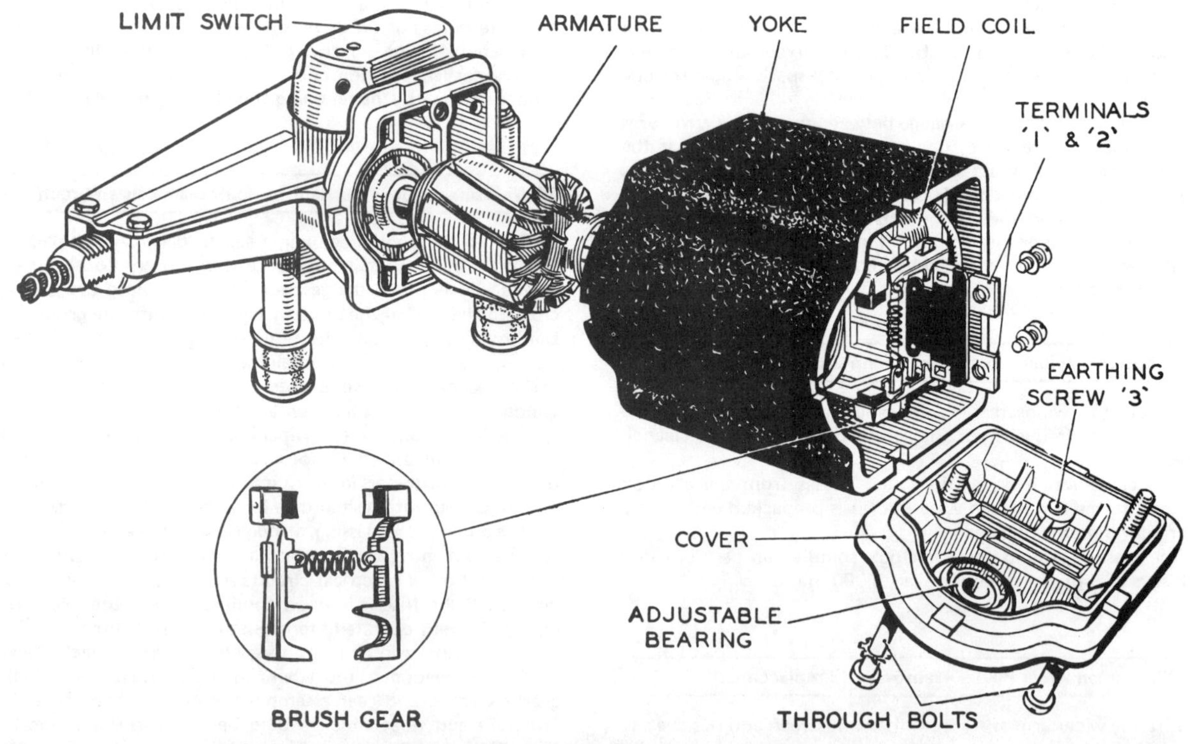

Fig 10.19 EXPLODED VIEW OF WINDSCREEN WIPER MOTOR (TYPE DR2)

2 Withdraw the terminal screws and the through bolts at the commutator end bracket.
3 Ease the commutator end bracket away from the end of the yoke.
4 Lift the brush gear clear of the commutator and remove as a complete unit. Whilst this is being done note the particular side occupied by each brush so that each may be replaced in its original setting on the commutator.
5 Access to the armature and field coils can be gained by withdrawing the yoke.
6 Should it be necessary to remove the field coil, unscrew the two screws securing the pole piece to the yoke. These screws should be marked so that they can be replaced in their original holes.
7 Drift out the pole pieces complete with field coil, marking the pole piece so that it can be replaced in its correct position inside the yoke. The pole piece can now be pressed out of the field coil.
8 To dismantle the gearbox first remove the circlip and washer from the crosshead connecting link pin and lift off the crosshead and cable rack assembly.
9 Remove the circlip and washer from the final gear shaft located on the underside of the gearbox casing.
10 With a scraper remove any burrs from the circlip groove before lifting out the final gear.
11 The armature and worm drive can next be lifted away from the gearbox.
12 Wipe out any grease and dirt from the motor and gearbox and also from the gears and crosshead. Inspect the gear teeth for signs of damage or wear, which, if evident, obtain new gears.
13 Inspect the bushes for wear and if evident, obtain a new bush gear set.
14 Reassembly is the reverse sequence to dismantling. The following points about lubrication should however be noted.

a) Lubricate the armature bearings with Castrol GTX. The self-aligning bearing should be immersed in oil for 24 hours before assembly if a new one is being fitted.
b) Lubricate the commutator end of the armature shaft with Castrol GTX.
c) Apply a little Castrol GTX to the felt lubricator in the gearbox.
d) The worm wheel bearings, crosshead, guide channel, connecting rod, crankpin, eccentric coupling assembly, worm and final gear shaft should be liberally coated with Castrol LM Grease.

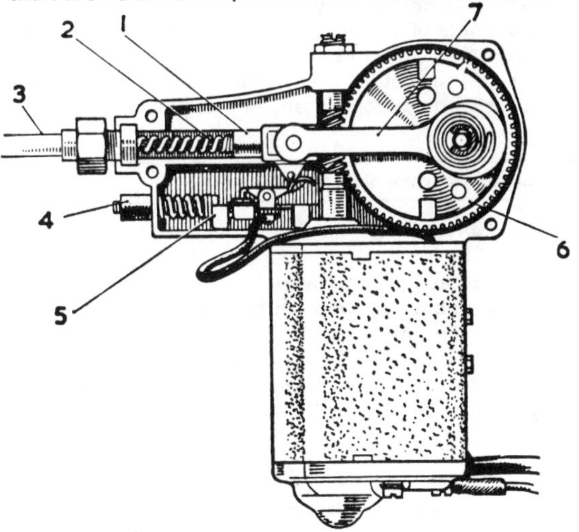

Fig 10.20 WINDSCREEN WIPER MOTOR (LATER TYPE)

1 Crosshead
2 Cable rack
3 Outer casing
4 Adjusting nut
5 Park switch
6 Final gear
7 Connecting rod

37 Windscreen wiper motor and gearbox - dismantling and reassembly (Type DR3A)

1 Undo the four screws holding the gearbox cover in place and remove the cover.
2 Undo and remove the two through bolts from the commutator end bracket. Pull out the connector and free the end bracket from the yoke.
3 Carefully remove the brush gear as a unit from the commutator and then withdraw the yoke.
4 Clean the commutator and brush gear and if worn, fit new brushes. The resistance between adjacent commutator segments should be 0.34 to 0.41 ohms.
5 Carefully examine the internal wiring for signs of chafing, breaks or charring which would lead to a short circuit. Insulate or replace any damaged wiring.
6 Measure the value of the field resistance which should be between 12.8 and 14 ohms. If a lower reading than this is obtained it is likely that there is a short circuit and a new field coil should be fitted.
7 Renew the gearbox gear teeth if they are damaged, chipped or worn.
8 Reassembly is a straightforward reversal of the dismantling sequence, but ensure the following items are lubricated:

a) Immerse the self-aligning armature bearing in SAE 20 engine oil for 24 hours before assembly.
b) Oil the armature bearings in SAE 20 engine oil.
c) Soak the felt lubricator in the gearbox with SAE 20 engine oil.
d) Grease generously the worm wheel bearings, crosshead, guide channel, connecting rod, crankpin, worm, cable rack and wheelboxes and the final gear shaft.

38 Windscreen wiper switch - removal and refitting

1 Switch off the battery master switch for safety reasons.
2 Using a small electrician's screwdriver push in the knob retaining plunger and pull the knob away from the instrument panel.
3 Unscrew the nut on the outside of the switch body.
4 Make a note of the electrical cable connections at the rear of the switch and disconnect the cables from the terminals.
5 The switch may now be withdrawn from the instrument panel.
6 Refitting the switch is the reverse sequence to removal.

39 Horn - fault tracing and rectification

1 If the horn works badly or fails completely, first check the wiring leading to it for short circuits and loose connections. Also check that the horn is firmly secured and that there is nothing lying on the horn body.
2 The horn should never be dismantled but it is possible to adjust it. This adjustment is to compensate for wear only and will not affect the tone.
3 At the rear of the horn is a small adjustment screw as shown in Fig 10.21. Do not confuse this screw with the coil securing screws.
4 Turn the adjustment screw anticlockwise until the horn just fails to sound. Then turn the screw a quarter of a turn clockwise which is the desirable optimum setting.
5 It is recommended that if the horn is to be reset in the car the fuse A1-A2 should be removed and replaced with a piece of wire otherwise the fuse will continually blow due to the continuous high current required for the horn in continual operation.

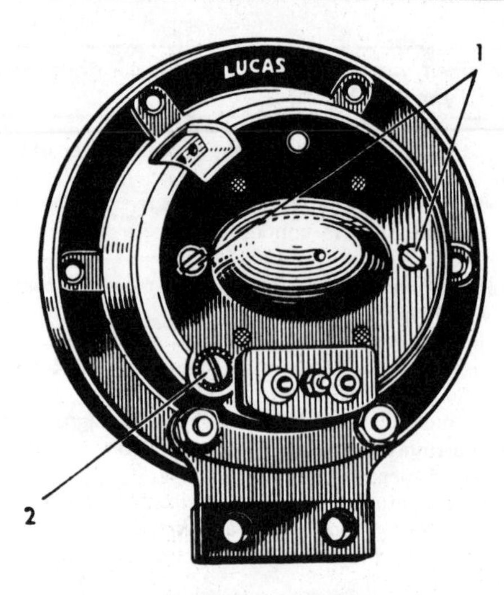

Fig 10.21 REAR VIEW OF HORN

1 Coil securing screws *2 Adjustment screw*

40 Horn push and direction indicator switch - removal and refitting

1 The horn push and direction indicator switch is mounted on the steering wheel hub and comprises a spring metal push covering the hub with the indicator switch lever positioned in its centre. The switch cables pass through a long tube down the steering column shaft which is secured by an olive in the base of the steering box.
2 If an adjustable steering column is fitted the stator tube is in two parts, the shorter piece being attached to the horn quadrant.
3 For safety reasons switch off the battery master switch.
4 Detach the horn and flasher light cables that protrude from the end of the stator tube at the nearest snap connectors. Note the cable colours for correct refitting.
5 Where an adjustable steering column is fitted, undo and remove the three grub screws in the steering wheel hub and withdraw the quadrant together with the short stator tube and cables. The longer part of the stator tube will remain in the steering column.
6 It is important to note that the short stator tube has an indentation in it which fits in a slot in the long stator tube. The horn quadrant must be withdrawn without any twisting motion to avoid enlarging the slot in the long stator tube. Any enlargement of this slot will result in excessive movement of the horn quadrant after refitting.
7 When a non-adjustable steering column is fitted, undo and remove the nut and olive at the bottom end of the steering box. This will free the stator tube which may be withdrawn as one piece with the horn push quadrant attached to the end. At the same time the cables will be drawn up from the outer column.
8 Using a piece of tapered wood, plug the hole left in the bottom of the steering box so as to prevent the oil draining out.
9 Refitting is the reverse sequence to removal. Top up the oil level in the steering gearbox with Castrol Hypoy.

41 Ignition switch - removal and refitting

1 For safety reasons switch off the battery master switch.
2 Unscrew the locknut which secures the switch to the instrument panel.
3 Note the electrical cable connections to the rear of the switch and disconnect the cables.
4 Draw the switch from the instrument panel.
5 Refitting is the reverse sequence to removal. Apply a little light machine oil to the barrel to ensure ease of movement.

42 Fuel gauge - removal and refitting

1 Switch off the battery master switch for safety reasons.
2 Note the locations of the T and B terminals at the rear of the gauge and disconnect the cables.
3 Unscrew and remove the centrally placed knurled securing nut at the rear of the instrument.
4 Draw the gauge forwards and away from the instrument panel.
5 Refitting is the reverse sequence to removal.

43 Switches - removal and refitting

1 In all cases switch off the battery master switch.
2 Refitting is the reverse sequence to removal.

Overdrive switch
1 Unscrew the locknut securing the switch to the instrument panel
2 Draw the switch from its location in the instrument panel.
3 Make a note of the electrical cables and disconnect from the instrument panel.

Panel lamps
1 Undo and remove the two screws that secure the panel lamps switch to the underside of the instrument panel.
2 Make a note of the electrical cables and disconnect from the switch terminals.
3 Withdraw the switch from the instrument panel.

Lighting switch
1 Using a small electrician's screwdriver, push in the knob retaining plunger and pull the knob away from the instrument panel.
2 Unscrew the nut on the outside of the switch body.
3 Make a note of the electrical cable connections at the rear of the switch and disconnect the cables from the terminals.
4 The switch may now be withdrawn from the instrument panel.

Headlamp dip switch
1 Undo and remove the two screws securing the switch to the bracket welded to the floor.
2 Carefully withdraw the switch from the bracket.
3 Make a note of the electrical cables and disconnect the three cables from the connectors.
4 Check the operation of the dip switch. Lightly smear the moving parts with petroleum jelly on reassembly.

44 Panel and warning light bulbs

In all cases refitting is the reverse sequence to removal.

Direction indicator warning lamp
1 Pull the bulb holder with the bulb in position from the rear of the warning lamp.
2 Unscrew the bulb and fit a new bulb.
3 Should it be necessary to release the green lens unscrew the chrome plated retaining ring situated on the front of the fascia panel.

Panel light
1 Pull the bulb holder with the bulb in position from the rear of the warning lamp.
2 Unscrew the bulb and fit a new bulb.

Ignition warning light bulb and headlamp main beam warning light
1 Pull the bulb holder with the bulb in position from the rear of the warning lamp.

2 Unscrew the bulb and fit a new bulb.
3 Should it be necessary to release the green lens unscrew the chrome plated retaining ring situated on the front of the fascia panel.

45 Headlamp bulb - removal and refitting

1 To remove the light unit to replace a bulb, unscrew the retaining screw at the bottom of the chromium plated rim and with a wide blade screwdriver gently prise the rim (1) (Fig 10.22) away from the rubber sealing ring (2).
2 Remove the rubber sealing ring and press the light unit in onto the springs of the adjusting screws. Turn the light unit in an anticlockwise direction until the heads of the screws come opposite the enlarged ends of the keyhole slots and withdraw the light unit.
3 The bulb holder is on the back of the light unit and is released by turning anticlockwise.
4 Replace the defective bulb, making sure that the new bulb is properly seated in the holder.
5 Replacement of the bulb holder and light unit is a reversal of the above procedure.

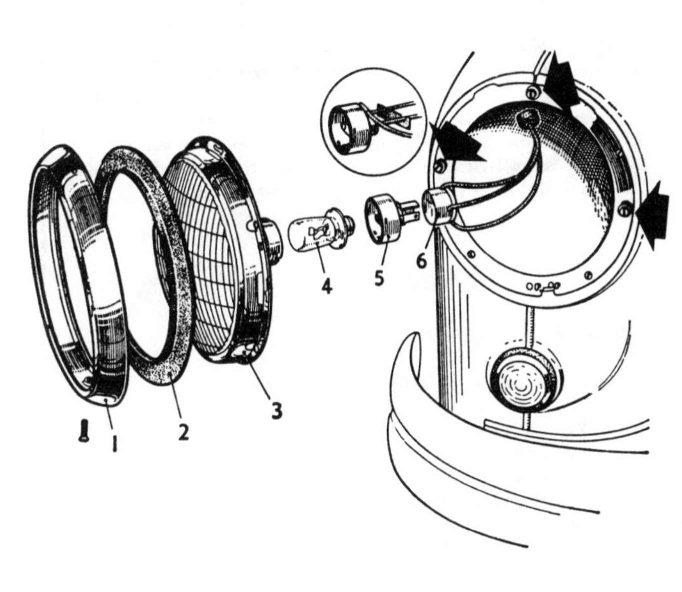

Fig 10.22 HEADLIGHT UNIT

1 Front rim
2 Rubber seal
3 Glass and reflector
4 Bulb
5 Bulb holder (USA)
6 Three pin socket

Inset shows bulb holder for all models except USA
Adjustment screws arrowed

46 Headlight light unit - removal and refitting

1 Remove the light unit as described in the previous section.
2 Undo and remove the three self-tapping screws from the unit rim and remove the sealing rim and unit rim from the light unit.
3 Place the replacement light unit between the unit rim and setting rim, taking care to see that the die cast projection at the edge of the light unit fits into the slot in the seating rim, and also check that the seating rim is correctly positioned.
4 Secure in position with the three self-tapping screws.

47 Headlamps - sealed beam unit replacement

1 On later produced models sealed beam light units of Lucas manufacture are fitted. To remove this unit start by removing the rim fixing screw at the base of the rim and lift the rim off the locating lugs at the top of the headlamp shell.
2 Remove the three Phillips screws holding the light unit rim, remove the rim and withdraw the light unit from its seating.
3 Disconnect the three pin socket from the lamp unit. Should a sealed beam fail, the whole unit will have to be replaced as owing to its sealed filaments no repairs are possible.
4 Replacement of a new unit is a direct reversal of the removal procedure.

48 Headlamps - adjustment

1 On earlier cars without sealed beam units there are three adjusting screws (Fig 10.22). The screw at the top controls the vertical adjustment and the two side screws the horizontal adjustment.
2 On models with sealed beam units there are only two adjusting screws. The top one being for vertical movement and the side one for horizontal movement.
3 Where possible adjustment should always be carried out using an optical type beam setter or similar equipment. If this is not available the lamps can be roughly aligned on a level piece of road at night, checking both in the dipped and undipped position.

49 Side, tail and flasher light bulbs - removal and refitting

1 Prise back the rubber lip and insert a screwdriver blade under the glass retaining collar.
2 Lever the collar out from the lamp body.
3 Remove the lamp glass and unscrew the bulb.
4 Fit a new bulb and replace the lamp glass and collar.

Combined stop, tail and flasher light bulbs
1 Prise back the rubber lip and insert a screwdriver blade under the glass retaining collar.
2 Lever the collar out from the lamp body.
3 Remove the lamp glass and unscrew the bulb.
4 Fit a new bulb and replace the lamp glass and collar.

Rear number plate light bulb
1 Remove the screw which secures the lamp cover and lift off the cover.
2 Remove the defective bulb and fit a new one.
3 Replace the cover and secure with the screw.

Side, stop, tail and flasher lamp bulbs (BN7, BT7, BJ7 and BJ8)
1 Unscrew the two screws securing the lens and lift off the lens.
2 Remove the defective bulb and fit a new one. It will be noted that the front side and direction indicator lamps have single filament bulbs and may be fitted either way round.
3 The tail and stop lamp bulbs have twin filaments and an offset bayonet fixing to ensure correct replacement.

WIRING DIAGRAM – AUSTIN HEALEY 100-6 SERIES BN4 and BN6

1 LH pilot lamp and front flasher
2 LH headlamp
3 RH headlamp
4 RH pilot lamp and front flasher
5 Direction switch and horn push
6 RH horn
7 Distributor
8 LH horn
9 Ignition coil
10 Stop lamp switch
11 Starter
12 Overdrive solenoid (when fitted)
13 Gearbox switch
14 Throttle switch (when fitted)
15 Flasher relay
16 Dynamo
17 Control box
18 Heater motor (when fitted)
19 Overdrive relay (when fitted)
20 Starter solenoid
21 Flasher unit
22 Ignition switch
23 Starter switch
24 Flasher warning lamp
25 Lighting switch
26 Heater switch (when fitted)
27 Wiper motor
28 Dip switch
29 Panel light
30 Panel light
31 Panel light
32 Fuel gauge
33 Petrol pump
34 Panel lamps switch
35 Wiper motor switch
36 High beam warning lamp
37 No charge warning lamp
38 Cigar lighter (when fitted)
39 Overdrive switch (when fitted)
40 12 volt battery
41 Tank unit
42 Battery master switch
44 LH stop and tail lamp and rear flasher
45 Number plate lamp
46 RH stop and tail lamp and rear flasher

CABLE COLOUR CODE

U	*BLUE*	*GN*	*GREEN with BROWN*
UR	*BLUE with RED*	*GB*	*GREEN with BLACK*
UW	*BLUE with WHITE*	*Y*	*YELLOW*
W	*WHITE*	*YG*	*YELLOW with GREEN*
WR	*WHITE with RED*	*N*	*BROWN*
WG	*WHITE with GREEN*	*NU*	*BROWN with BLUE*
WP	*WHITE with PURPLE*	*NG*	*BROWN with GREEN*
WN	*WHITE with BROWN*	*NP*	*BROWN with PURPLE*
WB	*WHITE with BLACK*	*NB*	*BROWN with BLACK*
G	*GREEN*	*R*	*RED*
GR	*GREEN with RED*	*RW*	*RED with WHITE*
GY	*GREEN with YELLOW*	*B*	*BLACK*
GU	*GREEN with BLUE*	*BG*	*BLACK with GREEN*
GW	*GREEN with WHITE*	*LG*	*LIGHT GREEN*
GP	*GREEN with PURPLE*	*WU*	*WHITE with BLUE*

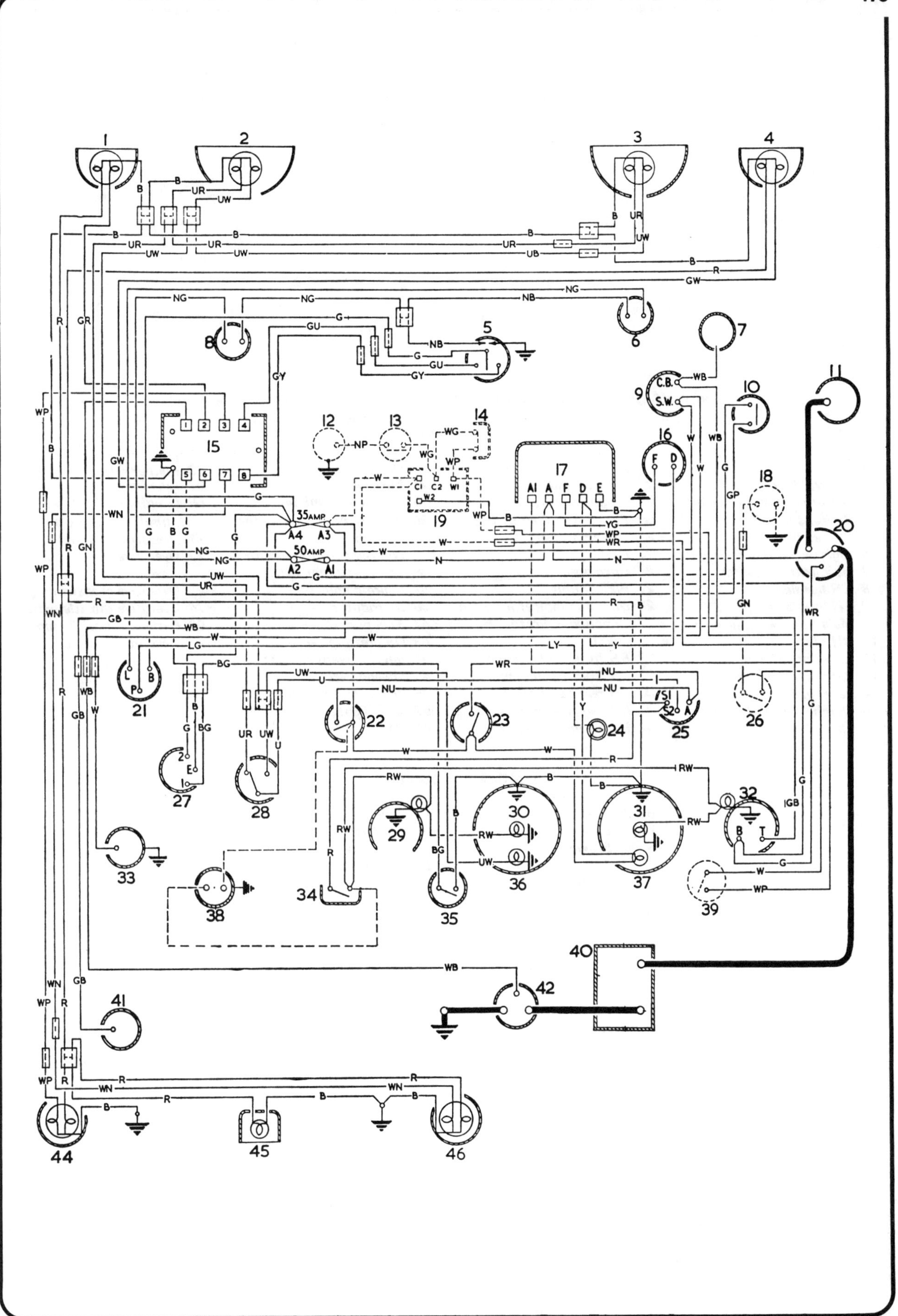

1
2
3
4
5
6
7
8
9
10
11
12
13
14
15
16
17
18
19
20
21
22
23
24
25
26
27
28
29
30
31
32
33
34
35
36
37
38
39
40
41
42
44
45
46
35AMP
A4
A3
50AMP
A2
A1
C.B.
S.W.
C1
C2
W1
W2
A1 A F D E
F D
L P B
B T

**WIRING DIAGRAM – AUSTIN HEALEY Mk I and Mk II (Series BN7 and BT7)
Mk II and Mk III (Series BJ7 and BJ8)
(Early Cars)**

1 Dynamo
2 Control box
3 Battery
4 Starter solenoid
5 Starter motor
6 Lighting switch
7 Headlight dip switch
8 RH headlamp
9 LH headlamp
10 Main beam warning light
11 RH pilot lamp
12 LH pilot lamp
13 Panel light switch
14 Panel lights
15 Number plate illumination lamp
16 RH stop and tail lamp
17 LH stop and tail lamp
18 Stop light switch
19 Fuse unit (50 amps 1-2, 35 amps 3-4)
23 Horns
24 Horn push
25 Flasher unit
26 Direction indicator switch
27 Direction indicator warning lights
28 RH front flasher lamp
29 LH front flasher lamp
30 RH rear flasher lamp
31 LH rear flasher lamp
32 Heater motor switch
33 Heater motor
34 Fuel gauge
35 Fuel gauge tank unit
36 (Windscreen wiper motor switch
37 Windscreen wiper motor
38 Ignition/starter switch
39 Ignition coil
40 Distributor
41 Fuel pump
43 Oil pressure gauge
44 Ignition warning light
45 Speedometer
46 Water temperature gauge
57 Cigar lighter
60 Radio
63 Flasher relay
67 Line fuse
68 Overdrive relay (25 amp fuse)
71 Overdrive solenoid
72 Overdrive manual control switch
73 Overdrive gear switch
74 Overdrive throttle switch
92 Battery cut off switch
95 Revolution counter

CABLE COLOUR CODE

B	*Black*	*R*	*Red*
U	*Blue*	*W*	*White*
N	*Brown*	*Y*	*Yellow*
G	*Green*	*LG*	*Light Green*
P	*Purple*		

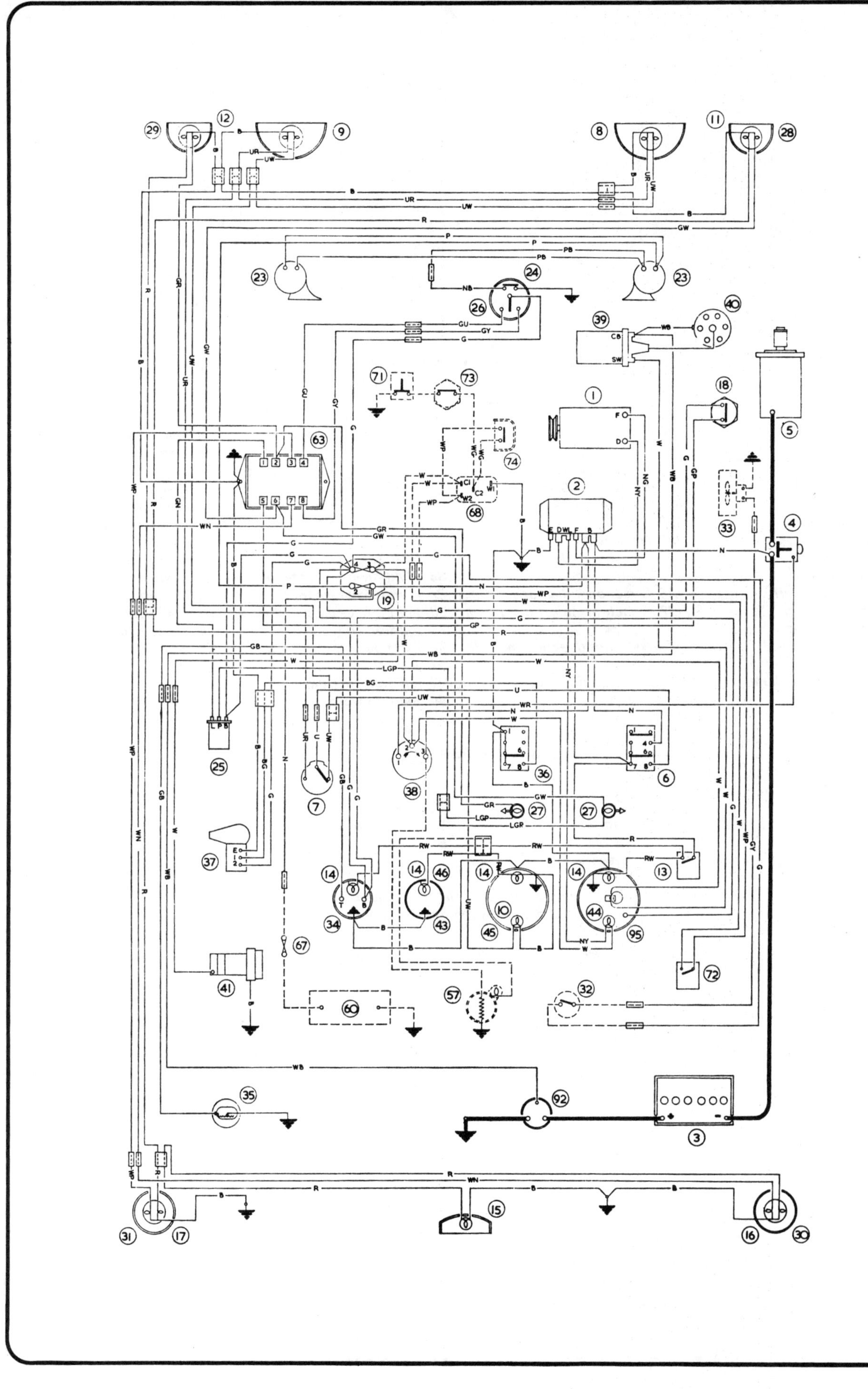

WIRING DIAGRAM – AUSTIN HEALEY Mk I and Mk II (Series BN7 and BT7)
Mk II and Mk III (Series BN7 and BJ8)
(Later Cars)

1 Dynamo
2 Control box
3 Battery
4 Starter solenoid
5 Starter motor
6 Lighting switch
7 Headlamp dip switch
8 RH headlamp
9 LH headlamp
10 Main beam warning lamp
11 RH pilot lamp
12 LH pilot lamp
13 Panel lamp switch
14 Panel lamps
15 Number plate illumination lamp
16 RH stop and tail lamp
17 LH stop and tail lamp
18 Stop lamp switch
19 Fuse unit (50 amps 1-2, 35 amps 3-4)
23 Horns
24 Horn push
25 Direction indicator flasher unit
26 Direction indicator switch
27 Direction indicator warning lamps
28 RH front direction indicator lamp
29 LH front direction indicator lamp
30 RH rear direction indicator lamp
31 LH rear direction indicator lamp
*32 Heater motor switch **
*33 Heater motor **
34 Fuel gauge
35 Fuel gauge tank unit
36 Windscreen wiper switch
37 Windscreen wiper motor
38 Ignition/starter switch
39 Ignition coil
40 Distributor
41 Fuel pump
42 Oil pressure gauge
44 Ignition warning light
45 Speedometer
46 Water temperature gauge
57 Cigar lighter
*60 Radio **
67 Line fuse
*68 Overdrive relay unit **
*71 Overdrive solenoid **
72 Overdrive manual control switch
*73 Overdrive gear switch **
74 Overdrive throttle switch
92 Battery cut-off switch
95 Revolution counter

CABLE COLOUR CODE

N	Brown	LG	Light green
U	Blue	W	White
R	Red	Y	Yellow
P	Purple	B	Black
G	Green		

When a cable has two colour code letters the first denotes the main colour and the second denotes the tracer colour

** Optional extra, circuit shown dotted*

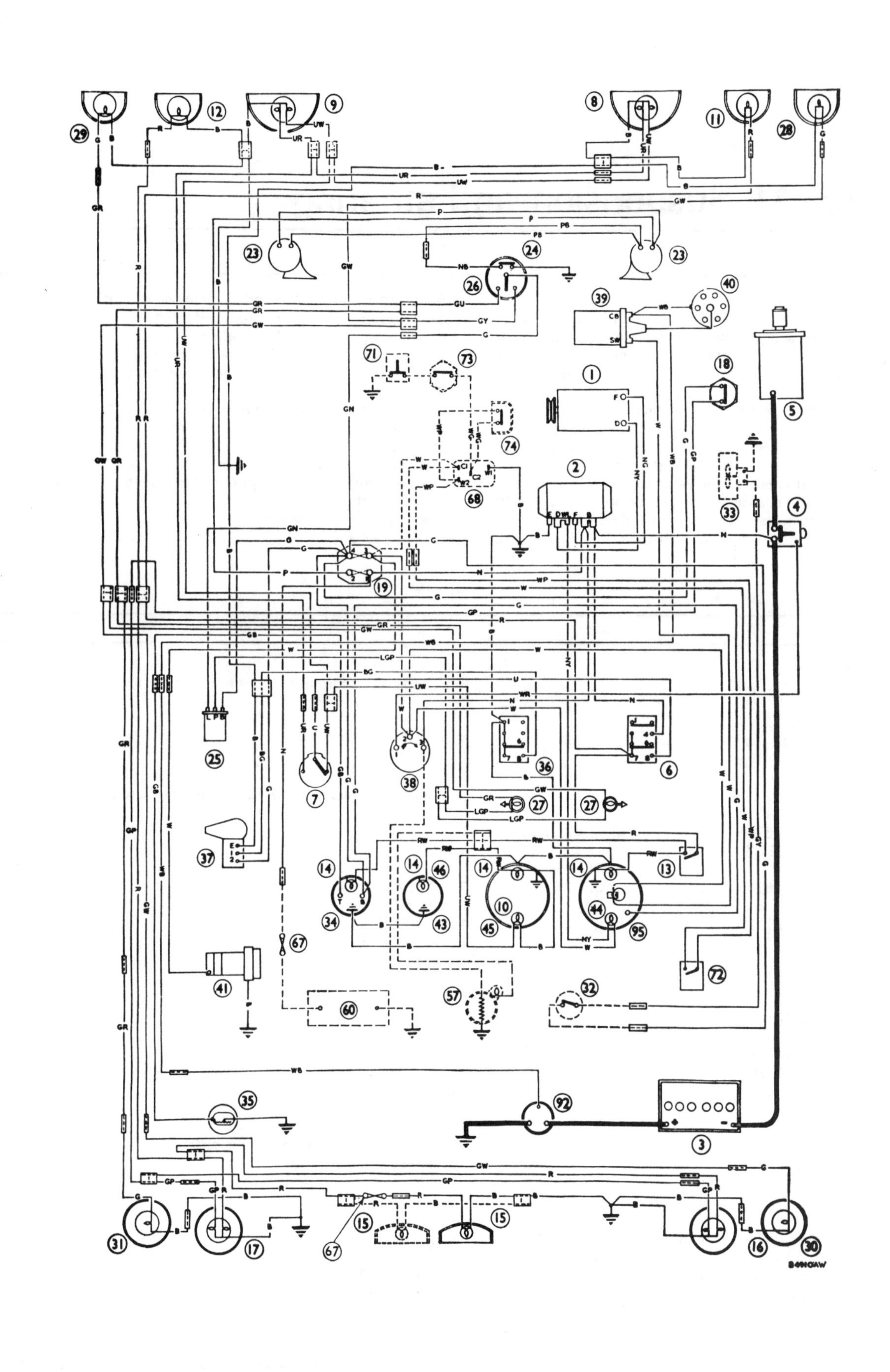
1
2
3
4
5
6
7
8
9
10
11
12
13
14
15
16
17
18
19
23
24
25
26
27
28
29
30
31
32
33
34
35
36
37
38
39
40
41
43
44
45
46
57
60
67
68
71
72
73
74
92
95

Chapter 11 Suspension and steering

Contents

Specifications

Front suspension

Type	Independent coil springs and wishbones
Castor angle	2°
Camber angle	1°
Swivel pin inclination	6½°

Rear suspension

Type	Semi-elliptic understrung leaf springs and panhard rod
Number of leaves:	
up to Mk III No 26705	7
from Mk III No 26706	6
Thickness of leaves:	
up to Mk III No 26705	6 at 0.1875 inch 1 at 0.15625 inch
from Mk III No 26706	4 at 0.1875 inch 2 at 0.15625 inch
Width	1¾ inch
Deflection:	
up to Mk III No 26705	4 ± ¼ inch
from Mk III No 26706	5¼ inch
Load camber:	
up to Mk III No 26705	½ ± 1/8 inch
from Mk III No 26706	1 inch
Zinc leaves	3 at 0.03125 inch

Shock absorbers

Type	Armstrong double-acting hydraulic piston

Steering

Type	Cam gear
Ratio:	
cars fitted with 2639 cc engines	14 : 1
all others	15 : 1
Toe-in	1/16 to 1/8 inch
Steering lock angle	Outer 20° Inner 21°

Wheels

Type	4J x 15 ventilated steel disc or 4J x 15 knock-on wire (optional)

Tyre sizes and pressures

Standard:	
cars fitted with 2639 cc engines	5.90 x 15 tubeless
all others	5.90 x 15 Road Speed (with tubes)
Pressures:	
* front	20 lb/sq in
* rear	25 lb/sq in

* These pressures are suitable for road speeds up to 110 mph

Note: For maximum performance increase all pressures by 5 lb/sq in

Torque wrench settings

Steering wheel nut	41 lb ft (5.76 kg m)
Wheel nuts (pressed steel wheels)	60-63 lb ft (8.3-8.64 kg m)

1 General description

The front suspension unit is of the independent wishbone type and is mounted onto the outer ends of a strong cross-member which is secured to the longitudinal body frame members by four studs and nuts. Large rubber mountings are used to dampen road noise.

Each unit comprises a coil spring, swivel axle unit, lower suspension arms and Armstrong double-acting hydraulic damper which is also part of the top of the suspension. The upper arms of the suspension are formed by the lever arms of the hydraulic damper, this being mounted on the top of the crossmember by four set bolts. The outer ends of the damper are secured to the swivel pin upper link by a trunnion pin and tapered rubber bushes.

The suspension lower arms are secured to the crossmember by a single spindle with Metalastik bushes, nuts and washers and by two screwed bushes and by a screwed fulcrum pin to the lower end of the swivel pin.

The coil spring is held in position at its lower end by a plate mounted on the suspension lower arms, whilst its upper end is located in the underside of the suspension crossmember.

A rubber bushed anti-roll bar is fitted.

The rear axle is of the three-quarter floating type incorporating hypoid final drive. It is possible to remove the axle shafts and differential unit from the axle without moving it from the car.

The rear axle wheel bearing outer races are located within the hubs and the inner races are mounted on the axle tubes and screwed in place with large diameter nuts and lockwashers.

Wheel studs, pressed into the hubs, pass through the brake drums and axle shaft driving flanges. The brake drums are accurately located on the hub flanges by two countersunk screws to each drum.

The rear suspension is of the semi-elliptic leaf spring design located fore and aft beneath the axle by anchor pins and flexible rubber mountings.

Double acting Armstrong piston type dampers are fitted to the rear between the body frame and each end of the axle. A Panhard rod is fitted to help locate the rear axle more positively.

The steering gear is of the cam and peg type in which the cam takes the form of a worm mounted on the ball bearings within the steering box and is an integral part of the steering inner column.

The rocker shaft is mounted in the steering gearbox in plain bearings and a lever which is integral with the shaft carries a conical peg. It is mounted in needle bearings and engages with the cam. As the peg does not bottom in the cam groove, the depth of engagement can be adjusted should wear occur. For this an adjuster screw is fitted to the steering gearbox top cover.

The steering arm is attached to the tapered splined end of the rocker shaft by a retaining nut and lockwasher and connected to the steering idler arm on the opposite side of the car by the track rod.

The steering arms are connected to the steering levers mounted on the swivel axles by non-adjustable tie rods, each end of both the track and tie rods being fitted with a spring loaded ball joint.

2 Front and rear suspension - maintenance

1 Lubrication nipples are located at the points listed below and should be lubricated with three or four strokes of the grease gun filled with Castrol LM Grease at the intervals recommended in Routine Maintenance:

Front suspension lower arm outer fulcrum pins (2 nipples)
Swivel pins (4 nipples)
Steering track and tie rod ball joints (6 nipples)

2 The rear springs are mounted in rubber and spraying with oil should be strictly avoided. The only attention required is an occasional tightening of the spring seat bolts to make sure they are not loose.
3 Periodically the hydraulic dampers should be examined for leaks and the fluid level checked and topped up if necessary to the level at the bottom of the filler plug hole. Before removing the filler level plug, wipe the area clean of dust as it is important that no dirt enters the damper. Only use ARMSTRONG SUPER (THIN) SHOCK ABSORBER FLUID 624.
4 At the same time as checking the damper fluid level make sure that their anchorages are tight.

3 Cam and peg steering and steering idler - maintenance

1 The steering gearbox and idler filler plug should be removed at the recommended intervals (see Routine Maintenance) and the oil level topped up with Castrol Hypoy Light. Wipe the area around the filler plugs before removal as it is important that no dust enters the steering gearbox or idler. The level of oil in the steering gearbox should be up to the bottom of the filler hole.
2 To top up the steering idler unscrew the square headed filler plug from the top of the steering idler and top up to the bottom of the filler plug orifice. It is important that the oil in the steering idler is kept at the correct level and not allowed to fall otherwise severe loading will be imposed on the steering gearbox.
3 Nylon seated ball joints, sealed during manufacture, were fitted on Healey 3000 Mk II cars from car number BT7 19191 and these do not require further lubrication.

4 Suspension, steering and shock absorbers - check for wear

1 To check for wear in the outer ball joints of the tie-rods, place the car over a pit, on a ramp or lie underneath the front of the car. For safety reasons do not forget to chock the rear wheels.

An assistant should rock the steering wheel from side to side whilst the ball joints are inspected. If wear is present this will be indicated by movement within the joints.

2 To check for wear in the rubber and metal bushes, jack up the front of the car until the wheels are clear of the ground. Hold each wheel in turn with the hands placed at the top and the bottom and try to rock it. If the wheel should rock continue the movement at the same time inspecting the upper trunnion link rubber bushes and the rubber bushes located at the inner ends of the wishbone for play.

3 Should the wheel rock and there is no side movement in the rubber bushes then the king pins and rubber bushes will be worn. Alternatively, if the movement occurs between the wheel and brake backplate, then the hub bearings require replacement.

4 The rubber bushes can be renewed but, if there is play between the lower end of the king pin and the wishbone, then it will be necessary to renew the fulcrum pin.

5 Side play or vertical or horizontal movement of the upper link or shock absorber arms relative to the shock absorber body is best checked with the outer end of the shock absorber arms free from the upper trunnion link. Should play be present the shock absorber bearings are worn and a replacement shock absorber should be obtained.

6 To check the correct action of the shock absorbers the car should be bounced at each corner. After each bounce the car should return to its normal ride position within one to one and a quarter up-and-down movements. If the car continues to move up and down in decreasing amounts it means that either the shock absorbers need topping up, or if they are already full, that the dampers are worn and must be renewed.

7 The shock absorbers cannot be adjusted without the use of special tools and therefore should not be dismantled but exchanged at your local BLMC agent.

8 Excessive play in the steering gear will lead to wheel wobble, and can be confirmed by checking if there is any loss to movement between the steering wheel and the steering lever at the base of the steering box. Provision is made for adjustment, but if all the adjustment has been taken up it will be necessary to fit a rebuilt steering box.

9 The ball joints are not prone to excessive wear providing that they have been regularly greased and the rubber gaiters renewed immediately if they are damaged.

5 Front hub - removal, inspection and refitting (disc wheel, drum brake)

1 Chock the rear wheels, apply the handbrake, jack up the front of the car and place on firmly based axle stands. Remove the road wheels.

2 With a wide blade screwdriver prise out the grease retaining cap (1) (Fig 11.2) from the centre of the hub.

3 Extract the split pin and undo and remove the castellated nut (2). Withdraw the flat washer (3) from the end of the stub axle.

4 Using a three legged puller and thrust pad withdraw the hub (7) complete with bearings and brake drum.

5 If the complete inner race (8) stayed on the stub axle instead of being withdrawn from the hub, remove the ball bearings and cage and with a universal three legged puller and thrust pad, draw the race inner track from the stub axle. Remove the oil seal (9) noting that the lip is facing inwards towards the centre of the hub.

6 Remove the bearing spacer (5) from the rear of the hub and with a soft metal drift carefully tap out the inner race, cage and ball bearings from the outer bearing (4).

7 The bearing outer races may next be drifted out of the hub using a soft metal drift. For this the hub should be placed on the top of a bench vice and with the jaws open and supporting the drum mounting flange or wheel stub flange depending on which outer track is being removed.

8 Thoroughly clean the bearings (4, 8) and the hub (7) in paraffin and then examine them for grooving, flat spots, chips, pitting or other damage. Renew the bearings if worn.

9 To reassemble, first fit the bearing spacer (5) with its larger diameter towards the inside of the hub.

10 Assemble the two bearings (4, 8) making sure that the marked edges of the inner and outer tracks and the ball cage are all the same side when the bearing is assembled.

11 Pack the bearings with Castrol LM Grease making sure that it is worked well into the bearing cage and insert them into the hub thrust adjacent to the bearing spacer.

12 Remove any excess grease so as to leave room for expansion otherwise the bearings could overheat caused by excessive churning of the grease.

13 Refit a new oil seal (9), lip facing inwards to the centre of the hub, into its recess using a drift of suitable diameter so as to locate on the rear of the oil seal. Carefully tap it into position making sure that it does not tilt.

14 Refit the assembled drum and hub onto the stub axle and using a tubular drift located on the outer bearing inner track, drive the hub into position.

15 Replace the plain washer (3) and castellated nut (2) and tighten fully before locking in position with a new split pin. Do not slacken the nut to align the slots with the split pin hole in the stub axle.

16 Do not put any grease in the grease retaining cap (1) but tap it into position with a soft faced hammer.

17 Replace the road wheel, remove the supports and lower the car to the ground.

6 Front hub - removal, inspection and refitting (wire wheels, drum brake)

1 Chock the rear wheels, apply the handbrake, jack up the front of the car and place on firmly based axle stands.

2 Remove the knock-on hub cap using a soft faced hammer and unscrewing in the normal direction of rotation. Pull the road wheel from the splines.

3 Slacken back the brake adjuster and undo and remove the four self-locking nuts and spring washers that secure the brake drum to the hub. Using a soft faced hammer gently tap the brake drum clear of the front hub assembly.

4 Supplied in every tool kit with the car is a special grease retaining cup extractor. This should be screwed onto the stud of the cup (1) (Fig 11.3) and the nut screwed down.

5 Straighten the split pin (3) locking the hub securing nut (2) and ease it out through the hole in the hub.

6 Using a box spanner and tommy bar or socket, undo and remove the hub securing nut (2) and flat washer (4) from the swivel axle (11).

7 Using a suitable size three legged puller and thrust pad draw the hub (7) from the swivel axle. The inner and outer bearings (5, 9) and oil seal (10) will also be drawn from the swivel axle at the same time as the hub.

8 With the hub away from the swivel axle drift the outer bearing track (6) using a soft metal drift through the inner bearing. The inner bearing track (8) and oil seal (10) may then be removed in a similar manner but working from the other end of the hub.

9 Thoroughly clean the bearings and the hub in paraffin and then examine them for grooving, flat spots, pitting or other damage. Obtain new bearings if the original ones are worn.

10 To reassemble, first drift the two bearing outer tracks (6, 8) into the hub, making sure that they are fitted the correct way round.

11 Insert the inner race and rollers (9) of the inner bearing and the bearing spacer (3) into the hub. The cavity between the bearings and the inner bearings should next be padded out with Castrol LM Grease making sure that it is worked well into the bearing cage.

12 Refit a new oil seal (10), lip facing inwards to the centre of the hub into its recess using a drift of suitable diameter so as to locate on the rear of the oil seal. Carefully tap it into position making sure that it does not tilt.

13 Pack the inner race and rollers (5) of the outer bearing with

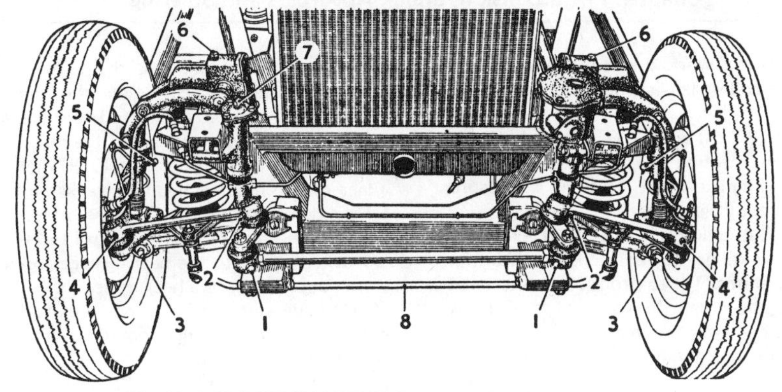

Fig 11.1 FRONT SUSPENSION AND STEERING ASSEMBLY

1 Cross tube connections
2 Side rod inner connections
3 Lower link
4 Side rod outer connections
5 Swivel pin
6 Shock absorber
7 Steering idler
8 Anti-roll bar (no lubrication required)

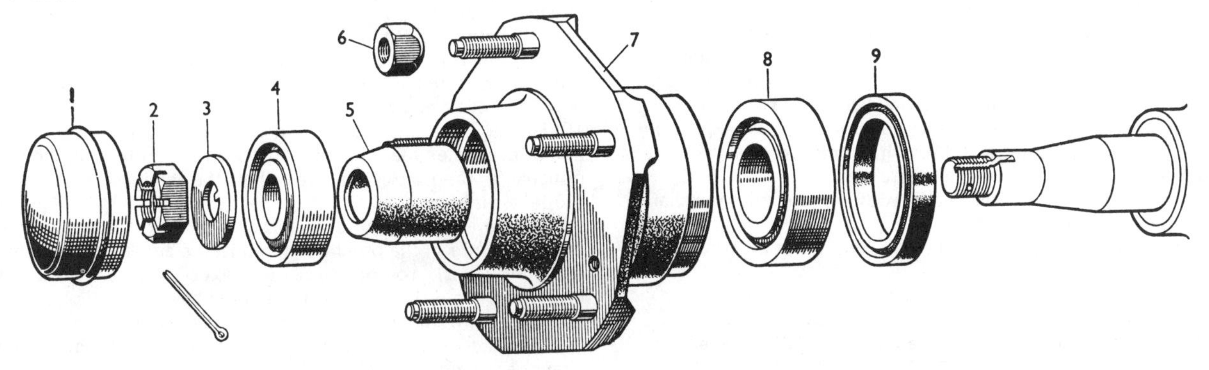

Fig 11.2 EXPLODED VIEW OF FRONT HUB (DISC WHEELS)

1 Hub cap
2 Castellated nut
3 Locating washer
4 Outer bearing
5 Distance piece
6 Wheel nut
7 Hub
8 Inner bearing
9 Oil seal

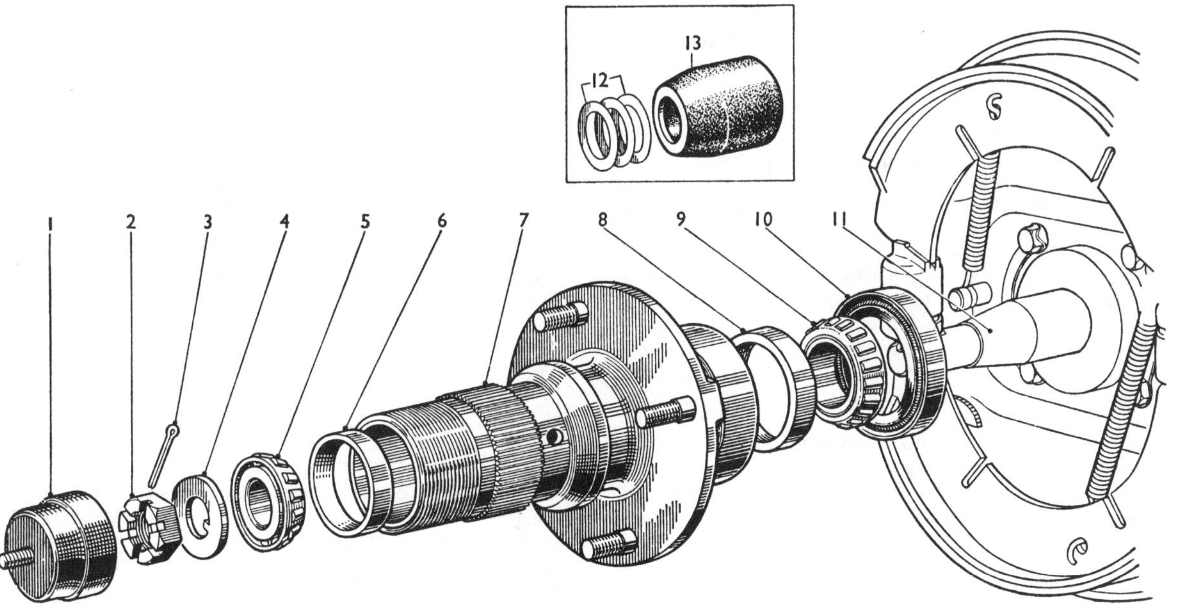

Fig 11.3 EXPLODED VIEW OF FRONT HUB (WIRE WHEELS)

1 Grease cup
2 Axle nut
3 Split pin
4 Washer
5 Outer bearing
6 Bearing outer race
7 Hub
8 Bearing outer race
9 Inner bearings
10 Oil seal
11 Swivel axle

Inset shows distance piece and shims

Castrol LM Grease making sure it is worked well into the bearing cage. Refit the inner race and rollers leaving out any shims (12) that were previously used.

14 Replace the stub axle nut (2) and washer (4) and tighten the nut whilst at the same time rotating the hub forwards and backwards until a noticeable drag is evident. This will make sure that the bearing cones are correctly seated.

15 Unscrew and remove the stub axle nut (2) and lift away the washer (4) and inner race and rollers (5) of the outer bearing. Insert a sufficient number of shims (12) to produce an excessive amount of end float. Note the total thickness of the shims inserted using feeler gauges to compare the thickness.

16 Refit the bearing centre (5) and washer (4) and tighten the stub axle nut (2).

17 With feeler gauges determine the amount of end float in the bearings.

18 Remove the stub axle nut (2), washer (4) and outer bearing centre (5) again and reduce the number of shims (12) so as to eliminate end float and yet still allow the hub to rotate freely.

19 Refit the outer bearing centre (5) and washer (4) and stub axle nut (2). Tighten the nut to a torque wrench setting of 40 to 70 lb ft. The reason for such a wide range is to allow the nut to be aligned with the split pin hole in the stub axle. Lock the nut (2) with a new split pin (3) and bend over the ends.

20 Remove any excess grease so as to leave room for expansion otherwise the bearings could overheat caused by excessive churning of the grease. Refit the grease cup.

21 Refit the brake drum and secure with the four spring washers and self-locking nuts.

22 Grease the wheel hub splines and refit the road wheel and replace the knock-on hub cap.

7 Front hub - removal, inspection and refitting (disc brake)

1 Chock the rear wheels, apply the handbrake, jack up the front of the car and place on firmly based stands. Remove the road wheel, the method depending on whether disc or wire wheels are fitted.

2 Ease back the locking tabs and remove the two bolts securing the calliper to the swivel axle. With a piece of string or wire support the weight of the calliper so that the flexible hose is not strained. Note any shims placed between the calliper and the swivel axle.

3 On cars fitted with wire wheels extract the split pin (3) and undo and remove the castellated nut (2). Withdraw the flat washer (4) from the end of the stub axle. See Fig 11.4.

4 Straighten the split pin locking the hub securing nut and ease it out of the axle stub and nut. On models with wire wheels a special hole is drilled in the hub which will allow the split pin to be removed.

5 Using a three legged puller and thrust pad withdraw the hub (7) complete with bearings and brake disc.

6 Should it be necessary to separate the disc (11) from the hub (7), clamp the disc between soft faces in a vice and with a scriber or file mark the disc and hub so that the two parts may be refitted in their original positions if new parts are not to be fitted.

7 Undo and remove the five bolts and self-locking nuts so as to release the wheel hub from the disc.

8 If the complete inner race (9) stayed on the stub axle instead of being withdrawn from the hub, remove the roller bearings and cage and with a universal three legged puller and thrust pad, draw the race inner track from the stub axle. Remove the oil seal (10) noting that the lip is facing inwards towards the centre of the hub.

9 Remove the bearing spacer (13) from the rear of the hub and with a soft metal drift carefully tap out the inner race, cage and roller bearings from the outer bearing (5).

10 The bearing outer races (6, 8) may next be drifted out of the hub using a soft metal drift. For this the hub should be placed on top of a bench vice and with the jaws open and supporting the disc brake mounting flange or wheel stub flange depending on which outer track is being removed.

11 Thoroughly clean the bearings (5, 9) and the hub (7) in paraffin and then examine them for grooving, flat spots, chips, pitting or other damage. Renew the bearings if worn.

12 To reassemble first fit the bearing spacer (13) with its larger diameter towards the inside of the hub.

13 Assemble the two bearings (5, 9) making sure that the marked edges of the inner and outer tracks and the ball cage are all the same side when the bearing is assembled.

14 Pack the bearings with Castrol LM Grease making sure that it is worked well into the bearing cage and insert them into the hub thrust adjacent to the bearing spacer.

15 Remove any excess grease so as to leave room for expansion otherwise the bearings could overheat caused by excessive churning of the grease.

16 Refit the new oil seal (10), lip facing inwards to the centre of the hub, into its recess using a drift of suitable diameter so as to locate on the rear of the oil seal. Carefully tap it into position making sure that it does not tilt.

17 If the disc (11) was removed it should next be refitted. Wipe the mating faces of the disc and hub (7), refit the disc to the hub and tighten the five bolts and nuts in a diagonal manner to a torque wrench setting of 28 to 30 lb ft.

18 Refit the assembled disc and hub onto the stub axle and using a tubular drift located on the outer bearing inner track, drive the hub into position.

19 Fit the stub axle nut (2) and washer (4). Tighten the nut, and at the same time rotate the hub back and forth until there is noticeable drag. This ensures that the bearing cones are properly seated.

20 Unscrew and remove the stub axle nut. Extract the washer and the centre of the outer bearing. Insert a sufficient thickness of shims (12) **to produce an excessive amount of endfloat.** Note the total thickness of shims used. Replace the bearing centre, the washer, and tighten the stub axle nut.

21 Measure accurately the total amount of endfloat in the bearings. Remove the stub axle nut, washer, and outer bearing centre. Reduce the number of shims to eliminate endfloat, while still allowing the hub to rotate freely, when the stub axle nut has been refitted and tightened to a torque wrench setting of 40 to 70 lb ft. A wide latitude is given so that the nut may be aligned with the split pin hole in the stub axle.

22 Insert a new split pin (3) through the hole provided in the hub and lock the stub axle nut.

23 Remove any surplus grease to allow room for expansion and, using a drift, tap the grease retaining cap gently but firmly up against the outer bearing. Do not put grease in the cap.

24 Refit the caliper to the swivel axle and secure with the two bolts and locking tab washer. The caliper securing bolts should be tightened to a torque wrench setting of 45 to 50 lb ft. Any shims noted when removing the caliper must be refitted in their original positions. Lock the two bolts by bending over the locking tab.

25 Operate the brake pedal several times to adjust the position of the pistons within the caliper.

26 Replace the road wheel, remove the supports and lower the car to the ground.

8 Front suspension coil spring - removal and refitting

1 Removal of one of the coil springs is not difficult but it requires the use of a special tool part number 18G37. If the special tool is not available, two 3/8 inch BSF high tensile steel bolts, 4 inches long and threaded along their entire length may be used instead.

2 Chock the rear wheels. Position a jack under the chassis front crossmember and raise the car until the front wheels are clear of the ground.

3 Remove the road wheel.

4 If the BLMC spring compressor is available fit it in place on the lower wishbone arms and adjust it to take the tension of the spring (Fig 11.5).

5 Undo and remove the four nuts, bolts and lockwashers (13, 15) (Fig 11.6) and carefully unscrew the compressor so as to free the spring tension.

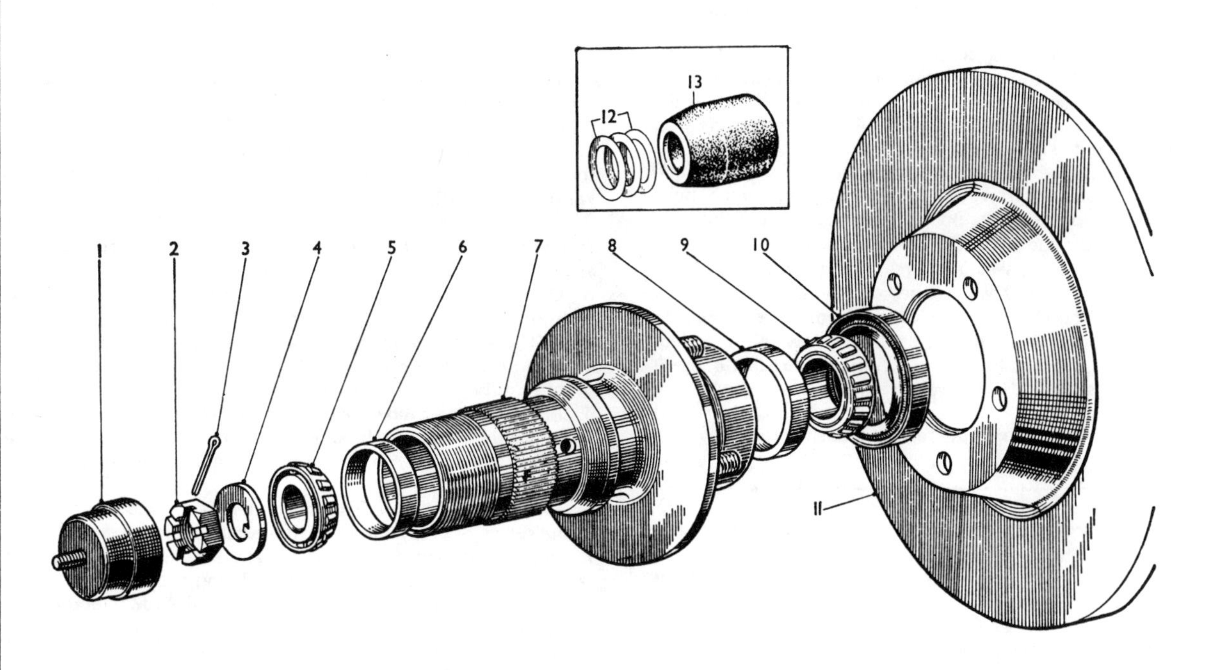

Fig 11.4 EXPLODED VIEW OF FRONT HUB (DISC BRAKE)

1 Grease cup
2 Axle nut
3 Split pin
4 Washer
5 Outer bearing
6 Bearing outer race
7 Hub
8 Bearing outer race
9 Inner bearing
10 Oil seal
11 Brake disc

Inset shows distance piece and shims

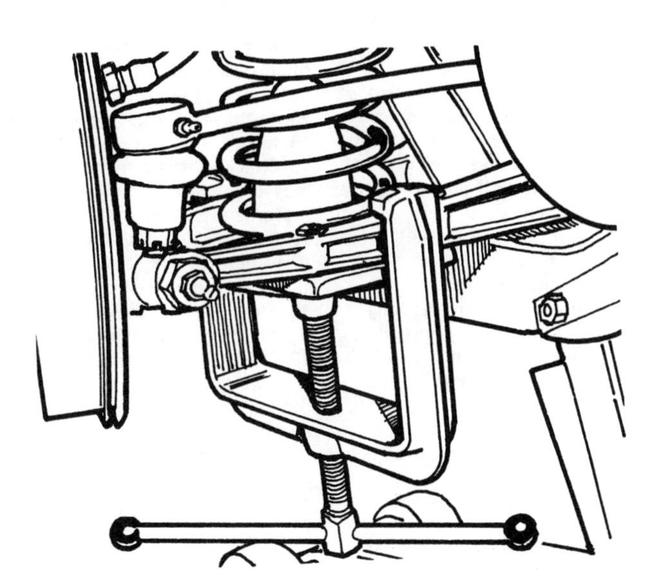

Fig 11.5 USE OF SPECIAL SPRING COMPRESSOR TO RELEASE TENSION OF THE SPRING

6 The spring (11) and the pan (16) may now be lifted away from the underside of the front suspension.
7 Should the BLMC spring compressor not be available free the two diagonally opposite nuts, bolts and lockwashers (13, 15) from the four which hold the lower wishbone arms (14) to the spring retaining pan. Substitute the two 4 inch high tensile steel bolts and tighten the nuts down fully.
8 Undo and remove the two remaining small nuts, bolts and lockwashers. Next unscrew the nuts from the 4 inch bolts equally until the spring is fully extended. Lift away the spring and pan from the underside of the front suspension.
9 Inspect the spring seat rubbers and if they have been contaminated with oil or have perished obtain new rubbers.
10 Clean the spring (11) and check for signs of excessive rusting or fracturing of the wire which, if evident, a new pair of springs should be fitted. Do not fit one new spring as the other one will probably have settled slightly which could cause inconsistent handling.
11 Refitting the coil spring is the reverse sequence to removal.

9 Front suspension unit - removal and refitting

1 Refer to Section 8 and remove the coil spring.
2 Extract the steering side rod to steering arm castellated nut split pin and unscrew the castellated nut. Using a ball joint remover to separate the ball joint from the steering arm.
3 Wipe the top of the brake master cylinder and unscrew the cap. Place some polythene sheet over the top of the master cylinder and refit the cap. This will stop hydraulic fluid syhponing out when the flexible brake pipe is disconnected.
4 Detach the brake flexible hose from the brake backplate or calliper as described in Chapter 9. Do not forget to clean the surrounding area first.
5 Using a garage hydraulic jack or other suitable means, support the weight of the suspension unit.
6 Extract the split pin locking the suspension arms inner fulcrum pins castellated nuts and unscrew the castellated nuts (21). Remove the fulcrum pins (18) special shaped washers (22) and rubber bearings (19, 20). These parts are shown in Fig 11.6.
7 Unscrew and remove the four set pins (3) and lockwashers that secure the shock absorber (1) to the top spring bracket.
8 The suspension unit may now be lifted away from the car.
9 To refit the suspension unit fit one rubber bearing (19) to each of the suspension lower links (14) on the side which corresponds to the small hole in each of the frame brackets.
10 Offer up the links to the frame brackets and insert the fulcrum pins (18) and slide the second bearing (20) and special washer (22) over the protruding end of each pin.
11 Refit the castellated nuts (21) but do not tighten fully yet.
12 Position the shock absorber (1) on its top bracket and partially tighten the four setscrews (3) and lockwashers.
13 Obtain a piece of metal or hardwood exactly 2 inches long and place between the shock absorber wishbone arm and the upper spring plate at a point opposite the rubber buffer. This is to enable the front suspension unit to be set in its normal loaded position.
14 Tighten the castellated nuts (21) (Fig 11.6) on the fulcrum pins (18) securing the lower wishbone arms (14) to the frame brackets. Lock the castellated nuts with new split pins.
15 Next tighten the four setscrews (3) that secure the shock absorber (1) to its bracket on the frame.
16 Tighten the upper trunnion fulcrum pin castellated nut (9) and secure with a new split pin.
17 Refit the coil spring which is the reverse sequence to removal as described in Section 8.
18 Reconnect the brake flexible hydraulic pipe and bleed the system as described in Chapter 9.
19 Reconnect the steering side tube to the steering arm.
20 Refit the road wheel and lower the car to the ground. Remove the distance piece that was used to retain the suspension in its normal loaded position.
21 It is recommended that whenever the front suspension has been removed or worked on in situ the front wheel alignment be checked. Further information on this subject will be found in Section 25.

10 Front suspension unit - dismantling and reassembly

1 With the front suspension unit away from the car, unscrew the nut from the clamping bolt (5) that connects the top wishbone arms (2, 4) together (Fig 11.6).
2 Extract the split pin and unscrew the nut (9) from the upper trunnion fulcrum pin (8) on the outer end of the top wishbone arms.
3 On the left hand suspension unit the forward arm of the top wishbone is secured to the shock absorber spindle by a clamping bolt. Slacken this clamping bolt and partially withdraw the arm.
4 The trunnion fulcrum pin (8) can now be withdrawn and the shock absorber (1) removed complete with the top wishbone arm.
5 Lift away the rubber bearing halves (7) from each end of the upper trunnion (6). It will be observed that the bearings fit into a groove in the swivel pin and must be taken out before the swivel pin is removed.
6 Extract the split pin from the castellated nut (43) on the top of the swivel pin (30). Unscrew the castellated nut.
7 Remove the upper trunnion (6) and the three thrust washers (44, 45) and lift off the swivel axle (42) and hub assembly.
8 Lift away the cork washer (38) from the lower end of the swivel pin (30).
9 Slacken the nut (23) on each of the half moon cotters (24) that are located in the ends of the lower wishbone arms (14). Screw out the two threaded bushes (24, 25) and detach the two arms (14).
10 Unscrew the nut (31) for the cotter (28) to be found in the centre of the lower trunnion and carefully drift out the cotter.
11 Withdraw the fulcrum pin (34) and remove the cork washer (32) from each end of the trunnion.
12 The front suspension unit is now completely dismantled and should be cleaned and checked for wear.
13 Examine the swivel pin (30) for wear by checking for ovality. Should the pin not show any signs of ovality renewal of the swivel bushes may effect a satisfactory cure for noisy movement.
14 The bushes may be easily drifted out and new ones fitted. When refitting the top bush the greasing hole must locate with the grease hole in the swivel housing. The second bush must be flush with the recessed housing and protrude about 1/8 inch above the lower housing upper face.
15 Inspect the dust covers and obtain new if they are damaged.
16 Check if it is possible for the screwed bushes (29, 35) to be moved backwards or forwards on the fulcrum pin thread. If movement is evident the fulcrum pin must be renewed.
17 Next inspect the shock absorber for up and down or sideways movement of the cross shaft and if this is evident a replacement shock absorber must be obtained.
18 Examine the shock absorbers for signs of leaks and again, if evident, replacement shock absorbers should be obtained.
19 Secure the shock absorber mounting plate in a vice and move the wishbone arms up and down through a complete stroke. A moderate amount of resistance should be felt through the whole stroke.
20 Should resistance be erratic it is an indication that the fluid level is too low for efficient operation causing air locks within the unit.
21 To rectify this condition, remove the shock absorber filler plug and top up the fluid with Armstrong Super (Thin) Shock Absorber Oil to the correct level whilst the arms are moved steadily up and down through complete strokes. However, if this operation does not rectify the erratic operation a replacement shock absorber must be obtained.
22 To reassemble the front suspension unit, first fit the screwed fulcrum pin (34) into the lower trunnion at the bottom end of the swivel pin (30) (Fig 11.6). Make sure that it is centralised with the cotter pin milled location lined up with the cotter pin hole.
23 Insert the cotter pin (28) and secure in position with the spring washer and nut (31).

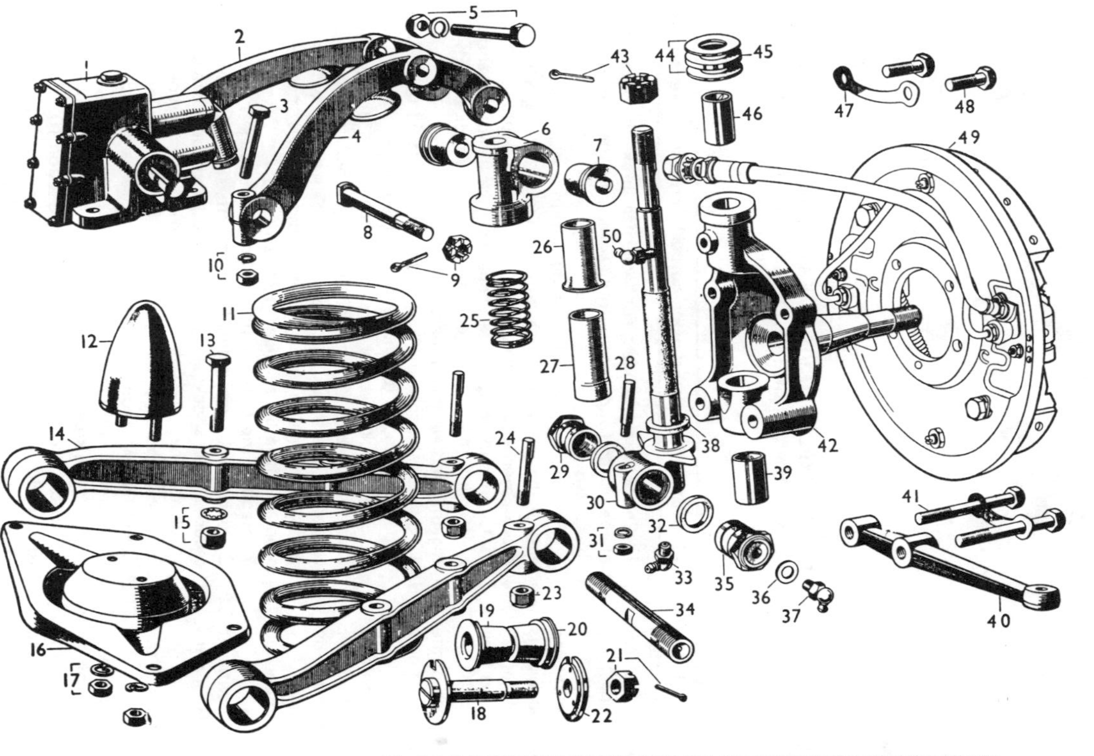

Fig 11.6 COMPONENT PARTS OF FRONT SUSPENSION UNIT

1 Shock absorber
2 Rear top wishbone arm
3 Clamping bolt for front wishbone arm
4 Front top wishbone arm
5 Joining bolt for top wishbone arms
6 Upper trunnion link
7 Trunnion rubber bearing
8 Upper trunnion fulcrum pin
9 Fulcrum locking nut and split pin
10 Nut and washer for clamping bolt
11 Coil spring
12 Rebound rubber bumper
13 Spring blade bolt
14 Rear lower wishbone arm
15 Simmonds nut and lockwasher
16 Spring pan
17 Rebound bumper nut and washer
18 Fulcrum pin for inner lower bearing
19 An inner lower rubber bearing
20 An outer lower rubber bearing
21 Fulcrum pin nut and split pin
22 Fulcrum pin special washer
23 Nut for bush cotter
24 Bush cotter
25 Swivel pin dust cover spring
26 Upper dust cover
27 Lower dust cover
28 Cotter for fulcrum pin
29 Rear screwed bush
30 Swivel pin and lower trunnion
31 Nut and washer
32 Cork ring
33 Trunnion oil nipple
34 Screwed fulcrum pin
35 Front screw bush
36 Flat washer
37 Oil nipple
38 Cork ring
39 Swivel axle lower bush
40 Steering arm
41 Steering arm setpin
42 Swivel axle
43 Swivel pin nut and split pin
44 Staybrite washers
45 Oilite washer
46 Swivel axle upper bush
47 Backplate setpin lockwasher
48 Backplate setpin
49 Backplate assembly
50 Swivel pin oil nipple

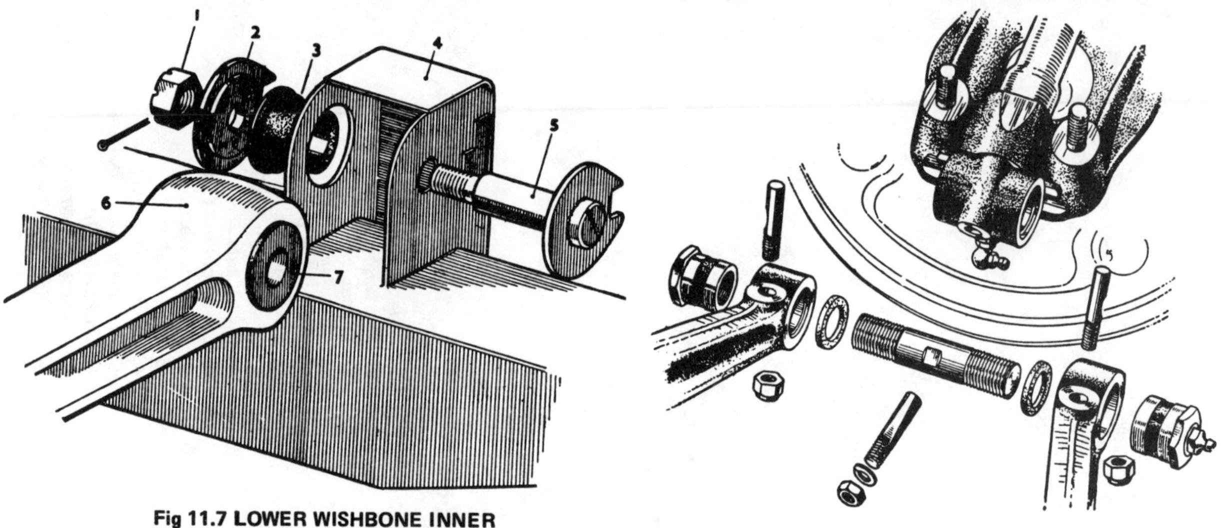

Fig 11.7 LOWER WISHBONE INNER BEARING ASSEMBLY

1 Castellated nut
2 Special washer
3 Bush
4 Mounting bracket
5 Fulcrum pin
6 Wishbone arm
7 Bush

Fig 11.8 EXPLODED VIEW OF THE LOWER END WISHBONE ARMS AND SCREWED BUSH HOUSING ASSEMBLY

24 Fit the cork ring (32) into the recess provided at each end of the lower trunnion and place the lower wishbone arms (14) in their respective positions. Make sure that the half moon cotters (24) are correctly positioned to receive the steel bushes (29, 35) which should be smeared with Castrol LM Grease and partially screwed home.
25 Position the lower spring pan (16) under the lower wishbone arms (14) and temporarily secure in position with the nuts, bolts and lockwashers (13, 15).
26 Screw the threaded bushes (29, 35) home evenly and then slacken them back one flat only.
27 Secure the bushes by tightening the nuts (21) on each of the half moon cotters. Do not however overtighten the cotter pin nuts as the bushes could become distorted.
28 If the reassembly has been carried out correctly it should be possible to insert a 0.002 inch feeler gauge between the inner shoulder of the bush and the outer face of the wishbone arm on each side.
29 It should also be seen that the lower trunnion assembly can now operate freely in the screwed bushes.
30 Position the cork washer (38) on the swivel pin with its chamfered face downwards and smear a little Castrol GTX onto the swivel pin.
31 Position the swivel axle and hub (42) assembly onto the swivel pin (30).
32 The thrust washers should now be refitted (44, 45). It will be observed that three thrust washers are made up of an Oilite washer interposed between two Staybrite washers. The Staybrite washers are supplied in varying thicknesses so as to permit adjustment as it is necessary to provide easy operation of the swivel axle with the minimum amount of lift. The maximum permissible amount of lift is 0.002 inch.
33 Refit the upper trunnion and swivel nut (43) and check the clearance. If it is in excess of 0.002 inch obtain new and thicker Staybrite washers and then slacken the swivel pin nut to permit further reassembly.
34 Lubricate the upper trunnion rubber bearings (7) with a little water and place them in position.
35 Place the trunnion complete with its bearings in position between the two upper wishbone arms.
36 Refit the fulcrum pin (8) and then reposition and tighten the previously slackened upper wishbone arm to the shock absorber arms. Do not tighten the swivel pin and upper trunnion fulcrum pin nuts yet as this will be done when the front suspension unit is being refitted.

11 Front suspension shock absorber - removal and refitting

1 Chock the rear wheels, apply the handbrake, jack up the front of the car and support on firmly based axle stands located under the chassis.
2 Remove the road wheel and position a jack under the outer end of the lower wishbone arm and then raise the jack until the shock absorber arms are clear of their rebound rubber.
3 Undo and remove the clamp bolt, securing nut and spring washer (5) (Fig 11.6) connecting the two shock absorber arms together and withdraw the clamp bolt.
4 Extract the split pin and undo the castellated nut (9) on the upper fulcrum pin (8) and withdraw the fulcrum pin.
5 Each shock absorber arm (34) is secured to the spindle by a clamp pinch bolt (3) and this bolt should next be removed. This will allow the shock absorber arm to be partially withdrawn so that the trunnion link and its rubber bushes may be easily separated fromthe shock absorber arms.
7 Recover the trunnion link half rubber bushes (7).
8 Undo and remove the four set bolts (3) and spring washers securing the shock absorber to the frame and lift away the shock absorber.
9 It is important that the jack is left in position under the suspension wishbone whilst the top link remains disconnected so that the coil spring is kept securely in position thereby avoiding straining the steering connections.
10 To test the shock absorber refer to Section 10, paragraphs 17 to 21 inclusive.
11 Refitting the shock absorber is the reverse sequence to removal but the following additional points should be noted.
12 Once the shock absorber has been refitted to the chassis frame and before the upper trunnion pin is reconnected, work the arms of the shock absorber at least four times through their full travel so as to expel any air which may have found its way into the shock absorber hydraulic system.
13 Should the rubber of the fulcrum pin bushes have perished or be contaminated with oil or if side movement is evident obtain and fit new rubber bushes.
14 Refit the trunnion with its bushes between the shock absorber arms and refit the fulcrum pin before pushing the loosened arm home on the shock absorber spindle and replacing the clamp bolt.
15 The fulcrum pin nut and clamp bolt that connects the two shock absorber arms may only be tightened when there is a load

on the suspension. This may be set by placing a 2 inch long metal or hardwood block between the shock absorber arm and the chassis frame.

12 Rear semi-elliptic spring - removal and refitting

1 Chock the front wheels, jack up the rear of the car and place on axle stands located under the chassis frame as near to the spring rear anchorage as possible. Remove the road wheel on the side which the spring is to be removed.
2 Place the jack under the centre of the spring so to relieve its tension.
3 Using a box spanner release the four self-locking nuts from the U bolts which secure the spring to the axle banjo.
4 Undo and remove the nut and spring washer on the inside of the upper rear shackle and the locknut, spring washer and nut on the inside of the lower rear shackle (Fig 11.9).

Fig 11.9 EXPLODED VIEW OF SPRING REAR SHACKLE

5 Remove the shackle inside connecting link and extract the top and bottom shackle pins together with the outside link.
6 Remove the nut and spring washer from the inside of the pin at the front end of the spring and with a soft metal drift drive the pin out.
7 Remove the jack from under the spring and lift the spring from the underside of the car.
8 Refitting the rear spring is the reverse sequence to removal.

13 Rear suspension shock absorber - removal and refitting

1 Undo and remove the nut and spring washer that secures the shock absorber lever to the link arm.
2 Undo and remove the two set pins that secure the shock absorber to the chassis bracket.
3 Lift away the shock absorber threading the lever over the link arm bolt.
4 Check the operation of the shock absorber as described in Section 10, paragraphs 17 to 21 inclusive.
5 It should be noted that the connecting link rubber bushes are integral with both ends of the connecting link which joins the shock absorber to the rear casing and they cannot be renewed. If these bushes are worn the complete arm must be renewed.
6 Refitting is the reverse sequence to removal but the following additional points should be noted.
7 Always keep the shock absorber assembly upright otherwise air may enter the hydraulic system so causing erratic resistance.
8 The shock absorber link must be above the arm when refitting the unit to the chassis bracket and axle.

14 Panhard rod assembly - removal and refitting

1 It will be seen that there is a tie rod which is mounted in rubber bushes between a bracket welded to the axle casing and a bracket welded to the chassis frame and its function is to prevent lateral motion between the axle and the chassis frame.
2 Should it be necessary to remove the tie rod to fit new rubber bushes, undo and remove the self-locking nuts, washers and outer bushes from the ends of the tie rod.
3 Refer to Chapter 8, Section 3, paragraphs 1 to 9 inclusive, so as to free the rear axle and lift away the tie rod from under the car.
4 Remove the inner rubber bushes.
5 Fit new inner rubber bushes and refit the tie rod assembly, this being the reverse sequence to removal.

15 Steering wheel - removal and refitting

1 The horn push and direction indicator switch must first be removed. It will be seen that this is mounted on the steering wheel hub and comprises a spring metal push covering the hub with the indicator switch lever positioned in its centre. The switch cables pass through a long tube down the steering column shaft which is secured by an olive in the base of the steering box.
2 If an adjustable steering column is fitted the stator tube is in two parts, the shorter piece being attached to the horn quadrant.
3 For safety reasons switch off the battery master switch.
4 Detach the horn and flasher light cables that protrude from the end of the stator tube at the nearest snap connectors. Note the cable colours for correct refitting.
5 Where an adjustable steering column is fitted undo and remove the three grub screws in the steering wheel hub and withdraw the quadrant together with the short stator tube and cables. The longer part of the stator tube will remain in the steering column.
6 It is important to note that the short stator tube has an indentation in it which fits in a slot in the long stator tube. The horn quadrant must be withdrawn without any twisting motion to avoid enlarging the slot in the long stator tube. Any enlargement of this slot will result in excessive movement of the horn quadrant after refitting.
7 When a non-adjustable steering column is fitted, undo and remove the nut and olive at the bottom end of the steering box. This will free the stator tube which may be withdrawn as one piece with the horn push quadrant attached to the end. At the same time the cables will be drawn up from the outer column.
8 Using a piece of taper wood plug the hole left in the bottom of the steering box so as to prevent the oil draining out.
9 Undo and remove the nut from the centre of the steering wheel hub.
10 With a scriber or pencil mark the relative position of the hub and inner column so that they may be refitted in their original position.
11 With the palms of the hands thump the underside of the spokes to separate the steering wheel from the inner column.
12 If an adjustable steering column is fitted, prise off the circlip from the end of the inner column and then release the locking ring behind the steering wheel hub.
13 Pull the steering wheel clear of the column and lift away the telescopic spring and locating collar.
14 Refitting the steering wheel is the reverse sequence to removal. It is recommended that in order to facilitate the threading of the horn and flasher cables through the stator tube, the ends of the cables are taped together.
15 If the colour coding of the cables is not too clear reference to the wiring diagrams given in Chapter 10 will facilitate this.
16 Tighten the steering wheel securing nut to a torque wrench setting of 41 lb ft.
17 Top up the steering gearbox with Castrol Hypoy.

16 Steering gear assembly - removal and refitting

1 Refer to Section 15 and remove the steering wheel.
2 Working behind the fascia release the two-piece clamping bracket supporting the top end of the steering column.
3 Refer to Chapter 2, Section 6 and remove the radiator.
4 Refer to Chapter 12, Section 20 and remove the front grille.
5 Located on each side of the scuttle are two sealing plates through which the steering column passes. Release each plate by undoing the four metal threaded screws.
6 Chock the rear wheels, apply the handbrake, jack up the front of the car and support on firmly based stands located under the main chassis frame. Remove the front road wheels.
7 Refer to Sections 21 and 22 of this Chapter and disconnect the steering cross tube and side rod from the steering lever.
8 Unscrew and remove the three nuts, bolts and spring washers that secure the steering box mounting bracket to the chassis.
9 Carefully manoeuvre the steering column together with the steering box downwards and forwards from the front grille aperture.
10 Refitting the steering gear is the reverse sequence to removal but the following additional points should be noted.
11 Carefully align the steering column so that there is no bending stress imposed upon it before tightening the support brackets.
12 When the side rod is being refitted make sure that the steering wheel is in the centre of its travel and the front wheels are in the straight ahead position.
13 Tighten the steering wheel securing nut to a torque wrench setting of 41 lb ft.
14 Do not forget to refill the steering gearbox with Castrol Hypoy.

17 Steering gearbox - dismantling and reassembly

1 Refer to Fig 11.10 and extract the split pin from the castellated nut (18).
2 Unscrew the castellated nut (18) from the base of the steering lever (16).
3 Using a two legged puller draw the steering lever (16) from the splines of the rocker shaft.
4 Remove the drain/filler plug (5) and allow the oil to drain out into a tin.
5 Undo and remove the four setscrews and spring washers (2) that secure the top cover (1) to the steering box (11). Lift away the top cover (1) and joint washer (7).
6 Invert the steering gearbox and suitably support the top face. Carefully tap out the rocker shaft (9) using a soft metal drift.
7 Undo and remove the nut (25) and olive (24) from the end cover (23).
8 Withdraw the long stator tube from the inside of the inner column (12).
9 Undo and remove the four set pins and spring washers (26) that secure the end cover (23) to the steering box. Lift away the end cover, adjusting shims (22) and joint washer (21).
10 Stand the steering box on the column with the steering box uppermost and by thumping the end of the inner shaft on a piece of softwood displace the worm and its two ball bearings.
11 Withdraw the complete inner column (12) from the casing via the open end of the steering box.
12 Remove the ball race from the top of the outer casing of the column by pulling upwards by hand. If it is tight ease it from the column with a screwdriver placed behind the protruding lip.
13 The gearbox and column are now dismantled for cleaning and inspection. Wash all parts in paraffin and allow to dry on a non-fluffy rag.
14 Examine the steering column shaft cam for signs of excessive wear in the grooves.
15 Check the rocker shaft, rocker shaft bush and splines for wear.
16 Carefully examine the steering lever for signs of cracks or other accidental damage.
17 Inspect the follower peg located in the rocker shaft, for wear. If this is evident as the peg is peened over at the top to ensure complete security carefully cut away the peening with a sharp cold chisel and with a soft metal drift drive out the follower peg. Drift in a new peg and peen over the top with a centre punch.
18 Reassembly is the reverse sequence to dismantling. However, before refitting the top cover plate (1) screw back the adjuster bolt (3).
19 Refer to Section 19 and adjust the steering gearbox as described in paragraphs 8 to 14 inclusive.

18 Steering gearbox - adjustment

1 To adjust the cam bearings with the steering gearbox in position, chock the rear wheels, apply the handbrake, jack up the front of the car and support on firmly based axle stands located under the chassis frame.
2 Extract the split pin, undo and remove the side rod ball joint to steering lever castellated nut. Using a universal ball joint separator disconnect the ball joint from the steering lever.
3 Switch off the battery master switch and disconnect the direction indicator switch and horn push cables at the snap connector behind the radiator grille. Note the cable colour coding to ensure correct reconnection.
4 Refer to Section 15, paragraphs 1 to 7 inclusive, but it is not necessary to completely withdraw the indicator and horn stator tube. Just draw it up the inner column so that the cables are protected from the steering gearbox oil.
5 Place a tin under the steering gearbox to catch any oil that may drain out.
6 Undo and remove the nut followed by the olive from the end cover.
7 Unscrew and remove the four end cover retaining bolts and spring washers and lift away the end cover and joint washer.
8 To obtain the correct adjustment add or subtract a number of shims from the end cover.
9 Adjustment is considered correct when the steering wheel rim is lightly held between the fingers and turned. No resistance to movement should be felt and there should be no perceptible end play.
10 Once the cam bearings have been adjusted the rocker shaft adjustment may now be carried out.
11 With the side rod still disconnected from the steering lever, slacken the adjusting screw locknut and screw in the adjusting screw.
12 Check for backlash by exerting a light pressure on the lower end of the steering lever alternatively in both directions whilst at the same time an assistant turns the steering wheel slowly from one lock to the other and back again.
13 It should be observed that the amount of slackness is not constant but there being less slackness in the centre than in the full lock position.
14 If slackness appears at all positions of the drop arm the adjusting screw should be screwed in further. Recheck this adjustment. The adjustment may be considered correct when there is a tight spot which is just evident as the steering wheel is moved past the centre position with no backlash at the steering drop arm. When this condition is met tighten the adjusting screw locknut.
15 Refit the stator tube and reconnect the electrical cable connections.
16 Reconnect the steering side rod and lock the castellated nut with a new split pin.
17 Refill the steering gearbox with Castrol Hypoy.
18 Refit the front grille and lower the car to the ground.

19 Steering idler - removal and refitting

1 Refer to Sections 21 and 22 and disconnect the steering side and cross tubes from their connections at the idler lever.
2 Undo and remove the three bolts and spring washers securing

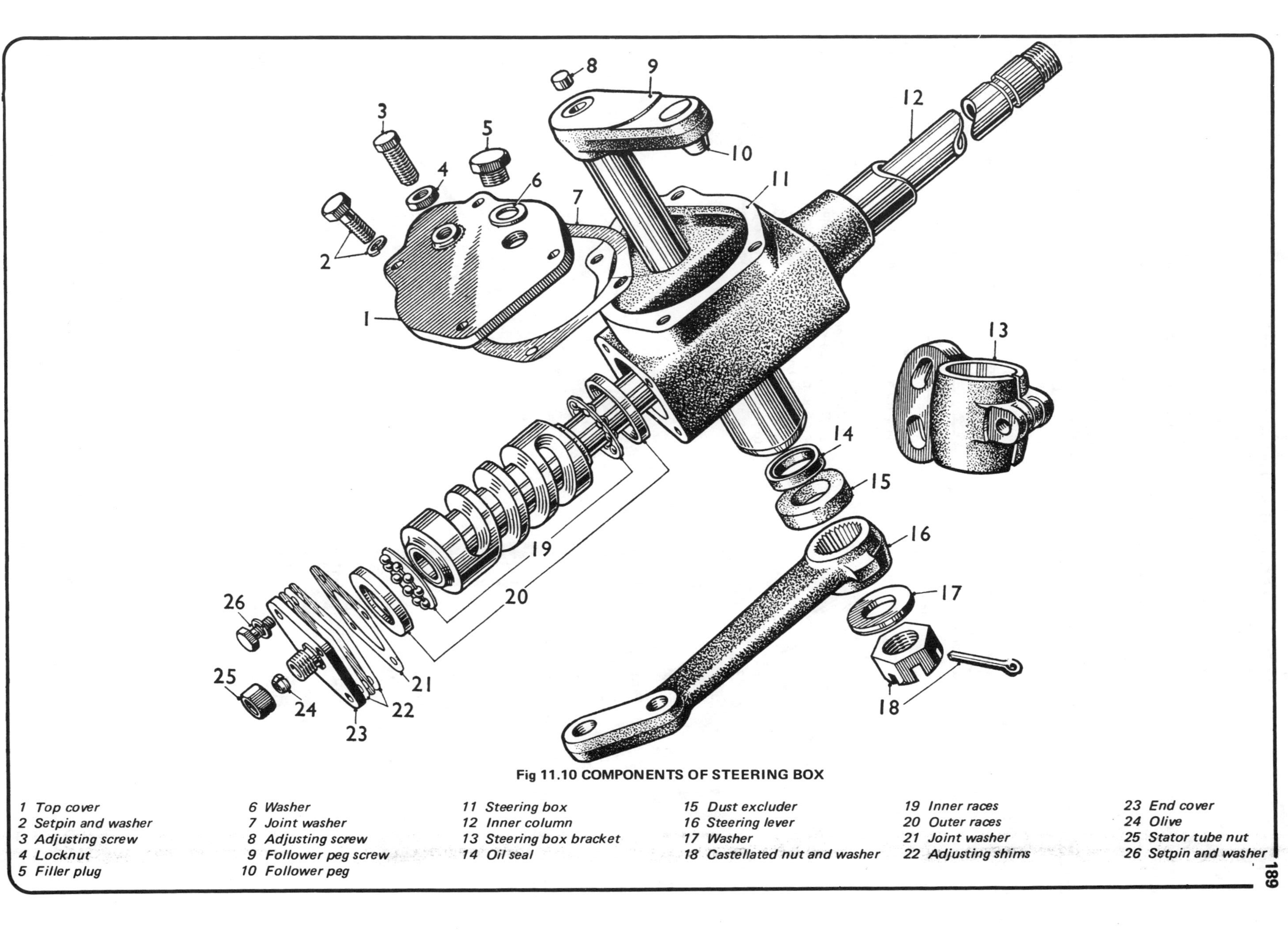

Fig 11.10 COMPONENTS OF STEERING BOX

1 Top cover
2 Setpin and washer
3 Adjusting screw
4 Locknut
5 Filler plug
6 Washer
7 Joint washer
8 Adjusting screw
9 Follower peg screw
10 Follower peg
11 Steering box
12 Inner column
13 Steering box bracket
14 Oil seal
15 Dust excluder
16 Steering lever
17 Washer
18 Castellated nut and washer
19 Inner races
20 Outer races
21 Joint washer
22 Adjusting shims
23 End cover
24 Olive
25 Stator tube nut
26 Setpin and washer

the idler to the mounting bracket.
3 Lift the idler and its lever away from the body.
4 Refitting the idler is the reverse sequence to removal. Make sure that the mounting bolts are tightened fully.
5 Whenever the idler has been removed it is recommended that the front wheel alignment is checked and further information on this subject will be found in Section 25 of this Chapter.

20 Steering idler - dismantling and reassembly

1 With the idler out of the car, undo and remove the three top cover bolts and spring washers (1) (Fig 11.11).
2 Lift off the cover (2) and its gasket (4).
3 Extract the split pin and remove the castellated nut (9) followed by the plain washer (10), idler lever (11) and dust excluder (8) from the base of the idler body.
4 Pull the idler shaft upwards through the idler body taking care not to damage the oil seal as the splines are being drawn through.
5 If the oil seal has been leaking carefully drift out the old oil seal (7).
6 Inspect the two bushes (12) for wear and, if evident, drift out the old bushes and fit new.
7 Reassembly is the reverse sequence to removal but the following additional points should be noted.
8 Fit a new oil seal with the lip facing inwards.
9 Take great care not to damage the oil seal when the idler shaft is being refitted.
10 When completely reassembled check that the idler shaft is able to turn freely without end float. If stiff or if end float is present add or subtract joint washers until the shaft fits correctly.
11 When the idler is back in the car remove the plug (3) in the top of the idler and fill up with Castrol Hypoy.

21 Steering side rods - removal and refitting

1 Extract the split pins and undo and remove the castellated nuts from the ball pins at the steering lever end and the swivel arm end of the side rod.
2 Using a universal ball joint separator detach the ball pin from the steering lever and swivel arm.
3 Should it be necessary to partially dismantle a side rod first remove the dust cover securing clip (6) (Fig 11.12) and prise off the rubber boot (7).
4 Check the ball pins for wear. They must be sufficiently tight to prevent end play yet loose enough to allow free movement. If wear is evident a complete new assembly must be obtained.

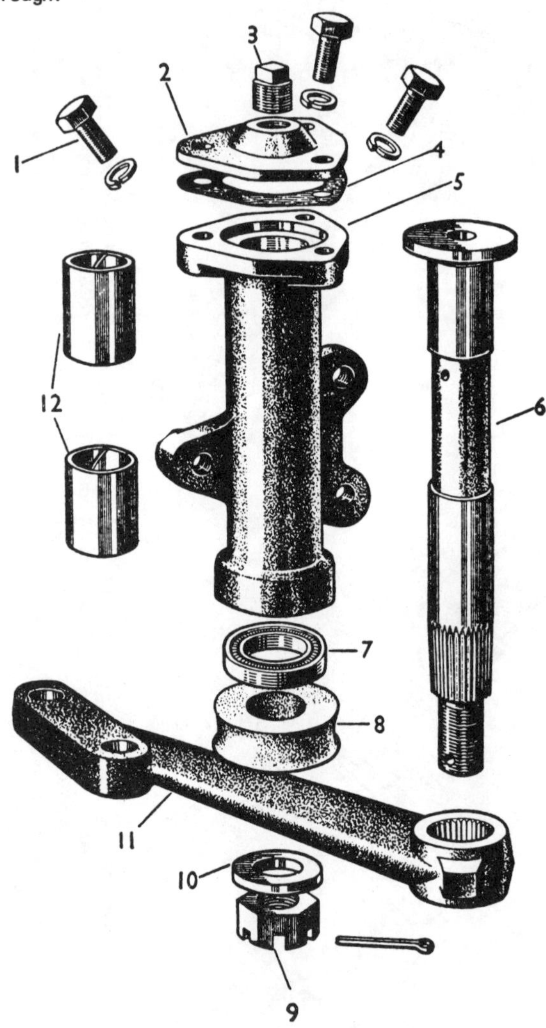

Fig 11.11 EXPLODED VIEW OF THE STEERING IDLER

1 Cap setpin
2 Idler cap
3 Oil plug
4 Joint washer
5 Idler body
6 Idler shaft
7 Oil seal
8 Dust excluder
9 Castellated nut
10 Plain washer
11 Steering lever
12 Bush bearings

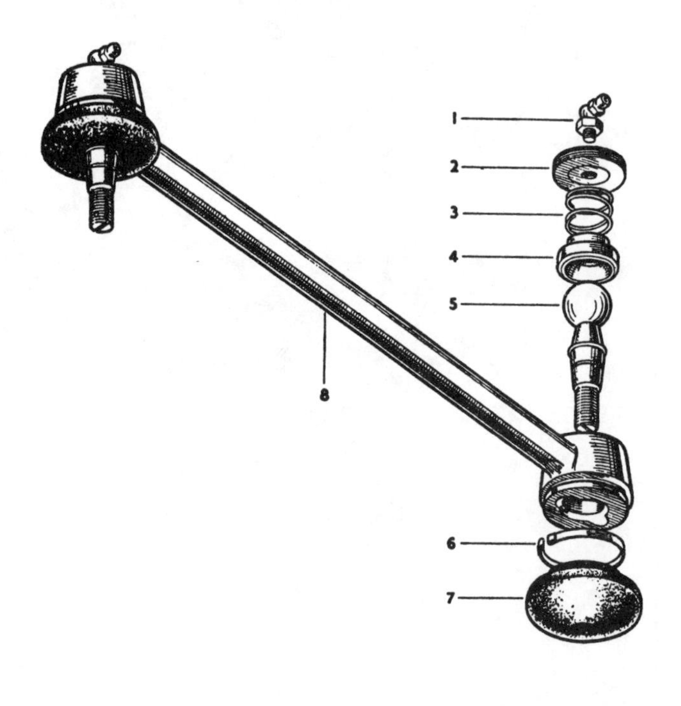

Fig 11.12 STEERING SIDE ROD, BALL AND SOCKET CONNECTION

1 Oil nipple
2 Screwed fixing plate
3 Spring
4 Socket
5 Ball
6 Spring clip
7 Rubber boot
8 Side rod

5 If the dust covers are damaged new covers must be obtained as otherwise wear will occur due to dirt ingress.
6 Carefully examine the side rod for damage as well as straightness. Fit new if this is evident.
7 Reassembly and refitting is the reverse sequence to removal.
8 Nylon seated ball joints which are sealed during manufacture were fitted on 3000 Mk II cars from BT7 19191 and it is very important that no dirt or abrasive matter enters the nylon ball joint. If the rubber boot has been accidentally damaged in service it is possible that the ball joint will have been subjected to road dust therefore the ball joint and boot must be renewed.
9 Should the rubber boot be damaged accidentally whilst the steering side rods or cross rod is being removed it is permissible for a new rubber boot to be fitted. Smear the area adjacent to the joint with a little grease before assembly the boot.
10 Whenever a steering rod has been removed it is recommended that the front wheel alignment be checked and further information on this subject will be found in Section 25 of this Chapter.

22 Steering cross tube - removal and refitting

1 The procedure for removing the steering cross tube from the steering lever and the idler lever is similar to that of the side rods. Full information will be found in Section 21, paragraphs 1 and 2.
2 To dismantle the cross tube, first slacken the socket locknuts at each end of the cross tube and unscrew the socket assemblies which have left hand and right hand threads respectively.
3 The procedure is now identical to that described in Section 21, paragraph 3 onwards.

23 Steering column lock and ignition starter switch - removal and refitting

1 On some models especially those destined for some export markets, a combined ignition and starter switch and steering column lock is mounted on the steering column.
2 A sleeve integral with the inner column is slitted so as to prevent engagement of the lock tongue. The outer column is also slotted to allow the tongue of the lock to pass through.
3 There is a hole drilled in the upper surface of the outer column which locates the steering lock bracket and this bracket is secured by two bolts each waisted below the head so as to permit removal of the bolt heads by shear action during assembly.
4 To remove the lock, first switch off the battery master switch and disconnect the ignition/starter switch connection. Note the cable colour coding so that the cables may be refitted in their original positions.
5 Turn the key to the Garage position to unlock the steering and then refer to Section 17 and remove the steering box and column from the car.
6 Remove the lock securing bolts using an easy-nut stud extractor.
7 Refitting is the reverse sequence to removal. Tighten the steering lock securing bolts fully and then one additional turn which should shear off the heads.

24 Steering geometry - general

Correct steering geometry is very important to the road holding and general control of the car and for any checks accurate wheel alignment equipment is required. This should be left to the local BLMC garage who will have the equipment as well as the specialist knowledge to interpret the readings and rectify any inaccuracies. Any car that has been in a front end accident or 'kerbed' should have these angles checked at the earliest opportunity. The angles are shown in Fig 11.13. For the owner's information a brief description is given below of the three major angles in question.

Camber angle

This is the outward tilt of the wheel and a rough check can be made by measuring the distance from the outside wall of the tyre, immediately below the hub, to a plumb line hanging from the outside wall of the tyre above the hub. The distance must be the same on both wheels. Before making this test, it is very important to ensure that the tyres are in a uniform condition and at the same pressure. Also that the car is unladen and on level ground.

Damage to the upper and lower wishbone arms may well affect the camber angle.

Castor angle

This is the tilt of the swivel pin when viewed from the side of the car. This also is only likely to be affected by damage to the upper and lower wishbone arms.

Swivel pin inclination

This is the tilt of the swivel pin when viewed from the front of the car and is again only likely to be affected by damage to the wishbone arms.

A useful tool which can be used for checking these settings is the Dunlop wheel camber, castor and swivel gauge. With the car standing on level ground this gauge will give readings enabling the castor, camber and swivel pin angles to be quickly verified.

25 Front wheel alignment

1 The front wheels are correctly aligned when they are turning in at the front between 1/16 to 1/8 inch as shown in Fig 11.14. It is important that this measurement is taken on a centre line drawn horizontally and parallel to the ground through the centre line of the hub. The exact point should be in the centre of the side wall of the tyre and not on the wheel rim which could be distorted and so give inaccurate readings.
2 The adjustment is made by rotating the cross tube once the locknuts have been slackened. This is achieved by left hand and right hand threads on the cross tube. The side rods are not adjustable.
3 This is a job best left to your local BLMC garage as accurate alignment requires the use of special equipment. If the wheels are not in alignment, tyre wear will be heavy and uneven and the steering will be stiff and unresponsive.

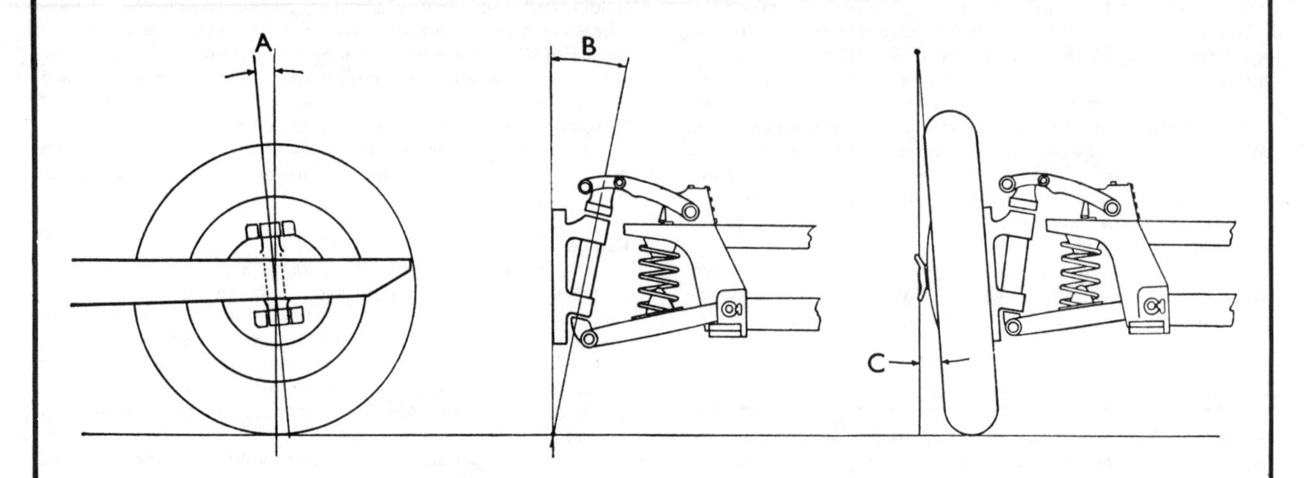

Fig 11.13 STEERING GEOMETRY ANGLES

A Castor angle 2º *B Swivel pin inclination 6½º* *C Camber angle 1º*

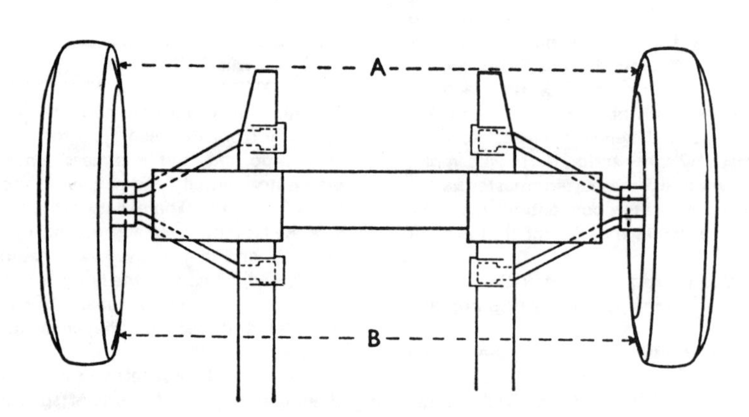

Fig 11.14 FRONT WHEEL ALIGNMENT

The toe-in must be adjusted so that A is 1/16 to 1/8 inch less than B

Chapter 12 Bodywork and underframe

Contents

1 General description

Two or four seater versions were built but all had hinged doors opening to nearly 90°. On the earlier models, with sliding windows, large open pockets are provided in the doors. Aluminium framed, detachable side screens with sliding perspex windows are fitted to all models up to Mk III. Thereafter wind-up windows and louvre ventilators are fitted.

The fixed curved windscreen has twin windscreen wipers. The vinyl treated folding hood, including a detachable transparent rear panel, may be stowed behind the rear seats when not required. A tonneau cover which completely covers the interior but which, by using a centre zip, may be opened for the driver only, was supplied as an optional extra.

The bonnet is hinged at the rear and incorporates a chromium plated air intake to assist engine cooling. It is supported in the open position by a pivoted rod. The bonnet lock is released from inside the car.

The boot has a lockable lid and contains the 12 volt battery and the spare wheel. However, on models without the occasional rear seats, two 6 volt batteries are located just forward of the rear axle and beneath the floor. For security, a battery master switch is also located within the boot.

The standard specification included adjustable steering wheel, interior driving mirror, adjustable bucket seats, passenger's grab handle, parcel shelf beneath the fasica, ashtray on transmission tunnel and twin horns. Externally, bumpers with overriders were fitted and the paint finish available in single or dual colours.

Optional extras included overdrive, heater and demister, radio, road speed tyres, wire spoke wheels, servo assisted brakes (some models) and a glass fibre hardtop.

2 Maintenance - interior

Many car owners leave interior cleaning to last and prefer to wash the exterior first. This is really working backwards because the dust created by removal of carpets will only settle on the clean exterior.

By regularly cleaning the interior, the upholstery will remain in good 'nearly new' condition, the carpets fresh and clean and the general appearance look smart and well cared for. When the carpeting is removed, any water leaks will be evident and the necessary corrective action can be taken before rust sets in.

First empty the under dash parcel tray and clear from under the front seats and boot all the bits and pieces that have collected over the last few months of motoring, and place in a large cardboard box ready for sorting out and replacing.

Lift out the rubber slip mats, the carpeting and underfelt. The rubber mats may be washed if very dirty or just shaken to remove loose dirt. The carpeting may be brushed, shaken or beaten to remove the dust and dirt. If badly marked they can be washed using a carpet shampoo and laid out to dry in the sun. Underfelt should be carefully shaken but not washed or beaten otherwise it will be difficult to dry and may start to break up. If the carpeting around the pedals is worn renew it.

Using a vacuum cleaner with a flexible hose, remove all traces of dust and grit that accumulate.

With a suitable upholstery cleaner diluted as recommended by the manufacturer wash down all upholstery on the seats, body trim and roof lining. Use a neat solution on stubborn stains. Wipe off all traces of cleaner or soap with a moistened cloth and finally rub dry with a clean non-fluffy rag. Do not use too much water as it will cause excessive condensation in the car, unless it is a hot day and the doors may be left open for a while.

The interior paintwork may next be cleaned using a cloth, and polished using a domestic aerosol polish and a clean non-fluffy rag. Door handles and chrome trim should be lightly rubbed with a moistened cloth.

To clean the interior glass, interior mirror and instrument cluster glass, add a little methylated spirits to water and wipe over with a soft cloth. Do not use ordinary domestic cleaners as they can cause smearing.

Inspect the seat belts for damage and make sure that the anchorages are still firm. The webbing may be washed in warm soapy water and wiped dry with an old towel.

Wash down the door, boot lid and bonnet apertures and also the edges of the doors, boot lid and bonnet. Remove all traces of lubricant with a paraffin moistened cloth. Take care to clean around the door hinges and locks as these are dust traps.

With a piece of wire probe the door drain holes to make sure that they are free of blockage. Inspect the floor pan for signs of rusting or leaking at the various seams. De-rust using Kurust and seal with a flexible sealing compound such as Seelastik.

Next go round all nuts, bolts and screws and make sure that all are tight. Then lubricate the door locks and hinges, choke control and front seat runners to ensure precise and free movement.

Turning to the boot, remove the complete contents including spare wheel, and vacuum out all the accumulated dust and dirt. Wipe the paintwork down with a damp cloth. If carpeting is fitted, clean this in a similar manner to the interior carpeting. Again look for water leaks especially in the corners and if necessary seal with Seelastik once the rust has been neutralised. Clear the drain holes using a piece of wire.

Using an oil can, lubricate the handbrake lever assembly and the pedal pivots. Inspect the pedal rubbers for signs of excessive wear and fit new ones if necessary. It is dangerous to drive with worn pedal rubbers.

Should you have a slight tear on one of the seats or trim panel, cut a piece of spare trim from the underside of one of the seats and apply a coat of impact adhesive such as clear Bostik. Insert the patch into the hole with the glue uppermost and then apply adhesive to the flap of the trim section. Allow the recommended drying time to pass and then press down the torn edges trying to get the edges as close together as possible which will make the repair less pronounced. Any large tears will have to be repaired using a piece of matching material.

3 Maintenance - exterior (underside)

It is recommended that if the car is in a dirty state it be taken to a garage equipped with steam cleaning equipment.

With the underside relatively clean it is an easy matter to keep it clean. Remove the interior carpeting and contents of the boot. Jack up the car as high as possible and remove the road wheels. With a garden hose, a stiff brush, tin of paraffin and scraper and, of course, with yourself suitably clad for a soaking, soak the dirt accumulated under the wheel arches and crevices, loosening where necessary with the hand scraper. This will require time and patience but, working systematically front to rear, remove dirt and oil.

Whilst the underside of the car is drying, check the seams for signs of leaking. Also generally check the tightness of all visible nuts and bolts and make sure the various pipes and wires are securely clipped to the underside of the body floor.

Inspect the underside for signs of rusting and, if evident, clear with a wire brush and neutralise with Kurust. When the underside is really dry seal any leaking seams with a flexible sealing compound. Wipe off the Kurust with a rag soaked in methylated spirits and apply a coat of suitable red oxide cellulose primer surfacer. Allow to dry and, if the part is visible, finish off with a coat of Holts Car Enamel Spray of the matching body paint colour.

Any underbody sealer requires regular inspection to make sure there are no loose flakes. If these are evident, scrape off the loose area and remove any rust as described in the last paragraph. Apply a coat of red oxide cellulose primer surfacer and allow to dry. Underbody sealer is available in a brush-on form, although when applied fresh at the garage it is sprayed on. A tin of this should be obtained and brushed on using a 2 inch paintbrush. On the wheel arches it is recommended that, because of stones being thrown up by the tyres, two coats are applied, with time allowed for drying between each coat.

Finally before lowering the car to the ground again check the exhaust system for leaks with the engine running.

4 Maintenance - exterior

It is recommended that once a week the exterior of the car be washed and wiped dry. For this job a flexibrush on the end of the garden hose is best, a sponge to assist wiping down and a leather to finish the operation off.

First make sure that all windows and doors are closed. Thoroughly wet the car with water using a gentle spray. Take care not to aim the jet of water directly at the windows or body seams which could start water leaks. Once the dirt has been loosened wipe down the panels using the brush with water still running through it as this way the paintwork should not be scratched by road grit.

Next apply wax car shampoo, or a little non-detergent washing up liquid, working from the roof downwards. Any stubborn dead flies, marks or tar may be removed using white spirit on a soft cloth. Do not forget to clean the front grille, the wiper blades and, of course, the wheels. A leather must not be used with a detergent or shampoo as it will cause it to rot.

Finally rinse off all the suds with plenty of clean water and wipe dry using a leather. Wipe all spots and smears from the windscreen and door glass using the leather. When the car is dry the glass may be polished with a rag soaked in a methylated spirits and water solution. Chromium plating requires regular cleaning with a damp cloth or leather. Occasionally one of the special polishes for chromium plating may be used but on no account use an ordinary metal polish.

Every six months it is recommended that the exterior be wax polished. There are, however, several important points to be noted before polish is used on a car:

1 If part of the paintwork has been resprayed allow at least two months for it to dry fully and harden.
2 Do not use a cutting paste to remove the dull film from cars sprayed with a metallic paint.
3 When purchasing a wax polish always make sure that it is suitable for the type of paintwork on the car.
4 Do not attempt to wax polish a car in the sun of when the body is still warm having been in the sun. It will bake on and have to be removed with petrol.
5 Do not wax polish a car which has just been washed because paintwork absorbs moisture slightly and the wax coating can hold this moisture so giving an effect called 'micro blistering' caused by minute rust spots under the paint film.

1 Don't dust down or polish a dusty car. Always wash.
2 Don't get polish or wax on any of the glass.
3 Don't neglect hidden parts of the doors when polishing.
4 Don't leave birdlime on the paintwork - it will cause stains.
5 Don't park under trees especially in the hot sun or when raining.
6 Don't use a cutting compound or haze remover on cars finished with an acrylic paint.
7 Don't use wax without cleaning the car first.

5 Bodywork - paint touch-up

On any car with a part or all-steel body, the greatest enemy of all is rust and this is most likely to start under the wings or along the sills because the road wheels will fling water, mud and grit onto the paint surface and it will only be a matter of time before the paint skin is penetrated and rusting starts.

It is for this reason that many new cars are given a thick coat of underseal, usually of a bituminous or rubber base, to guard against rusting. However, if this was the end of the story, paint maintenance would be relatively simple but unfortunately it is not, because chips appear at the front of the wings, along the outside of the wing panels and doors as well as the edges of the bonnet and boot lid. Whilst the car is being cleaned, these chip marks will become evident and it is important that they are attended to immediately otherwise rusting will occur and spread so that what was once a small chip will gradually turn into a large area requiring a great deal more renovation work.

Touch-up paint is usually available in either touch-up pencil, tin with a little brush in the lid, or aerosol form, and may be obtained as a good match to the original body colour. It must, however, be realised that some paint colours are more stable than others. Due to the action of sunlight on an older car an exact match may be difficult unless a tin of touch-up paint is mixed.

To prepare the surface for touching-up, first use a silicone solvent to remove all traces of polish which will not allow the paint to adhere properly. If there are signs of rusting or the paint beginning to lift, use a sharp penknife and carefully scrape away the loose paint and rust. Then neutralise the rust with a little Kurust and allow to dry. With a piece of rag soaked in methylated spirits wipe away the dry Kurust.

The prepared spot may now be touched in with the touch-up brush. Shake the tin vigorously for a few minutes to ensure that the paint is well mixed and withdraw the brush. Wipe the brush on the inside of the neck of the tin and then dip in the end of the brush until there is a little paint on it, just sufficient to touch in the area concerned. Very carefully apply a thin coat of paint only to the area concerned and allow to dry thoroughly. Apply a further thin coat so as to build up the paint to the original paint thickness. This will take time and patience but with care the touch-up should be indistinguishable from the surrounding area.

If there is a scratch on the paintwork which has penetrated the top coat of paint and the red primer is showing through, the basic procedure is the same. First remove all traces of polish with a solvent and then, with the knife, lift off any loose paint. Neutralise any rust and finally touch in the scratch preferably with one sweep of the brush. For this, a fine brush may be better than the brush provided with the tin. Build up the paint to the original paint thickness.

The edges of doors and boot lid seem to suffer very much and small areas of rust frequently appear. In this case an aerosol tin of primer and enamel top coat will be required. Again use the silicone solvent to remove any polish from the area concerned. Rub down the paint around the area with a little wet or dry paper grade 400, until the area is smooth. As the name implies, the paper can be used either dry or with water; the latter method keeps the grit of the paper clear of dust and also acts as a lubricant. Neutralise any rust with Kurust and, when dry, wipe with a piece of cloth soaked in methylated spirits.

Before spraying, make sure the car is sheltered from wind and dust. Shake the aerosol tin of primer for a few minutes to ensure that there is no sediment in the bottom. Usually the manufacturer drops in a ball bearing to assist agitation of the paint. If this is the first time that an aerosol tin is being used, try it on a piece of metal such as an old tin to get the 'feel' of the spray and then proceed to spray the prepared surface. Remember success of this work lies in the preparation. The smoother the prepared surface the better will be the finish. Hold the jet about six inches away from the area to be sprayed and work from the centre outwards keeping the centre moist and the outside lightly sprayed and dry.

When dry, very lightly rub the primer with wet or dry paper to roughen up the surface and inspect it for blemishes caused by dust or bad preparation. Rectify any faults by rubbing down again and applying a further coat of primer. It is only when the surface under repair is perfect that the final top coat may be applied. Again experiment on a piece of metal if this is your first time, and when you are confident apply the top coat to the primer. Remember, it is like ordinary household painting - two thin coats are better than one thick coat.

Should runs occur, it is an indication that either too much paint has been applied at one go or the nozzle was too near to the surface being sprayed. Rub down the area concerned and start again.

With all touching up, be it a small spot or a larger rusted area, allow the paint to dry thoroughly, at least overnight, and then use a little rubbing compound to blend in the edges of the paint and remove any dry spray.

If the rusted area is near to a piece of chrome trim, there is no need to remove it but mask up the chrome trim with a little sellotape or proper masking tape. This may be removed once the paint is half dry, leaving no paint overspray marks on the trim. Take care when sticking down the tape and use a knife to push the tape around any curved areas.

Should the scratch be only a minor one without penetration through to the undercoat, it may be removed using a rubbing compound.

6 Bodywork - deep scratching, dent or crease removal

This type of repair requires a little more work but is well within the do-it-yourself motorist's capabilities, provided that care is taken and the job is not rushed. Again preparation is the secret of good results. The method of approach will depend on the location of the damage, but in all cases if it is possible to push the dent or crease out from behind, so much the better. This may mean removal of a piece of interior trim. Should, however, this present problems, then do not worry too much unless the original shape cannot be achieved even with building up with a filler.

On this next operation wear a pair of goggles or a pair of glasses to protect the eyes. Using an electric sander with an abrasive disc on the rubber pad, remove all the paint right down to the bare metal from the area surrounding the damage as well as the damaged area itself. Work the area until all traces of paint including undercoat and primer have been removed, and an area of bare bright metal is obtained.

Next coat the area of bare metal with a special zinc primer, such as Holts Zinc Plate, to give additional protection against future corrosion as well as to provide a key for the body filler. Allow to dry thoroughly.

The body filler must next be prepared according to the manufacturer's instructions. Usually this comes in two parts, a tin of filler in paste form and a hardener. Read through the mixing instructions and, when fully conversant, mix only enough for immediate use to guard against waste. As well as being expensive, the hardener has a very limited working time of a few minutes. It is best to mix the hardener using a piece of plastic or very stiff cardboard on a piece of hardboard or plywood.

The filler should be applied to the damaged area and about one inch either side of it so as to allow for preparing the surface for final finishing. Do not apply the filler to paintwork as it will not adhere properly. Carefully smooth the filler to the contour of the body panel, but do not try to work the filler once it has started to harden.

When the filler has hardened it should be rubbed down using a coarse wet or dry paper, grade 120. Do not use an electric sander for this or subsequent operations as its action is too fierce. Carefully rub the surface smooth until the contour matches the rest of the panel and is also relatively smooth. Use paper either wet or dry. Wash down the area being worked upon and inspect for imperfections and small air holes or areas requiring further building up. On flat panels use a sanding block but on curved areas just use the paper by itself.

Mix some more filler and apply where necessary to make good any defects found. When the filler is dry, blend into the rest of the area using wet and dry paper.

Now using wet 280 grade paper rub down the complete area taking care to blend the filler edges to the bare metal. This may take time but remember the preparation determines the quality of finish.

Wipe the complete area dry and inspect again for any blemishes. These must be rectified at this stage. With the palm of the hand, feel the surface for any high or low spots caused by over-ambitious rubbing down and again rectify if evident.

When you are entirely satisfied that the area is perfect, the next stage is to mask over any adjoining panels or chrome trim with sellotape or masking tape and newspaper.

Apply a coat of Holts Zinc Plate to give a good key for the primer as well as to give additional rust protection. Allow to dry.

The primer may now be applied with a good quality paintbrush which will not moult. Paint the whole of the area under repair and allow to dry. Very lightly rub down the surface with wet 400 grade paper and inspect for any imperfections. Then wash down and allow the moisture to evaporate. Apply a second coat of primer and again lightly rub down and wash.

The repair is now ready for receiving the top coat. Holding the nozzle about six inches away from the surface spray behind any catches or fittings first and then work from the centre of the panel outwards until the repaired section of the panel is covered. Make sure that the part overlaps the existing panel by a couple of inches to allow for feathering and allow to dry. Lightly rub down the surface with wet 400 grade paper and allow to dry.

Now spray on a second top coat and, if necessary, a third coat until the depth of colour matches the original paintwork.

When the final coat is completely dry, rub the surface with a soft cloth and some rubbing compound, paying particular attention to the areas where the new paint overlaps the original. Wipe off the compound and inspect the finished result for signs of any blemishes which should be corrected by rubbing down and respraying.

Finally remove the masking tape and paper and lightly polish with a clean soft cloth. DO NOT apply a polish for at least two months to allow the paint to harden.

Provided that care was taken in the selection of the materials and the instructions followed, the results should be satisfactory, but if something has gone wrong the following table should be of assistance:

Fault	Cause
Blotchy finish	Insufficient number of primer or top coats
Paint runs	Spray nozzle too near panel during spraying. Too much paint applied.
Rippling (called 'Orange Peel')	Too thick a coat application
Matt finish	Spray nozzle too far away from panel. Not all dust from previous flatting operation removed.
Creasing	Unsuitable materials used for primer or top coat.
Overspray	Insufficient masking - use cutting compound to remove it.
Rough finish	Spraying in dusty or windy conditions.
Faded patches of top coat (called 'Blooming')	Usually caused by spraying in damp conditions. Most pronounced with dark colours.

7 Bodywork - serious corrosion

Should a removable body panel or part of it, such as a lower door panel, leading edge of a wing or sill, as opposed to a main structural member, be badly corroded, it is within the capabilities of a more experienced do-it-yourself owner to effect a repair. If the panel is badly corroded then it should be removed and replaced as whole. Some steel panels may no longer be available, fibreglass panels may have to replace them. If not badly corroded proceed as follows.

The first thing to do is to sand down the affected area for a further inspection. Do not forget to wear goggles or glasses to protect the eyes. Use an electric drill with sanding attachment and a coarse disc to remove the paint from the rusted section as well as the immediate area surrounding the more visible part affected.

Next hammer or cut away all affected metal until sound metal is reached and then treat with Holts Zinc Plate paint to check subsequent corrosion.

Obtain a piece of perforated zinc plate, as found in old meat safes, and cut off a section larger than the hole produced by the removal of affected metal, and insert into the hole. If necessary, mould it to the shape of the panel. Use paper clips, or small self-tapping screws, to keep it in place.

Mix sufficient filler according to the maker's instructions to apply a thin coat to the zinc plate and immediate surrounding area and work it well in so as to provide a good key for subsequent layers. Allow it to dry.

Continue building up, a layer at a time, until the contour of the body panel has been reproduced and then allow to dry.

Thereafter follow the instructions given in the last Section, shaping and blending the filler to the existing body contour, and then finally paint.

8 Major chassis and body repairs

1 Because the body is bolted to a separate chassis, major damage is easier to repair than that for an integral body where the chassis and body are designed as one unit. This is because the body may be separated from the chassis frame and the two parts repaired individually. However, it is recommended that any major repair is still left to a competent body repair specialist.

2 If the damage is serious it is important that the body alignment is checked for, otherwise, the handling of the car will suffer causing excessive tyre wear, as well as wear in the transmission and steering.

3 To check the chassis alignment of a car which has been damaged, a system of diagonal and measurement checks from parts projected from the underframe onto a level floor is used.

4 To ensure that the alignment check is performed accurately the chassis must first be raised so that its datum line O (Fig 12.1) is parallel with the floor. Use the comparative measurements given in the table accompanying Fig 12.1 to achieve this condition.

5 Raise the rear of the chassis to a convenient working height and then adjust the height of the front of the chassis, until the points given in the table accompanying Fig 12.1 for the front and rear on both sides of the chassis frame are in the correct relative vertical position to each other.

6 As an example of this, if the rear dimension is 36 inches from the floor and is quoted as 2 inches above the chassis datum line the front point quoted as 1 inch below the datum line must be 33 inches above the floor.

7 At the same time check the relative heights of all the intermediate points given in the table accompanying Fig 12.1 so that any distortion of the car in the vertical will be ascertained.

8 Chalk over the area of the floor directly below the points shown in Fig 12.3. Using a plumb line, project the points from the chassis to the floor, marking the positions with a pencilled cross.

9 The centre between each pair of points can be established by means of a large pair of compasses and the central points marked on the floor.

10 In addition diagonals can be determined between any two pairs of points, and the points of intersection marked on the floor. At this stage a length of thin cord covered in chalk can be held by two persons in such a position that it passes through as many of the central points and intersections marked, as possible.

11 While the cord is held taut, a third person raises the centre of it, and then allows it to spring back smartly to the floor. If the resulting white line passes through all the points, the chassis alignment is satisfactory. Any points through which the white line does not pass will be in a position where the underframe is out of alignment.

12 Considerable deviations in the transverse and longitudinal measurements given in the table in Fig 12.3 confirm chassis misalignment. It should however be appreciated that allowance must be made for normal manufacturing tolerances and that a reasonable departure from the nominal dimensions may be permitted without detriment to road performance and safety.

9 Maintenance - hinges and locks

Regularly lubricate the door, bonnet, boot hinges and locks

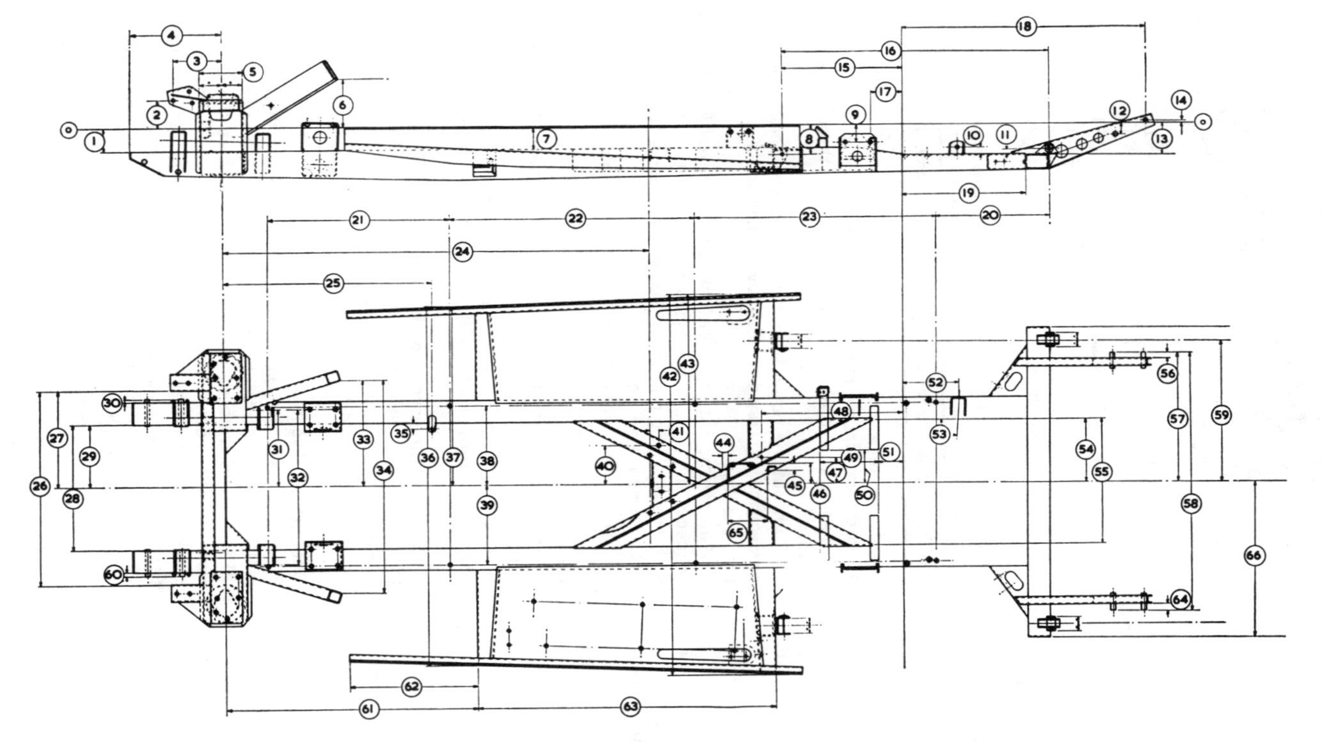

Fig 12.1 CHASSIS ALIGNMENT DIAGRAM (SERIES BN4, BN6, BN7, BT7 and BJ7 CARS)

1 3¼ in (8.26 cm) parallel
2 3.723 to 3.754 in (9.46 to 9.54 cm)
3 6.196 to 6.226 in (15.74 to 15.81 cm)
4 12 in (30.48 cm)
5 5¾ in (14.6 cm)
6 6½ in (16.51 cm)
7 3 in (7.62 cm)
8 4 1/8 in (10.48 cm)
9 2½ in (6.35 cm)
10 1 in (2.54 cm)
11 ¾ in (1.91 cm)
12 1.55 in (3.94 cm)
13 4¼ in (10.8 cm)
14 11/32 in (.873 cm)
15 16 in (40.64 cm)
16 35½ in (90.13 cm)
17 4 1/8 in (10.48 cm)
18 42 11/32 in (82.15 cm)
19 16 3/8 in (41.59 cm)
20 15 in (38.1 cm)
21 24 in (61 cm)
22 32½ in (82.55 cm)
23 32 in (76.2 cm)
24 56 9/32 in (143 cm)
25 27½ in (69.89 cm)
26 26¼ in (66.72 cm)
27 13 1/8 in (33.36 cm)
28 17 in (43.18 cm)
29 8½ in (21.59 cm)
30 ½ in (1.27 cm)
31 10½ in (26.67 cm)
32 21 in (53.34 cm)
33 14 13/32 in (36.6 cm)
34 28 13/16 in (73.2 cm)
35 ¾ in (1.91 cm)
36 48¾ ± 1/16 in (123.9 ± .16 cm)
37 24 3/8 ± 1/32 in (61.95 ± .08 cm)
38 10¾ in (27.31 cm)
39 10¾ in (27.31 cm)
40 5¼ in (13.34 cm)
41 1¼ in (3.18 cm)
42 51½ ± 1/16 in (130.89 ± .16 cm)
43 25¾ ± 1/32 in (65.45 ± .08 cm)
44 ¾ in (1.91 cm)
45 7/8 in (2.22 cm)
46 2 7/8 in (7.3 cm)
47 3½ in (8.89 cm)
48 18¾ in (47.61 cm)
49 7 13/16 in (19.84 cm))
50 4½ in (11.43 cm)) Series BN6 and BN7 cars only
51 3 5/16 in (8.41 cm))
52 7 3/8 in (18.73 cm)
53 87°
54 8½ in (21.59 cm)
55 17 in (43.18 cm)
56 7/8 in (2.22 cm)
57 17 9/16 in (44.61 cm)
58 35 1/8 in (89.22 cm)
59 19¼ in (48.88 cm)
60 ½ in (1.27 cm)
61 33 3/8 in (84.77 cm)
62 17 in (43.18 cm)
63 39 7/16 in (100.13 cm)
64 1/8 in (2.22 cm)
65 5 5/16 in (13.49 cm)
66 21 in (53.32 cm)

with a little Castrol GTX from an oil can. The door striker plates can be given a thin smear of grease to reduce wear and ensure free movement.

10 Door rattles - tracing and rectification

1 The commonest cause of door rattles is a misaligned, loose or worn striker plate, but other causes may be:

a) Loose door handles, window winder handles (later models) or door hinges.
b) Loose, worn or misaligned door lock components.
c) Loose or worn remote control mechanism.

2 It is quite possible for door rattles to be a result of a combination of the above faults, so a careful examination must be made to determine the cause of the fault.
3 If the nose of the striker plate is worn as a result of door rattles, it should be renewed.
4 If the nose of the door lock wedge is badly worn and the door rattles as a result, then fit a new lock.
5 Should the hinges be badly worn they must be renewed.

11 Door - removal and refitting

1 Upon inspection it will be observed that both the upper and lower hinge of each door are secured to the door post by four crosshead screws and one hexagonal head set pin.
2 At each hinge four crosshead setscrews are used to secure it to the door frame.
3 To remove the door, first undo the two set pins securing the check strap coupling bracket to the door pillar. This will allow the door to be opened fully. Take care however as if the door is opened too far the wing or door panel could be dented.
4 Using a pencil, accurately mark the outline of the hinge relative to the pillar to assist refitting. It is desirable to have an assistant to take the weight of the door once the two hinges have been released.

12 Door (early type) - dismantling and reassembly

Interior trim

To remove the door casing, undo and remove the sixteen crosshead screws located on the casing perimeter. Lift away the interior trim panel.

Top moulding

The aluminium moulding on the inner top edge of the door is secured by three crosshead screws. To remove the moulding undo and remove the three securing screws and lift away the moulding.

Outer handle

The outer door handle is secured by a nut which is accessible from inside the door, and a Phillips screw accessible from the outside when the handle is raised.

Inner handle

To remove the door operating handle, push the chrome escutcheon behind the handle inwards against spring pressure. A dowel which passes through the handle stem will now be visible and this should be removed with a small diameter parallel pin punch or electrician's screwdriver. Lift away the handle, escutcheon and spring.

In all cases refitting is the reverse sequence to removal.

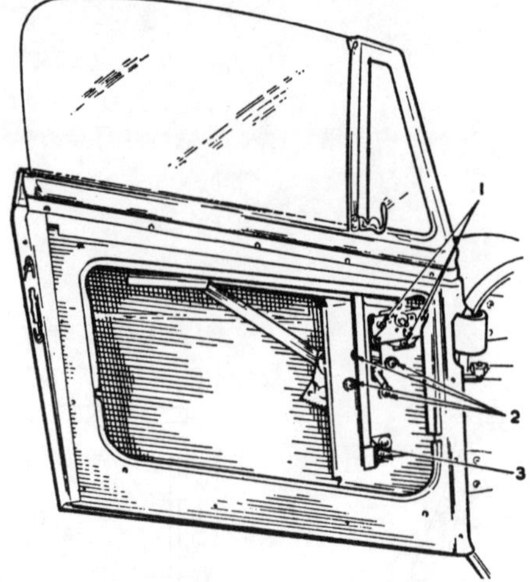

Fig 12.2 INTERIOR OF LATER TYPE DOOR

1 Door lock remote control screws
2 Window regulator screws
3 Window guide channels

13 Door (later type) - dismantling and reassembly

Interior handles

1 Push the window regulator handle escutcheon away from the handle so as to expose the dowel which passes through the handle stem.
2 Push out the dowel with a small diameter parallel pin punch or electrician's screwdriver. Lift away the handle, escutcheon and spring.
3 When removing the door lock remote control handle it may be necessary to turn the escutcheon until its slots are aligned with the retaining pin. By so doing it will allow the escutcheon to be pushed away from the handle.
4 Remove the dowel as described in paragraph 2.

Interior trim

5 Unscrew the self-tapping screws and remove the chromium plated door pull.
6 Unscrew the self-tapping screws located around the perimeter of the outer trim panel and lift away the outer trim panel.
7 Unscrew the self-tapping screws located around the perimeter of the inner trim panel and lift away the inner trim panel.
8 It will have been observed that upon removal of the outer trim panel, the self-tapping screws which hold the top trim panel will be exposed.

Door glass and ventilator

9 Unclip the door glass inner weather strip and the door waist plate moulding, complete with the attached rubber seal.
10 Undo and remove the four screws that secure the ventilator assembly to the door upper panel.
11 Unscrew the nuts adjacent to the window regulator mechanism to release the lower end of the window guide channel (Fig 12.2).
12 Temporarily refit the window regulator handle and lower the glass as far as possible.
13 Release the regulator arm from its guide channel in the glass frame.
14 Pull the ventilator and door glass upwards, as one assembly, from the door.

Window regulator

15 To remove the regulator, undo the four screws securing the regulator to the door and lift away the regulator complete with arm.

Door lock (early cars)

16 Remove the three screws securing the door lock remote control assembly and the further three screws that hold the door lock and guide plate to the door.

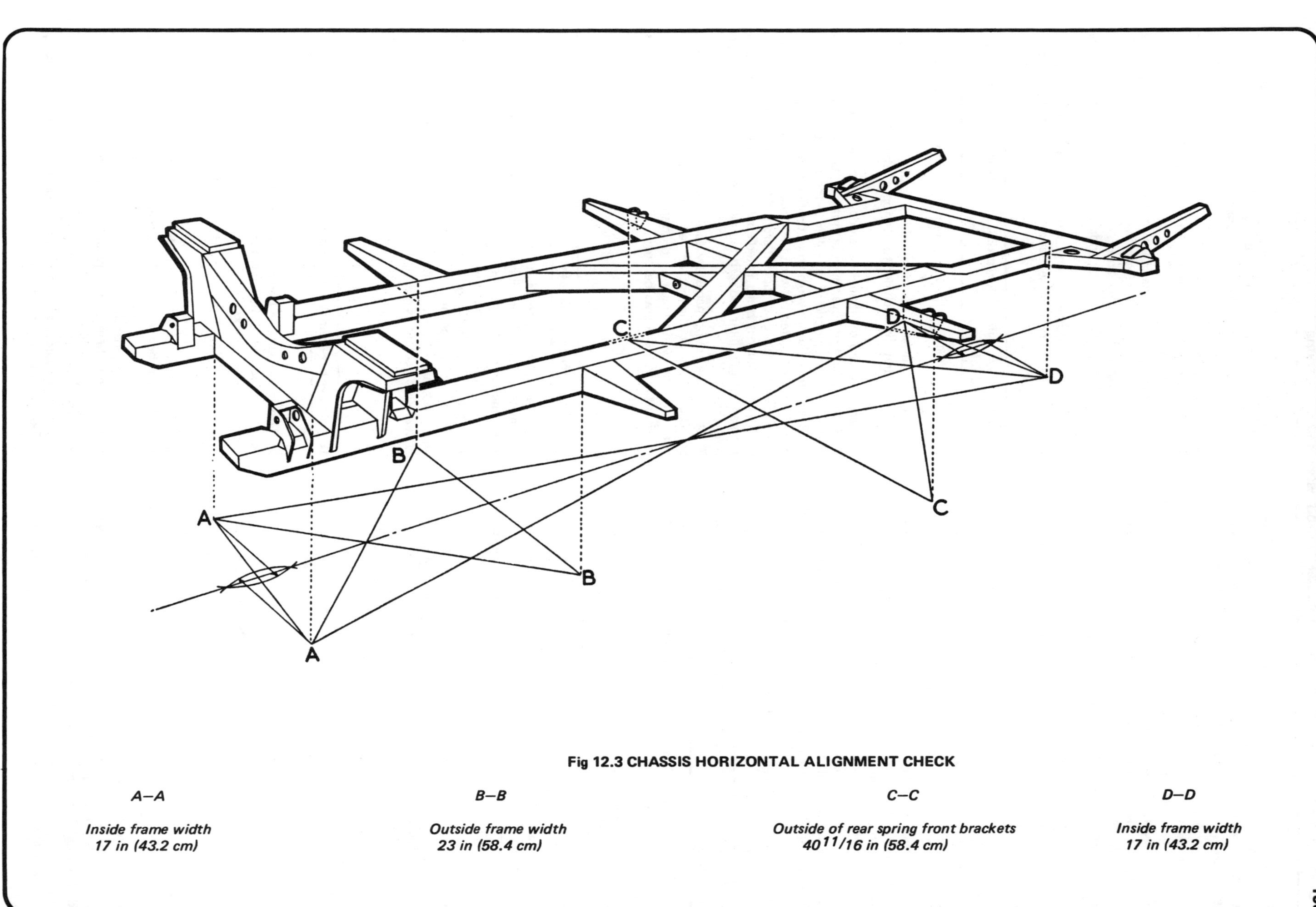

Fig 12.3 CHASSIS HORIZONTAL ALIGNMENT CHECK

A–A	*B–B*	*C–C*	*D–D*
Inside frame width 17 in (43.2 cm)	*Outside frame width 23 in (58.4 cm)*	*Outside of rear spring front brackets 40 11/16 in (58.4 cm)*	*Inside frame width 17 in (43.2 cm)*

19 Push the catch of the lock downwards into the locked position and push it through the aperture in the door panel.
18 Allow the remote control to drop downwards and withdraw the complete assembly through the door inner panel.

Door lock (modified type)
19 With the door trim removed, undo the three screws that retain the door lock and guide plate to the door. Also remove the three screws which secure the door lock remote control assembly to the door.
20 Undo and remove the two inside screws securing the door lock and remove the door lock.

Before final assembly, lubricate all moving parts with Castrol GTX. Reassembly is the reverse sequence to removal in all cases.

14 Windscreen and frame (early models) - removal and refitting

The windscreen frame is secured to the scuttle at each side by two nuts, bolts and one set pin. The nuts and set pin are accessible from within the cockpit behind the fascia. The bolt heads are visible at the door pillars once the doors are open.

With an assistant to take the weight, undo and remove the fixings mentioned in the last paragraph and lift away the frame.

When refitting, tighten all fixings in a progressive manner so as to ensure that the frame seats correctly without any strain which could cause the glass to fracture.

15 Windscreen and frame (Convertible Mk II and Mk III) - removal and refitting

1 Refer to Chapter 10 and remove the windscreen wiper arms from their spindles.
2 Undo and remove the two self-tapping screws from each corner of the fascia panel and the two screws that secure the driving mirror to the scuttle.
3 Carefully lift the scuttle top liner assembly from the scuttle, making a note of the six locating holes in the scuttle for the demister dust bezel pegs.
4 Undo and remove the one set bolt that holds the bracket at the centre of the windscreen bottom rail to the scuttle top.
5 Remove the four screws from the rear of each windscreen pillar.
6 The windscreen and frame assembly may now be drawn forwards from the pillars.
7 Inspect the rubber weatherstrip which is located in a channel in the windscreen lower frame member for signs of hardening, perishing or splitting and fit a new one if necessary.
8 Before refitting the windshield assembly make sure that the aperture at the base of each pillar is sealed with either Dum-Dum putty or Glasticon Compound.
9 When refitting, tighten all fixings in a progressive manner so as to ensure that the frame seats correctly without any strain which could cause the glass to fracture.

16 Windscreen glass (Convertible models Mk II and Mk III) - removal and refitting

1 Refer to Section 15 and remove the windscreen glass and frame assembly from the car.
2 Undo and remove the two screws to be found at each corner of the frame (top and bottom). The four sections of the frame can now be separated.
3 Upon inspection it will be seen that the windscreen glass glazing rubber is in two parts. One is in the frame top member and the other is in the frame lower member and the two side members.
4 Inspect the glazing rubber for signs of hardening, perishing or cracking and if evident obtain a complete new glazing rubber.
5 When the glazing rubber is being refitted it is important that it does not overlap at the corner brackets. No additional sealing rubber is necessary.
6 Insert the shorter section of the glazing rubber into the frame top member and position the larger section centrally in the lower member.
7 Assemble the frame members to the glass and then insert the glazing rubber protruding from the lower member of the frame into the two side members, and assemble them to the glass.
8 Refit the screws at both ends of the top and lower frame members.
9 Refit the windscreen frame assembly to the car, this being the reverse to the removal sequence as described in Section 15.

17 Windscreen pillar (Convertible models Mk II and Mk III) - removal and refitting

1 Refer to Section 15 and remove the windscreen glass and frame assembly from the car.
2 Using a suitable diameter drill, remove the rivet which secures the draught excluder to the top of the pillar which is to be removed.
3 Carefully ease the draught excluder from the pillar.
4 Unscrew and remove the four nuts, bolts and spring washers that hold the lower end of the pillar to the scuttle.
5 Draw the pillar upwards from the body. Recover the packing piece which is located between the pillar and the body. Note that there is some sealing compound around the base of the pillar.
6 If a new pillar is being fitted apply some Dum-Dum putty to the base of the pillar and refit the pillar in the reverse manner to removal.
7 It may be found necessary to insert or remove some of the packing pieces at the bottom of the pillar so as to prevent straining the windscreen and frame assembly. Fit all the windscreen frame screws before finally tightening the pillar nuts and bolts.

18 Heater - removal and refitting

Heaters were fitted as standard equipment to BN4 models commencing from car number 68960 and to BN6 models commencing from car number 501.

Heater removal and refitment is a simple process provided you study Fig 12.4 carefully and note where all the water, electrical and bracket connections fit. Once removed and inspected, repaired and then replaced, do not forget to refill the system with coolant as described in Chapter 2.

19 Bumper bar - removal and refitting

Front
1 The front bumper bar is mounted on two support brackets which are secured to the chassis frame by two set pins and spring washers to each bracket.
2 To assist removal, soak these set pins with a little penetrating oil.
3 Undo and remove the four set pins with spring washers and draw the front bumper forwards from the front of the car.
4 Should it be necessary to separate one of the mounting brackets, undo the two nuts and spring washers. This should release the overrider as well, if fitted.
5 Refitting is the reverse sequence to removal.

Rear
The procedure for removal of the rear bumper and mounting brackets is basically identical to that for the front bumper.

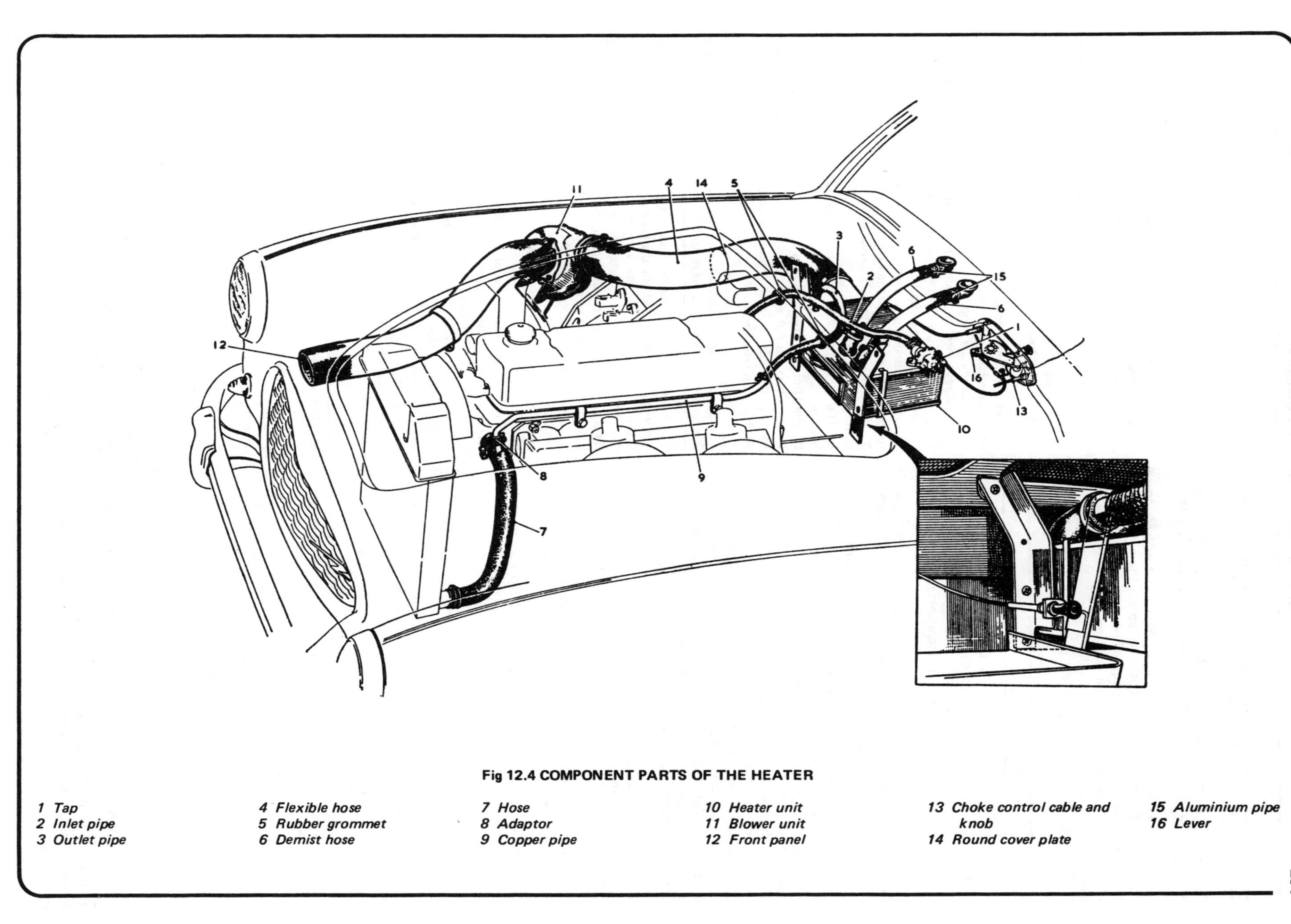

Fig 12.4 COMPONENT PARTS OF THE HEATER

1 Tap
2 Inlet pipe
3 Outlet pipe
4 Flexible hose
5 Rubber grommet
6 Demist hose
7 Hose
8 Adaptor
9 Copper pipe
10 Heater unit
11 Blower unit
12 Front panel
13 Choke control cable and knob
14 Round cover plate
15 Aluminium pipe
16 Lever

20 Front grille - removal and refitting

1 The grille is secured to the front of the shroud by eight fixing set pins which are all easily accessible from the underside of the car.
2 Undo and remove the three ¼ inch UNF set pins with spring washers which will be found at both the top and the bottom of the grille. A further 3/16 inch set pin will be found on either side of the grille. These should next be removed together with their spring washers. Lift away the front grille.
3 Refitting the front grille is the reverse sequence to removal.

21 Bonnet - removal and refitting

1 Open the bonnet and hold open using the bonnet stay.
2 It is recommended that an old blanket is placed under the top edge of the lid and spread over the area of the rear of the wings and under the windscreen to avoid scratching the paintwork during removal and replacement.
3 The bonnet, at its rear edge, has two brackets which form part of the hinges. A leg from the bulkhead is secured to each bracket by two nuts and bolts.
4 To act as a datum for refitting, mark the position of the hinges on their mounting brackets using a soft pencil.
5 The services of a second person should be enlisted to assist in taking the weight of the bonnet and lifting it away over the engine compartment.
6 Undo the bonnet securing bolts from each bracket and lift away the bonnet.
7 Refitting the bonnet is the reverse sequence to removal. Any adjustment necessary can be made at the brackets, but provided the bonnet is fitted correctly before removal and the outlines of the brackets were marked in pencil, the bonnet should be a good fit if the brackets are re-attached in their original positions.

22 Boot lid - removal and refitting

1 Open the boot lid. It is recommended that an old blanket is placed under the top edge of the boot lid and spread over the wing panels to prevent scratching the paintwork during removal and refitting.
2 To act as a datum for refitting, mark the position of the hinges on their mountings on the boot lid using a soft pencil.
3 The services of a second person should be enlisted to assist in taking the weight of the boot lid.
4 Release the limit cable from its connection under the boot lid.
5 Undo and remove the two hinge nuts and spring washers from the underside of the boot lid and lift away the boot lid from over the rear of the car.
6 Refitting is the reverse sequence to removal. Any adjustment necessary can be made at the hinges, but provided the boot lid was fitted correctly before removal, and the outlines of the hinges were marked in pencil the lid should be a good fit if the hinges are re-attached in their original positions.

23 Boot lid locking handle and lock - removal and refitting

1 To remove the locking handle from the boot lid, locate the two round head securing screws which will be seen on the underside of the lid. Undo and remove these two screws.
2 Lift away the locking handle through the lock and boot lid.
3 Undo and remove the four crosshead screws, nuts and spring washers that secure the lock assembly to the supporting bracket which is riveted and welded to the lid. Lift away the lock assembly.
4 Refitting is the reverse sequence to removal. It is beneficial to well lubricate the spring that is visible at the closing edge of the boot lid.

24 Front wing - removal and refitting

1 Refer to Section 11 and remove the door.
2 Refer to Chapter 10 and remove the headlight and sidelight from the front wing. It will also be necessary to remove the outer case of the headlamp. To do this undo and remove the four bolts and brass nuts which are accessible beneath the wing. The sidelights are secured by three crosshead bolts and nuts.
3 Upon inspection, it will be seen that there are three bolts beneath the headlight aperture securing the wing to the cowl. These bolts screw into spring clip type nuts. Undo and remove the three bolts.
4 A further four bolts positioned along the top edge of the wing flange and forward of the scuttle secure the wing to the bonnet surround. Undo and remove these four bolts which screw into spring clip type nuts.
5 Inside the car, and under the fascia, are a further three bolts which screw into clip nuts on the wing and secure the lower flange of the wing to the underside of the scuttle. Undo and remove these three bolts.
6 Using a small drill, remove the two rivets that secure the rubber water channel to the rear section of the wing.
7 Undo the metal thread screws securing the wing to the inside of the door pillar.
8 The wing may now be lifted away together with the plastic beading.
9 Refitting the front wing is the reverse sequence to removal. If pop rivets are not easily available, use two self-tapping screws to refix the rubber water channel to the wing. Protect all inner fittings with underseal.

25 Rear wing - removal and refitting

1 Chock the front wheels, jack up the rear of the car and support on firmly based stands. Remove the road wheel on the side of the car fromwhich the wing is to be removed.
2 The rear wing is secured to the main bodywork structure by six square headed bolts with spiral clip nuts, these being located over the wheel arch and round the rear curve of the wing.
3 Located at the top of the wheel arch is a plain nut and bolt with spring and plain washer. The head of the bolt is accessible from within the luggage compartment. Undo and remove this bolt.
5 At the lower front edge of the wing there is a flange which is secured to the chassis by two nuts and bolts and a metal drive screw. Undo and remove these fixings.
6 Undo and remove the eight 3/16 inch countersunk crosshead bolts and nuts, together with their plain and spring washers that secure the wing leading edge to the door pillar.
7 Refitting is the reverse sequence to removal. It is recommended that all fixings likely to be subjected to moisture are coated with Castrol LM Grease to protect against corrosion.

26 Shroud - removal and refitting

1 Refer to Section 24 and remove the two front wings.
2 Refer to Section 21 and remove the bonnet.
3 Refer to Section 19 and remove the front bumper.
4 Refer to Section 20 and remove the grille.
5 Refer to Section 14 or 15 and remove the windscreen.
6 Remove the fascia and cockpit moulding.
7 There are five drive screws that secure the shroud to the scuttle. These should next be undone and removed.
8 At the front of the bonnet opening there are three crosshead bolts and nuts which secure the shroud to the front cross bracing to the bodywork. Undo and remove these fixings.
9 Located at each side of the bonnet opening are two countersunk crosshead bolts which fix the shroud to the upright braces projected upwards from the chassis frame (Fig 12.5). Undo and remove these crosshead bolts.

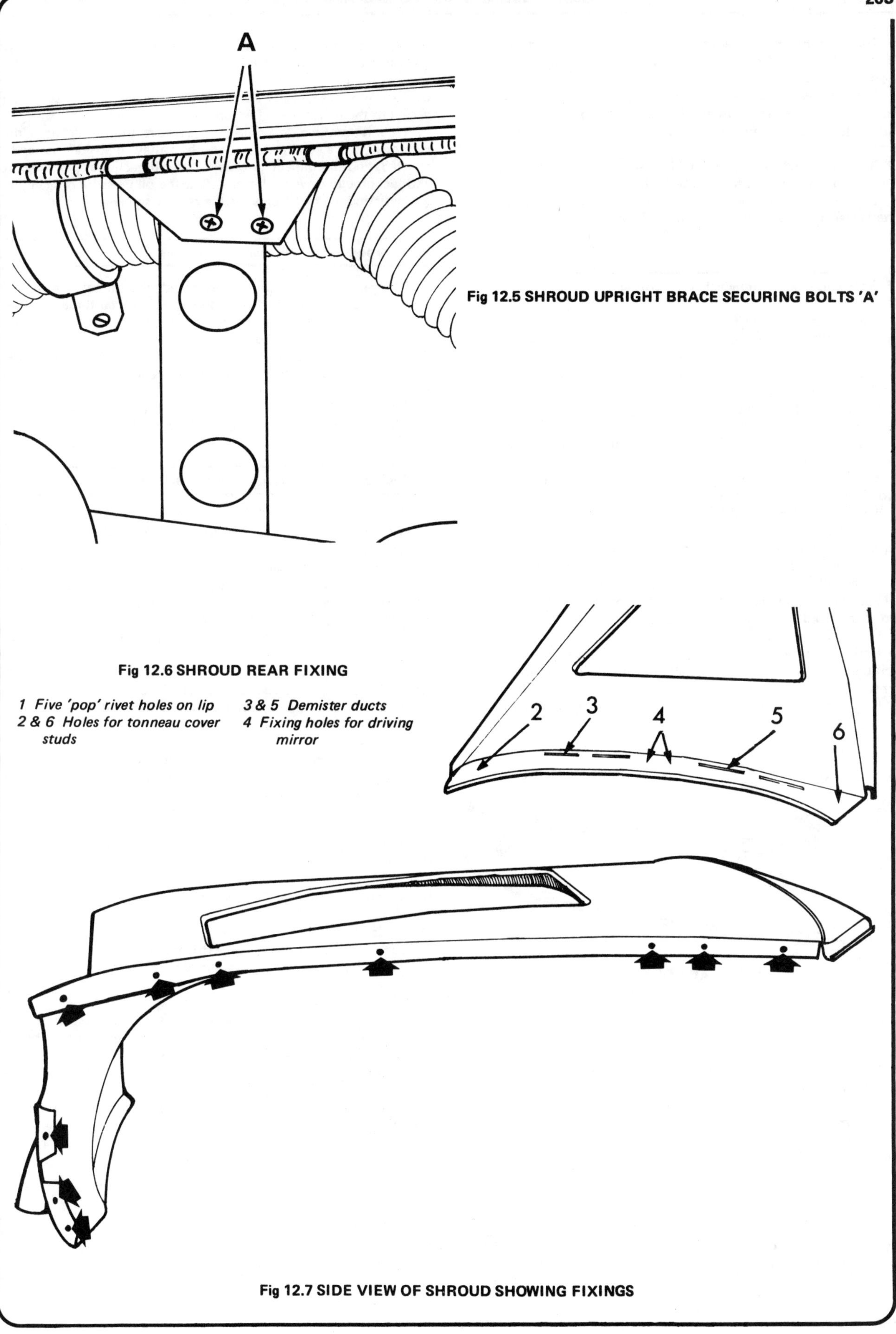

Fig 12.5 SHROUD UPRIGHT BRACE SECURING BOLTS 'A'

Fig 12.6 SHROUD REAR FIXING

1 Five 'pop' rivet holes on lip
2 & 6 Holes for tonneau cover studs
3 & 5 Demister ducts
4 Fixing holes for driving mirror

Fig 12.7 SIDE VIEW OF SHROUD SHOWING FIXINGS

10 There are two plate brackets which secure the shroud to each wheel arch panel. Next undo the two nuts and bolts to each bracket.
11 The cowl which is part of the shroud has two brackets which secure the body member to the frame dumb irons and two nuts and bolts act as attachments which should now be undone and removed.
12 Refer to Fig 12.6 where it will be seen that the rear end of the shroud is secured to the scuttle by five pop rivets.
13 The complete shroud as shown in Fig 12.7 may now be lifted clear of the chassis frame and bodywork.
14 Refitting the shroud is the reverse sequence to removal.

27 Rear body panel - removal and refitting

The rear body panel forms part of the lower rear part of the luggage compartment and it is usually this panel which suffers most if the rear of the car is subjected to minor body damage. Although it is welded at each side it is easy to break the weld with a cold chisel, and when a new panel is fitted small nuts and bolts may be used instead.
1 Refer to Section 22 and remove the boot lid.
2 Refer to Section 19 and remove the rear bumper bar and mounting brackets.
3 Each side of the panel is secured by two pop rivets and by two of the rear wing nuts and bolts. Drill out the pop rivets with a suitable diameter drill and remove the nuts and bolts.
4 The top edge of the panel has thirteen rivets holding it to the luggage compartment frame, and the lower lip has nineteen rivets. Drill out these pop rivets with a suitable diameter drill.
4 Refer to Chapter 10 and remove the rear lights.
5 Feel for the welded seam at each side of the rear body panel and with a cold chisel break open the weld.
6 Lift away the old panel.
7 Refitting is the reverse sequence to removal. If welding facilities are not available, use small nuts and bolts to remake the joint.
8 Refer to Section 6 of this Chapter and make good any minor panel defects and paintwork.

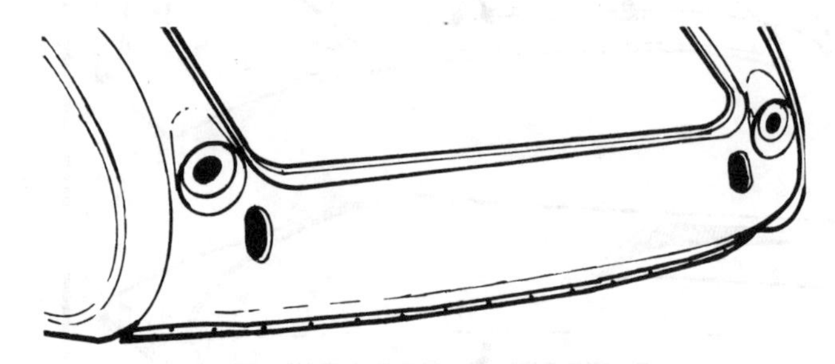

Fig 12.8 REAR BODY PANEL

28 Fascia panel

1 Refer to Chapter 11 and remove the steering wheel.
2 Switch off the battery master switch.
3 Undo and remove the two round head bolts, nuts and spring washers that secure the heater control panel to the fascia panel.
4 Locate the five screws, the heads of which are under the fascia, which pass through the fascia panel into tapped holes of brackets behind the fascia. There is also one further screw adjacent to the instrument panel. Undo and remove all six screws.
5 Carefully ease the fascia forwards into the cockpit so giving access to the rear of each instrument.
6 If it is required to completely remove the fascia, note the electrical cable connections so that they may be correctly refitted, and disconnect the cables. Also detach the speedometer and tachometer drive cables as well as the oil pressure gauge pipe.
7 Refitting the fascia panel is the reverse sequence to removal.

29 Gearbox cover - removal and refitting

The gearbox cover, sometimes called 'the tunnel' is attached at each side flange to the floor panels by six metal thread screws. To gain access to these screws, the carpeting must be peeled back. Lift away the gearbox cover.

If work is to be done on the clutch or gearbox, the carpet covered bulkhead plate can also be removed once the six self-tapping screws have been removed.

In both cases refitting is the reverse sequence to removal. Use an impact adhesive to stick the carpeting back into place.

30 Sidescreens and sockets - removal and refitting

1 The sidescreens each have one locating dowel at their base and these are a snug push fit into the sockets located in the top of each door.
2 Should it be necessary to remove one of the sockets because of wear, it may be screwed out using a wide bladed screwdriver.
3 Refitting a socket is the reverse sequence to removal. It is recommended that the threads be liberally coated with Castrol LM Grease to ensure that at a later date they may be easily unscrewed.

31 Front seats - removal, refitting and adjustment

Passenger's seat
1 Lift out the cushion from the passenger's seat.
2 It will be seen that there are two set pins on each side of the seat frame which secure it to the body floor panels.
3 Undo and remove the four set pins and lift away the seat frame.
4 If it is required to adjust the position of the seat there are four alternative holes for adjustment on each side of the seat frame.
5 Refitting is the reverse sequence to removal.

Driver's seat
1 Lift out the cushion from the driver's seat.
2 It will be seen that there are six nuts which secure the seat frame to the runners. Undo and remove these six nuts and spring washers and lift away the seat.
3 The seat runners, together with their packing pieces, are bolted to the floor, each runner being held by three nuts and bolts. Access to the nuts is gained under the floor panel.
4 Refitting is the reverse sequence to removal. Well grease the runners to ensure ease of operation.
5 To adjust the position of the seat once it has been refitted, push the lever located beneath the seat towards the runner, and then slide the seat forwards or backwards until the required position has been obtained. Release the lever.

Index

Printed by
J H Haynes & Co Ltd
Sparkford Nr Yeovil
Somerset BA22 7JJ England